Advances in Molecular Plant Nematology

NATO ASI Series

Advanced Science Institutes Series

A series presenting the results of activities sponsored by the NATO Science Committee, which aims at the dissemination of advanced scientific and technological knowledge, with a view to strengthening links between scientific communities.

The series is published by an international board of publishers in conjunction with the NATO Scientific Affairs Division

A	**Life Sciences**	Plenum Publishing Corporation
B	**Physics**	New York and London
C	**Mathematical and Physical Sciences**	Kluwer Academic Publishers
D	**Behavioral and Social Sciences**	Dordrecht, Boston, and London
E	**Applied Sciences**	
F	**Computer and Systems Sciences**	Springer-Verlag
G	**Ecological Sciences**	Berlin, Heidelberg, New York, London,
H	**Cell Biology**	Paris, Tokyo, Hong Kong, and Barcelona
I	**Global Environmental Change**	

Recent Volumes in this Series

Volume 265 — Basic Mechanisms of Physiologic and Aberrant Lymphoproliferation in the Skin
edited by W. Clark Lambert, Benvenuto Giannotti, and Willem A. van Vloten

Volume 266 — Esterases, Lipases, and Phospholipases: From Structure to Clinical Significance
edited by M.I. Mackness and M. Clerc

Volume 267 — Bioelectrochemistry IV: Nerve Muscle Function— Bioelectrochemistry, Mechanisms, Bioenergetics, and Control
edited by Bruno Andrea Melandri, Giulio Milazzo, and Martin Blank

Volume 268 — Advances in Molecular Plant Nematology
edited by F. Lamberti, C. De Giorgi, and David McK. Bird

Volume 269 — Ascomycete Systematics: Problems and Perspectives in the Nineties
edited by David L. Hawksworth

Volume 270 — Standardization of Epidemiologic Studies of Host Susceptibility
edited by Janice Dorman

Volume 271 — Oscillatory Event-Related Brain Dynamics
edited by Christo Pantev, Thomas Elbert, and Bernd Lütkenhöner

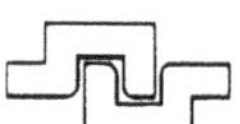

Series A: Life Sciences

Advances in Molecular Plant Nematology

Edited by

F. Lamberti
Institute of Agricultural Nematology of the CNR
Bari, Italy

C. De Giorgi
University of Bari
Bari, Italy

and

David McK. Bird
University of California
Riverside, California

Springer Science+Business Media, LLC

Proceedings of a NATO Advanced Research Workshop on
Molecular Plant Nematology,
held November 20–27, 1993,
in Martina Franca, Italy

NATO-PCO-DATA BASE

The electronic index to the NATO ASI Series provides full bibliographical references (with keywords and/or abstracts) to more than 30,000 contributions from international scientists published in all sections of the NATO ASI Series. Access to the NATO-PCO-DATA BASE is possible in two ways:

—via online FILE 128 (NATO-PCO-DATA BASE) hosted by ESRIN, Via Galileo Galilei, I-00044 Frascati, Italy

—via CD-ROM "NATO Science and Technology Disk" with user-friendly retrieval software in English, French, and German (©WTV GmbH and DATAWARE Technologies, Inc. 1989). The CD-ROM also contains the AGARD Aerospace Database.

The CD-ROM can be ordered through any member of the Board of Publishers or through NATO-PCO, Overijse, Belgium.

Library of Congress Cataloging-in-Publication Data

Advances in molecular plant nematology / edited by F. Lamberti, C. De Giorgi, and David McK. Bird.
p. cm. -- (NATO ASI series. Series A, Life sciences ; v. 268)
"Proceedings of a NATO Advanced Research Workshop on Molecular Plant Nematology, held November 20-27, 1993, in Martina Franca, Italy"--T.p. verso.
"Published in cooperation with NATO Scientific Affairs Division."
Includes bibliographical references and index.

1. Plant nematodes--Congresses. 2. Plant nematodes--Molecular aspects--Congresses. 3. Plant nematodes--Molecular genetics--Congresses. 4. Nematode diseases of plants--Congresses.
I. Lamberti, F. (Franco), 1937- . II. De Giorgi, C. (Carla)
III. Bird, David McK. IV. North Atlantic Treaty Organization. Scientific Affairs Division. V. NATO Advanced Research Workshop on Molecular Plant Nematology (1993 : Martina Franca, Italy)
VI. Series.
SB998.N4A38 1994
595.1'820488--dc20 94-36829
CIP

DOI 10.1007/978-1-4757-9080-1

Originally published by Plenum Press, New York in 1994.
MyCopy version of the original edition 1994

PREFACE

Plant parasitic nematodes are a main pest to crops. For example, the root-knot nematodes belonging to the genus *Meloidogyne* are worldwide in their distribution and attack almost every type of crop, causing considerable losses of yield and affecting quality of produce. The cyst nematodes within the genera *Globodera* and *Heterodera* constitute a major group of plant pathogens in many countries throughout the world, suppressing yields of potato, sugar beet, soybean and cereals. Several nematodes such as longidorids and trichodorids are implicated in the transmission of numerous plant viruses. Many others cause constraints to agricultural production either locally or on large areas.

However, despite their economic importance (they account for worldwide crop reduction in excess of 10%), plant parasitic nematodes are still poorly understood, because most of them are obligate parasites of roots. Environmental concerns over the agricultural use of pesticides demand the development of alternative measures to control them. To achieve environmentally sound control, knowledge of the basic biology of nematodes must be expanded. Important research areas include understanding the molecular bases for pathogenicity, the molecular mechanisms of the host-parasite interactions and the genetic bases for population fluctuations.

The workshop has, for the first time, brought together an international group of researchers using molecular approaches to study plant parasitic nematodes and their host responses.

Nematode genetics and genome organization were discussed in the frame of the highly advanced *Caenorhabditis elegans* project. Genetic transposition in *C. elegans* and plant parasitic nematodes and progress towards developing genetic maps of *Globodera* and *Heterodera* were presented. Work on the application of molecular systematics for examining relationships between nematode species and PCR strategies for identification and population dynamics studies of plant parasitic nematodes were illustrated in detail. A group of speakers described the induction of plant genes following nematode infection and presented analysis of a nematode responsive plant promoter. Molecular aspects of nematode resistance in tomato were also discussed. Finally, a section dealt with nematode functions that contribute to pathogenesis including stylet-exudate proteins, virulence genes, chemoreception in nematodes and molecular determinants of nematode transmission of plant viruses.

The NATO Advanced Research Workshop held at Martina Franca, Italy, during 20 to 27 November 1993 was attended by 50 participants from twelve different countries. They all contributed greatly to the success of the workshop with their interest and enthusiasm. We are grateful to them, to Mr. F. Elia, who was responsible for technical services, and to the management of the Park Hotel S. Michele, where the workshop was held, which created a friendly and warm atmosphere appropriate for a scientific meeting. However, special thanks are due on behalf of all those who participated and to the NATO Scientific Affairs Division for its support, which made this workshop possible.

F. Lamberti
C. De Giorgi
D. McK. Bird

February 1994

CONTENTS

PART I NEMATODE GENETICS

PART II IDENTIFICATION OF NEMATODES

PART IV HOST PARASITE INTERACTIONS: THE NEMATODE

PART I

NEMATODE GENETICS

THE *CAENORHABDITIS ELEGANS* GENOME PROJECT

Sean R. Eddy

MRC Laboratory of Molecular Biology
Hills Road
Cambridge CB2 2QH
England
sre@mrc-lmb.cam.ac.uk

INTRODUCTION

In the next five years, molecular biology will get its first look at the complete genetic code of a multicellular animal. The *Caenorhabditis elegans* genome sequencing project, a collaboration between Robert Waterston's group in St. Louis and John Sulston's group in Cambridge, is currently on schedule towards its goal of obtaining the complete sequence of this organism and all its estimated 15,000 to 20,000 genes by 1998 (Sulston et al., 1992). By that time, we should also know the complete genome sequence of a few other organisms as well, including the prokaryote *Escherichia coli* (Daniels et al., 1992; Plunkett et al., 1993) and the single-celled eukaryote *Saccharomyces cerevisiae* (Oliver et al., 1992).

Sydney Brenner became interested in *C. elegans* in the 1960s as a model genetic system for the study of nervous system development and animal behavior (Brenner, 1974). Since that time, a growing community of *C. elegans* researchers has extended Brenner's already ambitious vision, and *C. elegans* has become a major experimental organism for the study of eukaryotic molecular, cellular, and developmental biology (Wood, 1988). The community's goal is to understand the biology of this particular animal in the minutest detail possible. This unashamedly naive philosophy has been one of the driving forces behind a set of large descriptive projects that have supported the *C. elegans* community with a remarkably deep knowledge base. These include the elucidation of the somatic developmental lineage by direct observation of embryos (Sulston et al., 1983), the reconstruction of the connectivity and anatomy of the 302-cell nervous system from serial EM micrographs (White et al., 1986), and the cloning and physical mapping of the genome in cosmid and yeast artificial chromosome (YAC) vectors (Coulson et al., 1986; Coulson et al., 1988; Coulson et al., 1991). The genome sequencing project is the latest endeavor in this program of accumulating basic knowledge about the nematode.

As of this summer (1993), 2.2 million bases of DNA sequence have been obtained

Advances in Molecular Plant Nematology
Edited by F. Lamberti *et al.*, Plenum Press, New York, 1994

from the middle of chromosome III (Wilson et al., 1993) (Figure 1). This region was known to be relatively gene-dense, like all the centers of *C. elegans* autosomes. It was also known to contain some quite interesting genetically identified loci such as a cluster of homeobox genes involved in determination of anterior/posterior pattern (Burglin et al., 1991; Kenyon and Wang, 1991; Wang et al., 1993). Although the current data comprise only 2% of the genome, we have a taste of what is to come.

Here I describe the overall strategy and goals of the genome sequencing effort. I emphasize the mechanisms being used to make data and clones available to the rest of the research community, both via the international sequence databases and via the freely distributable *C. elegans* database, ACeDB (R. Durbin and J. Thierry-Mieg, unpublished). Finally, I describe the first exploratory efforts by the computational biology groups at St. Louis and Cambridge to predict genes and other features of the sequence, using both conventional similarity searching techniques and more recent developments in biological sequence pattern recognition.

STRATEGY

Physical map

The physical map originally consisted of overlapping 40-50 kb cosmid clones. 17,500 cosmids were mapped into about 700 contigs, covering perhaps 90% of the genome (Coulson et al., 1986; Coulson et al., 1988). The remaining gaps proved impossible to clone into cosmids and are thought probably to be repetitive sequence of some sort. Yeast artificial chromosome (YAC) vectors, which accommodate large inserts of about 200-400 kb (or more; megabase inserts are possible) became available by the late 1980s. YACs have an additional advantage in that DNA that is difficult to maintain stably in cosmids seems more easily cloned in YACs (Coulson et al., 1988; Coulson et al., 1991). Over 2,300 YAC clones have now been mapped, closing most of the remaining gaps on the physical map (Coulson et al., 1988; Coulson et al., 1991).

The physical map is currently eight gaps away from closure (Alan Coulson, personal communication). No special effort has yet been made to clone the ends of the chromosomes. Although the size of the remaining gaps and telomeric regions is unknown, the current physical map is thought to be about 98% complete because 98% of the current cDNA expressed sequence tags (McCombie et al., 1992; Waterston et al., 1992) can be mapped to existing clones (Alan Coulson, personal communication). About 95 million bases have been cloned and mapped.

Most new sequences can be rapidly physically mapped to a resolution of about 100 kilobases by hybridization to YAC grids dubbed polytene filters (Coulson et al., 1988; Coulson et al., 1991). The origin of "polytene" is a play on words, derived from the *Drosophila* technique of physical mapping by *in situ* hybridization to polytene chromosomes. Spotted on each filter are 958 representative YACs, covering most of the physical map with an average coverage of two-fold. Polytene filters are available on request from Cambridge.

Genome sequencing

The collaboration that was established between Robert Waterston's group in St. Louis and John Sulston's group in Cambridge for physical mapping was continued to sequence the genome (Sulston et al., 1992; Wilson et al., 1993). They began at a point in the middle of the chromosome III and sequenced outwards. The strategy is to sequence

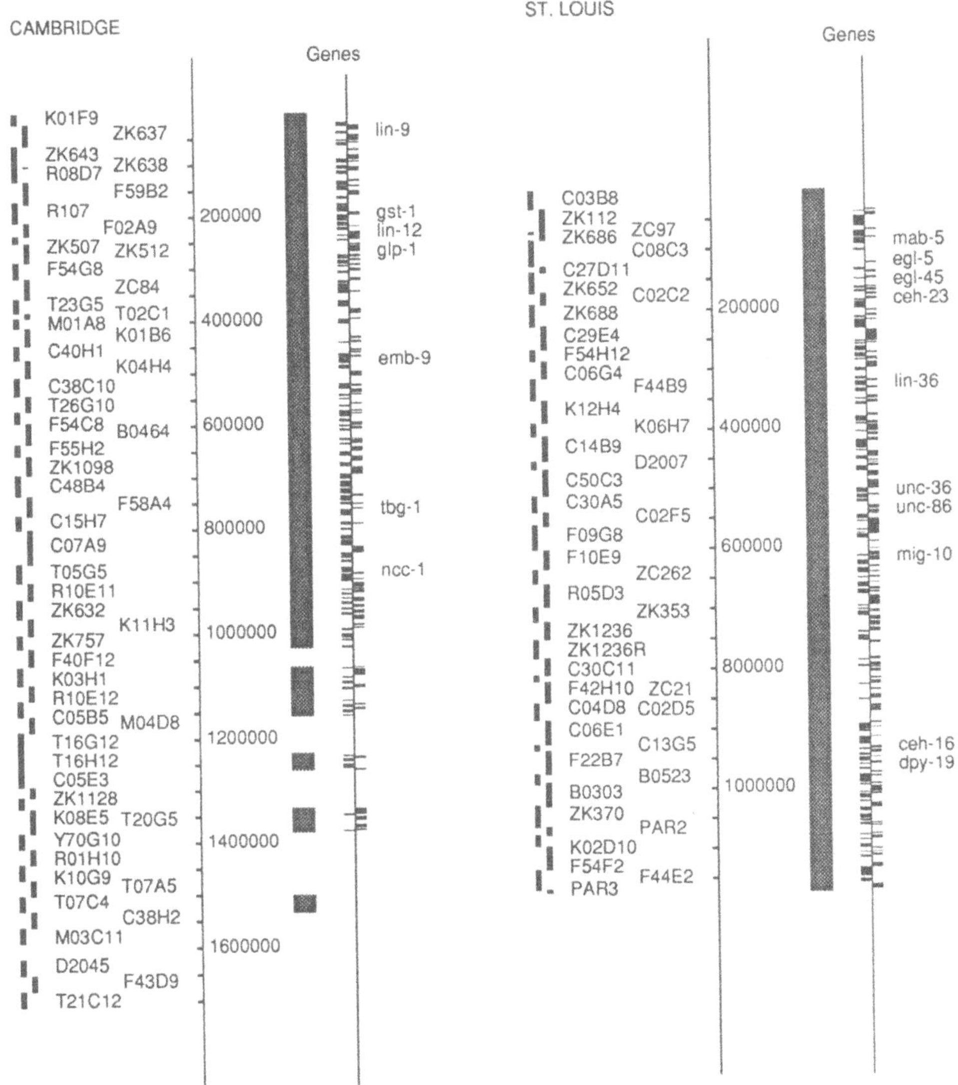

Figure 1: Summary of the current state of the sequencing project. Finished or in progress cosmids are indicated to the left of a linear scale in nucleotides. Gray bars indicate the extent of finished sequence. Positions of predicted genes on each strand are indicated as black bars to the left or right of a central line, and genetically identified genes are named to the right of this. More cosmids appear here than the 77 discussed in the text. The figure was generated from the Cambridge database, which includes Cambridge sequence in progress as well as the finished sequence from both groups.

the central gene clusters first. The center of chromosome II will be targeted when the center of III is completed.

Although there is much interest in new technological developments for large-scale genome sequencing, the strategy currently employed is one of scaling up and refining existing automated sequencing technology (Sulston et al., 1992; Wilson et al., 1993). Overlapping cosmid clones are selected from the physical map, random fragments of each cosmid are subcloned into M13 (1-3 kb insert size) and phagemid (6-9 kb insert size) vectors, and a large number of these "shotgun clones" are sequenced on automated machines using standard fluorescent primer extension methods. These sequence fragments, consisting of about 400 nucleotides per gel read, are assembled into contigs using software developed by Rodger Staden and his group (Dear and Staden, 1991). The remaining gaps are closed by primer walking from custom-synthesized oligonucleotides. A cosmid sequence is finished when it has been sequenced at least once on both strands and any ambiguities resolved by inspection of the original traces or by further sequence reads. Very roughly, 600–800 shotgun reads and about 10–40 custom oligos are used per cosmid, giving about 4- to 5-fold redundancy on average.

Through both improved automation – for instance, robotic DNA template preparation (Watson et al., 1993) – and additional man- and sequencing machine-power, the pace continues to accelerate. The groups plan to finish another ten million bases of sequence in 1994 (John Sulston, personal communication).

Preliminary computational analysis

The program GENEFINDER (Phil Green and LaDeana Hillier, unpublished) is used to predict coding regions in the sequence. In the nematode genome, which is gene-dense with (usually) short introns and relatively little repetitive DNA, gene-finding is not too difficult. The predictions of GENEFINDER are manually refined using ACeDB's sequence displays.

The DNA sequence is conceptually translated in all six frames and compared against the protein sequence databases, using the program BLASTX from the BLAST suite of similarity searching programs (Altschul et al., 1990). BLASTX output is filtered to remove hits resulting from biased composition, and to detect and assign higher significance to multiple BLAST matches consistent with a single gapped alignment (Sonnhammer and Durbin, 1993). Significant similarities are recorded and displayed in ACeDB, and become one of the criteria by which GENEFINDER predictions are manually revised.

Other features in the DNA or predicted protein sequence that are noticed by various sorts of programs (large inverted repeats, members of repeat sequence families) are annotated in ACeDB as well.

Data organization and distribution

Complete cosmid sequences are annotated with predicted and known genes, detected similarities, and other features. The annotated sequences are deposited in the EMBL and GenBank sequence databases (Benson et al., 1993; Rice et al., 1993). Release 35 of the EMBL database (June 1993) contained data for 53 cosmids, 1.6 million bases of slightly overlapping genome sequence.

A great deal of information about *C. elegans* is available through the database program ACeDB (A *C. elegans* Data Base), written by Jean Thierry-Mieg and Richard Durbin (unpublished). Among other things, ACeDB organizes and displays the genetic map, physical map, and growing genomic sequence data. The database is gathered with the help of the entire *C. elegans* community. The ACeDB software is organism-independent; different databases can be created for other purposes, and the ACeDB software has been adopted by a number of other genome projects. The program is built to accommodate new extensions and data types flexibly. The versions in use in Cambridge are in constant flux as new features are explored.

Released stable versions of ACeDB and the *C. elegans* database are freely available by anonymous ftp from a number of sites on the Internet, including `ncbi.nlm.nih.gov` and `cele.mrc-lmb.cam.ac.uk`. The software and the databases are discussed on the Usenet newsgroup `bionet.software.acedb`.

ANALYSIS

Gene predictions and database similarities

From the sequence of 77 cosmids (2.2 Mb), GENEFINDER predicts 470 coding regions. The predictions imply an average of 1 gene every 5,000 bp. About 29% of the genome appears to be coding. Seventeen genes which had been previously identified genetically have been localized in the sequence (some of which had already been sequenced individually): *ceh-16*, *dpy-19*, *emb-9*, *egl-5*, *egl-45*, *gst-1*, *glp-1*, *lin-9*, *lin-12*, *lin-36*, *mab-5*, *mig-10*, *ncc-1*, *sup-5*, *tbg-1*, *unc-36*, and *unc-86* (Wilson et al., 1993).

An estimate for the total number of genes in the worm that corrects for biased gene density along the chromosomes can be obtained using data from the expressed sequence tag (cDNA) sequencing projects (McCombie et al., 1992; Waterston et al., 1992). Although the cDNAs are biased towards highly-expressed genes, one assumes that highly expressed genes are not significantly clustered in the genome. 129 out of 4,615 (2.8%) of the current cDNA tags map into this 2.2 Mb interval. The extrapolation 470 divided by 2.8% implies there are about 17,000 total genes.

How many of the predicted genes are similar to something in the existing protein databases? I used the BLAST algorithm (Altschul et al., 1990) to scan the SWISS-PROT and PIR databases (Bairoch and Boeckmann, 1993; Barker et al., 1993) for proteins similar to these 470 coding regions. High-scoring matches were further checked with the Smith/Waterman dynamic programming alignment algorithm (which allows gaps) (Smith and Waterman, 1981) to ascertain how much of the two proteins could be aligned (Appendix).

Of the 470 predicted coding regions, 159 (34%) are significantly similar to one or more proteins in the database. Of these, 102 (22%) have alignments which span more than half of both the predicted and the database sequences, suggesting that they might be true functional homologues. The cutoff of 50% alignment was an arbitrary choice. The remaining 57 coding regions with database similarities may be more divergent homologues, or may share functional domains with each other. They may also reflect errant GENEFINDER predictions; we know that some GENEFINDER predictions are missing exons or are fused to inappropriate exons. The putative homologues and similarities are listed in the Appendix.

There are genes for multiple members of some protein families, such as protein kinases (10 proteins), homeodomain proteins (7 proteins), and RNA helicases (6 proteins). Although extrapolation from these small numbers is dangerous, particularly since related genes may occur in clusters, it appears that there could be hundreds of members of genes encoding these protein families in the nematode genome. The homeodomain estimate is probably biased because the sequence includes a known cluster of homeodomain genes that are involved in determining anterior-posterior pattern (Burglin et al., 1991; Kenyon and Wang, 1991; Wang et al., 1993). It has been estimated from hybridization and PCR data that there are about 60 homeodomain genes (Chalfie, 1993).

It is the common experience of genome projects in several different organisms that about a third of randomly selected inferred protein sequences match something in the databases (Bork et al., 1992; Green et al., 1993). To some extent, this reflects the incompleteness of the databases, but there is also a deeper lesson. Green *et. al.* performed extensive pairwise comparisons of unselected protein data sets from the human, yeast, and *C. elegans* genome projects to the protein sequence database (Green et al., 1993). About 30-40% of each dataset matched sequences in the database. Green *et. al.* refer to matches between sequences from different phyla as ancient conserved regions (ACRs). ACRs represent fundamental protein components of animal life – proteins that arose before the major radiation of metazoan phyla 580 to 540 million years ago (Knoll, 1992) and have been conserved since then. The surprising result of Green *et. al.* is that few new ACRs are detected by comparing the new datasets to *each other.* In other words, newly sequenced worm proteins that match a newly sequenced human protein are very likely (>90%) to match an existing database sequence as well. This implies that the database already contains examples of most ACR's, and that the number of ACR families is relatively small (Green *et. al.* estimate on the order of 1,000). The remaining 60-70% of sequences have either diverged too far to be detected by sequence-based comparisons, or they have arisen fairly recently in evolution.

In this regard, it is interesting that there are many similarities between pairs of the 470 predicted worm genes with no similarities to existing database sequences. Using approximately the same procedure as Green et. al. (1993), I found 29 ACR (encoding kinases, RNA helicases, collagens, etc.) and 23 non-ACR families in pairwise comparisons between the 470 predicted proteins (119/470 have similarity to at least one other predicted protein in the set). Many of the non-ACR similarities are very highly similar and sometimes nearly identical; one match, between two predicted genes (F22B7.5 and C38C10.4) about 750 kb apart, shows 95% identity over almost 500 deduced amino acids and must be the result of a recent duplication. There are a number of scenarios that could explain the presence of a high proportion of gene families within the nematode that are not already in the database, given the ACR story of Green et. al. (1993). Two likely scenarios are: 1) The matches may be to gene families that emerged late in evolution and are playing functional roles specific to all or part of the phylum *Nematoda* (a strong Darwinist viewpoint). 2) The matches may reflect the fact that the genome is fluidly evolving by gene duplication and divergence, and parts of the genome are being copied around at a high rate regardless of function (a more neutral, evolutionary drift viewpoint). Either way, it appears likely that a significant fraction of nematode genes radiated fairly recently.

Transposons

Two different sorts of transposons are known in *C. elegans* (Collins et al., 1989; Dreyfus and Emmons, 1991; Levitt and Emmons, 1989; Wood, 1988; Yuan et al., 1991). The first type, exemplified by Tc1, Tc2, and Tc3 elements, are 1.6 to 2.1 kilobases in length with short inverted terminal repeats, and open reading frames that may encode transposases (Collins et al., 1989; Levitt and Emmons, 1989; Wood, 1988). They are members of a large family of eukaryotic transposable elements that includes, for instance, *Drosophila* mariner and P elements (Robertson, 1993). A Tc3 element has been detected in the sequence of cosmid B0303, and two elements nearly identical to each other and significantly similar to *Drosophila* mariner elements have also been found (in C30A5 and ZK370). The other known type of *C. elegans* transposon, exemplified by Tc4 and Tc6 elements, are composed entirely of long inverted repeats and appear to encode no protein; their structure is analogous to that of *Drosophila* foldback elements (Dreyfus and Emmons, 1991; Yuan et al., 1991). Two Tc4-related elements have been detected in the genome sequence (in C27D11 and ZK686).

Another major class of eukaryotic transposable elements are retrotransposons (Singer and Berg, 1991). Retrotransposons encode proteins with similarity to reverse transcriptases and are thought to transpose through an RNA intermediate. Retrotransposons come in two distinct families. Some have long terminal direct repeats of several hundred nucleotides (LTR's) and have structural similarity to retroviruses. A second class, the non-LTR retrotransposons, lacks the direct repeats and is less understood. Until recently, neither form of retrotransposon had been seen in *C. elegans*, though examples of LTR-containing retrotransposons have been found in the nematodes *Panagrellus redivivus* and *Ascaris lumbricoides* (Wood, 1988).

Six regions with significant similarity to reverse transcriptases have been detected in the genome sequence. None of these appear to be associated with long terminal repeats. Five of the six are clearly related to each other, though divergent (less than 60% identity), and their closest similarities are to insect non-LTR retrotransposons, so they are likely to represent a family of nematode non-LTR retrotransposons. The sixth element (F44E2.1) is very diverged from the other five, and is more closely related to RVT's of the gypsy-like LTR-containing retrotransposons, though I can find no LTR's around it. It is impossible to tell whether any of the elements encode functional reverse transcriptase. One element, in C07A9, almost certainly is nonfunctional because a small inverted repeat element has apparently inserted itself into the retrotransposon and disrupted the RVT reading frame. Much more analysis needs to be done on these regions, but for now it appears that there are at least two families of retrotransposons in the nematode genome sequence, and members of the one family may be quite numerous.

SENSITIVE FEATURE RECOGNITION

There are many interesting parallels between linguistic problems – for instance, parsing a sentence, or recognizing words spoken by speakers with different accents – and sequence recognition problems – for instance, parsing sequence into exons and introns, or recognizing divergent sequence family members (Searls, 1992). A direction that I have been involved in is to use adaptive statistical models and other theoretical techniques borrowed from the fields of speech recognition and formal linguistics to recognize features in the genome sequence.

These methods build probabilistic models of sequence families from multiple alignments, similar to sequence "profiles" (Barton, 1990; Gribskov et al., 1990; Krogh et al., 1993). The models can be used for sensitive recognition of more members of a family. Unlike profiles, these models are "adaptive," meaning that they can be learned automatically from a set of initially unaligned example sequences (Krogh et al., 1993). For analysis of many families in large amounts of genome sequence, it is quite advantageous to have methods which learn on their own, bypassing the laborious step of constructing a trustworthy multiple sequence alignment.

Two kinds of adaptive statistical models are in use. The first, called hidden Markov models (HMM's), model primary sequence information only. They are good for recognition of protein and DNA sequence family members (Krogh et al., 1993). The second, called covariance models (CM's), additionally capture base-pairing secondary structure consensus for RNA sequence families (Eddy and Durbin, 1993; Sakakibara et al., 1993).

Work with both sorts of models is at a preliminary stage. One of our short-term goals is to build a library of models for many known protein, DNA, and RNA families and motifs, and automate the process of screening new sequence using these sensitive models. So far, we have used HMM's and covariance models on some trial problems.

An immunoglobulin superfamily member detected by HMMs

A number of labs use *C. elegans* as a model system for neural development (Wood, 1988). Many nematode genes are known which disrupt proper axonal guidance (Hedgecock et al., 1987; Hedgecock et al., 1990; Wadsworth and Hedgecock, 1992). In the development of mammalian nervous systems, a family of molecules containing multiple repeats of an immunoglobulin (Ig) superfamily domain, including the neural cell adhesion molecule (NCAM), are thought to be some of the key players (Bixby and Harris, 1991; Hynes and Lander, 1992). Indeed, one of the *C. elegans* axon guidance genes, *unc-5*, contains two NCAM-like Ig motifs; mutations in *unc-5* disrupt a subset of dorsal-wards circumferential pioneer axon migrations (Hedgecock et al., 1990). It would be very interesting to detect additional superfamily members in the genome sequence.

Unfortunately, the Ig superfamily is one of the most divergent sequence families. It can be difficult or impossible to detect Ig superfamily members by routine database search techniques. I trained an HMM to recognize NCAM-like Ig domains, using a training set of domains taken from the SWISSPROT protein database, and used that model to search through the 470 predicted proteins. A single protein was detected, ZC262.3, which has one Ig superfamily domain (Figure 2). BLAST searches had missed the similarity. A closer examination of ZC262.3 shows that it has a putative transmembrane region. The Ig domain would presumably be extracellular (although a few examples of intracellular Ig domains are known). The overall structure of the predicted ZC262.3 protein is not similar to any known members of the NCAM family, however, and its function is unclear. Perhaps the next logical step would be to subclone a piece of ZC262.3 and ask whether its expression is tissue-specific, for instance if it is localized to neurons (by either *in situ* RNA hybridization or construction of transgenic animals carrying ZC262.3::*lacZ* fusions).

tRNAs detected by covariance models

Obviously, much of the interest in the genome sequence focuses on the protein coding regions. But there are also RNA genes, such as the well-known tRNAs, snRNAs, and rRNAs, and it will not be at all surprising to find other functional RNA genes and motifs as well. Many RNA sequence families are difficult to detect by conventional primary sequence analysis techniques, because it is in general much more satisfactory to represent RNA families as consensus secondary structures rather than consensus primary sequences. From work by Richard Durbin and myself in Cambridge and work by David Haussler and coworkers at UC-Santa Cruz, it has recently become possible to construct fully probabilistic models of RNA secondary structure consensus, analogous to the HMM's we use for protein and DNA families, and search very sensitively for members of known RNA families (Eddy and Durbin, 1993; Sakakibara et al., 1993) . We call these models "covariance models."

There are thought to be about 300 tRNA genes in the nematode (Wood, 1988). Figure 3 shows the results of a search over the genomic sequence using a covariance model trained on tRNA example sequences. Fourteen tRNA genes have been detected.

```
NRG_DROME  DNPFIIEC EADGQP  EPE YSWIKN  GKKFDWQAYDNRMLRQP                       GRGTLVITIPKDEDR    GHYQCFASNEFG
NRG_DROME  GEPFMLKCAAPDGFP  SPT VNWMIQ   ESIDGSIKSINNSRMTLD                     PEGNLWFSNVTREDASSDFYYACSATSVFR
NRG_DROME  GKRMELFC IYGGTP  LPQ TVWSKD  GQRIQWSDRITQG H                         YGKSLVIRQTNFDDA    GTYTCDVSNGVG
NRG_DROME  DEEVVFEC RAAGVP  EPK ISWIHN  GKPIEQSTPNPRRTV                         TDNTIRIINLVKGDT    GNYGCNATNSLG
NRG_DROME  GRNVTIKC RVNGSP  KPL VKWLRA  SNWLTGGRYNVQ                            ANGDLEIQDVTFSDA    GKYTCYAQNKFG
NRG_DROME  GQSATFRC NEAHDDTLEIE IDWWKD  GQSIDFEAQPRFVKT                        NDNSLTIAKTMELDS    GEYTCVARTRLD
CAML_MOUSE TDDISLKC EARGRP  QVE FRWTKD  GIHFKPKEELGVVVHEAP                      YSGSFTIEGNNSFAQRFQGIYRCYASNKLG
CAML_MOUSE GESVVLPCNPPPSAA  PPR IYW MN  SKIFDIKQDERVSMG                         QNGDLYFANVLTSDNH   SDYICNAHFPGT
CAML_MOUSE GQSLILEC IAEGFP  TPT IKWLHP  SDPMPTDRVIYQN                           HNKTLQLLNVGEEDD    GEYTCLAENSLG
CAML_MOUSE GETARLDC QVQGRP  QPE ITWRIN  GMSMETVNKDQKYRI                         EQGSLILSNVQPTDT    MVTQCEARNQHG
CAML_MOUSE GSTAYLLC KAFGAP  VPS VQWLDEE GTTVLQDERFFPY                           ANGTLSIRDLQANDT    GRYFCQAANDQN
CAML_MOUSE GARVTFTC QASFDPSLQAS ITWRGD  GRDLQERGDSDKYFI                         EDGKLVIQSLDYSDQ    GNYSCVASTELD
NCA2_HUMAN GESKFFLC QVAGDA  KDKDISWFSPN GEKLTPNQQRISVVWNDD                      SSSTLTIYNANIDDA    GIYKCVVTGEDG
NCA2_HUMAN GEDAVIVC DVVSSL  PPT IIWKHK  GRDVILKKDVRFIVL                         SNNYLQIRGIKKTDE    GTYRCEGRILAR
NCA2_HUMAN GQSVTLVC DAEGFP  EPT MSWTKD  GEQIEQEEDDEKYIFSD                       DSSQLTIKKVDKNDE    AEYICIAENKAG
NCA2_HUMAN EEQVTLTC EASGDP  IPS ITWRTS  TRNISSEEKTLDGHMVVRSHA                   RVSSLTLKSIQYTDA    GEYICTASNTIG
NCA2_HUMAN GNQVNITC EVFAYP  SAT ISWFRD  GQLLPSSNYSNIKIYNTP                      SASYLEVTPDSENDF    GNYNCTAVNRIG
ZC262.3    EKPTAISC FSYGVP  SPK ISWWRFRPAEKLGSYDPITDEISYTNVSETMKESYEIQSGGSLLIRSPNRSHV    ERYVCVVENEYG
```

Figure 2: Alignment of Ig-like domains from three neural cell adhesion molecules with the Ig-like domain detected in ZC262.3. NRG_DROME is *Drosophila* neuroglian; CAML_DROME is mouse N-CAM L1; NCA2_HUMAN is a human N-CAM. The alignment was produced automatically by an HMM trained to recognize N-CAM sequences.

Two of the genes contain introns in their anticodon loop. These genes were detected despite the fact that this particular model was trained entirely on intronless sequences, which is a nice example of the model's flexibility. Three of these tRNA genes turn out to be in introns of predicted protein-coding genes.

Discussion

The complete cloning and sequencing of *C. elegans* will change how its molecular biology is studied. Much time and effort is still devoted to the cloning of genes, but soon *everything* will be cloned and sequenced from *C. elegans*, and molecular cloning will often be as simple as an electronic database retrieval of the sequence and a letter to Cambridge or St. Louis for the clone. The burden of work can shift to the daunting task of identifying the functions of the estimated 17,000 genes.

Clearly, some hundreds of "interesting" genes will be attacked immediately by conventional molecular biology. But it is neither wise nor possible to devote this kind of effort to every one of the genes. A challenge will be to get at the functions of the rest by scaling up rapid screening techniques, particularly ones that can be done in parallel and/or partially automated. To ask if and where a predicted gene is expressed, one might use *in situ* RNA hybridization to PCR-produced probes, or transgenic animals carrying reporter gene fusions to the predicted promoter (Hope, 1991), or hybridization to tissue-specific or even cell-specific cDNA populations. Gene knockout by PCR-based selection of transposon insertions and systematic determination of null phenotypes is feasible though a bit laborious (Plasterk, 1992). Various sorts of molecular genetic screens could be scaled up, such as enhancer trapping (Bellen et al., 1989; Bier et al., 1989; Wilson et al., 1989) and insertional mutagenesis, since one will be able to immediately identify a candidate gene from a bit of flanking sequence around an insert.

A more computer-reliant style of molecular biology will continue to develop. Computational analysis will become an initial exploratory step in many projects, rather than just the familiar final step of "what is my sequence related to?" For instance, computational probing techniques for particular sequence motifs and sequence families are far more sensitive than molecular hybridization and degenerate PCR cloning approaches, and will rapidly supercede them where genomic sequence is available.

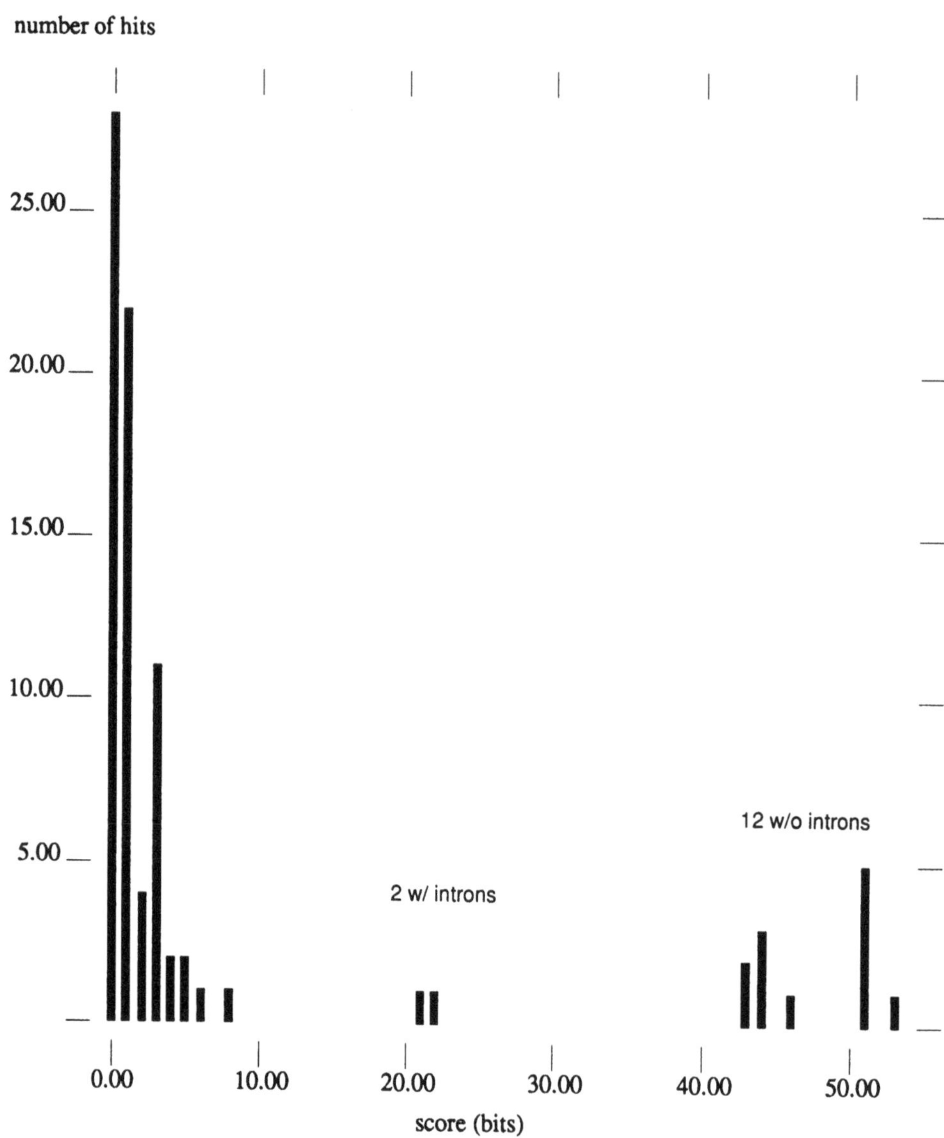

Figure 3: tRNA's detected in the genomic sequence. The tail of non-tRNA scores is shown as well to illustrate the good separation of signals from noise. The scores are in bits, which are log base 2 probabilities.

Genomic computational biology itself is still an immature and somewhat awkward science. At present, the data are limited and we remain confined to something much like the compilation of a dictionary. But however fascinating the etymology of our "words" may be, we should not be contented with merely cataloging genes and motifs into tidy Linnaean trees with some elaborate nomenclature. The future challenges for computational biology are to find ways to ask questions about the larger context of complex eukaryotic genomes. What genomic changes or new gene families correlate with the relatively explosive major evolutionary radiations, such as the diversification of multicellular eukaryotes about 1,200-1,000 million years ago, or the radiation of the animal phyla about 580-540 million years ago (Knoll, 1992)? What is occupying the spaces between genes – is it functional, or junk? Are there sequences determining higher order chromosome structure? What sequences control the regulation of individual genes and groups of coordinately regulated genes? How much sequence is created by autonomously replicating "selfish" elements? What are the mechanisms responsible for the evolution of genes and whole genomes? These sorts of questions can and will be tentatively explored in the *C. elegans* genome sequence, but they will be attacked most powerfully by comparative analysis of multiple divergent genome sequences.

Acknowledgements

Many thanks to the members of the nematode genome project and the St. Louis and Cambridge genome informatics groups, particularly John Sulston, Alan Coulson, Richard Durbin, Erik Sonnhammer, and Graeme Mitchison on the Cambridge side, for their continuing help and advice. I am supported by a Human Frontier Science Program long-term postdoctoral fellowship.

REFERENCES

Altschul, S. F., Gish, W., Miller, W., Myers, E. W., and Lipman, D. J., 1990, Basic local alignment search tool, *J. Mol. Biol.* 215:403.

Bairoch, A. and Boeckmann, B., 1993, The SWISS-PROT protein sequence data bank, recent developments, *Nucl. Acids Res.* 21: 3093.

Barker, W. C., George, D. G., Mewes, H.-W., Pfeiffer, F., and Tsugita, A, 1993, The PIR-international databases, *Nucl. Acids Res.* 21:3089.

Barton, G. J., 1990, Protein multiple sequence alignment and flexible pattern matching, *Meth. Enzymol.* 183:403.

Bellen, H. J., O'Kane, C. J., Wilson, C., Grossniklaus, U., Pearson, R. K., and Gehring, W. J., 1989, P-element-mediated enhancer detection: a versatile method to study development in *Drosophila*, *Genes Dev.* 3:1288.

Benson, D., Lipman, D. J., and Ostell, J., 1993, GenBank, *Nucl. Acids Res.* 21:2963.

Bier, E., Vaessin, H., Shepherd, S., Lee, K., McCall, K., Barbel, S., Ackerman, L., Carretto, R., Uemura, T., Grell, E., Jan, L. Y., and Jan, Y. N., 1989, Searching for pattern and mutation in the *Drosophila* genome with a P-lacZ vector, *Genes Dev.* 3:1273.

Bixby, J. L. and Harris, W. A., 1991, Molecular mechanisms of axon growth and guidance, *Ann. Rev. Cell Biol.* 7:117.

Bork, P., Ouzounis, C., Sander, C., Scharf, M., Schneider, R., and Sonnhammer, E., 1992, Comprehensive sequence analysis of the 182 predicted open reading frames of yeast chromosome III, *Protein Science* 1:1677.

Brenner, S., 1974, The genetics of *Caenorhabditis elegans*, *Genetics* 77:71.

Burglin, T., Ruvkun, G., Coulson, A., Hawkins, N., McGhee, J., Schaller, D., Wittmann, C., Muller, F., and Waterston, R., 1991, Nematode homeobox cluster, *Nature* 351:703.

Chalfie, M., 1993, Homeobox genes in *Caenorhabditis elegans*, *Curr. Opin. Genet. Dev.* 3:275.

Collins, J., Forbes, E., and Anderson, P., 1989, The Tc3 family of transposable genetic elements in *Caenorhabditis elegans*, *Genetics* 121:47.

Coulson, A., Kozono, Y., Lutterbach, B., Shownkeen, R., Sulston, J., and Waterston, R., 1991, YACs and the *C. elegans* genome, *BioEssays* 13:413.

Coulson, A., Sulston, J., Brenner, S., and Karn, J., 1986, Toward a physical map of the genome of the nematode *Caenorhabditis elegans*, *Proc. Natl. Acad. Sci. USA* 83:7821.

Coulson, A., Waterston, R., Kiff, J., Sulston, J., and Kohara, Y., 1988, Genome linking with yeast artificial chromosomes, *Nature* 335:184.

Daniels, D. L., Plunkett, G., Burland, V., and Blattner, F. R., 1992, Analysis of the *Escherichia coli* genome: DNA sequence of the region from 84.5 to 86.5 minutes, *Science* 257:771.

Dear, S. and Staden, R., 1991, A sequence assembly and editing program for efficient management of large projects, *Nucl. Acids Res.* 19:3907.

Dreyfus, D. H. and Emmons, S. W., 1991, A transposon-related palindromic repetitive sequence from *C. elegans*, *Nucl. Acids Res.* 19:1871.

Eddy, S. R. and Durbin, R., 1993, Analysis of RNA sequence families using adaptive statistical models, unpublished manuscript.

Green, P., Lipman, D., Hillier, L., Waterston, R., States, D., and Claverie, J.-M., 1993, Ancient conserved regions in new gene sequences and the protein databases, *Science* 259:1711.

Gribskov, M., Luthy, R., and Eisenberg, D., 1990, Profile analysis, *Meth. Enzymol.* 183:146.

Hedgecock, E. M., Culotti, J. G., and Hall, D. H., 1990, The *unc-5*, *unc-6*, and *unc-40* genes guide circumferential migrations of pioneer axons and mesodermal cells on the epidermis in *C. elegans*, *Neuron* 4:61.

Hedgecock, E. M., Culotti, J. G., Hall, D. H., and Stern, B. D., 1987, Genetics of cell and axon migrations in *Caenorhabditis elegans*, *Development* 100:365.

Hope, I., 1991, 'Promoter trapping' in *Caenorhabditis elegans*, *Development* 113:399.

Hynes, R. O. and Lander, A. D., 1992, Contact and adhesive specificities in the associations, migrations, and targeting of cells and axons, *Cell* 68:303.

Kenyon, C. and Wang, B., 1991, A cluster of antennapedia-class homeobox genes in a nonsegmented animal, *Science* 253:516.

Knoll, A. H., 1992, The early evolution of eukaryotes: a geological perspective, *Science* 256:622.

Krogh, A., Brown, M., Mian, I., Sjolander, K., and Haussler, D., 1993, Hidden Markov models in computational biology: applications to protein modeling, unpublished manuscript.

Levitt, A. and Emmons, S. W., 1989, The Tc2 transposon in *Caenorhabditis elegans*, *Proc. Natl. Acad. Sci. USA* 86:3232.

McCombie, W. R., Adams, M. D., Kelly, J. M., Fitzgerald, M. G., Utterback, T. R., Khan, M., Dubnick, M., Kerlavage, A. R., Venter, J. C., and Fields, C., 1992, *Caenorhabditis elegans* expressed sequence tags identify gene families and potential disease gene homologues, *Nature Genet.* 1:124.

Oliver, S., van der Aart, Q., Agostoni-Carbone, M., Aigle, M., et al., 1992, The complete DNA sequence of yeast chromosome III, *Nature* 357:38.

Plasterk, R. H., 1992, Reverse genetics of *Caenorhabditis elegans*, *BioEssays* 14:629.

Plunkett, G., Burland, V., Daniels, D. L., and Blattner, F. R., 1993, Analysis of the *Escherichia coli* genome. III. DNA sequence of the region from 87.2 to 89.2 minutes, *Nucl. Acids Res.* 21:3391.

Rice, C. M., Fuchs, R., Higgins, D. G., Stoehr, P. J., and Cameron, G. N., 1993, The EMBL data library, *Nucl. Acids Res.* 21:2967.

Robertson, H. M., 1993, The mariner transposable element is widespread in insects, *Nature* 362:241.

Sakakibara, Y., Brown, M., Hughey, R., Mian, I. S., Sjölander, K., Underwood, R. C., and Haussler, D., 1993, The application of stochastic context-free grammars to folding, aligning and modeling homologous RNA sequences, unpublished manuscript.

Searls, D. B., 1992, The linguistics of DNA, *American Scientist* 80:579.

Singer, M. and Berg, P., 1991, *Genes and Genomes*, University Science Books, Mill Valley, CA.

Smith, T. and Waterman, M., 1981, Identification of common molecular subsequences, *J. Mol. Biol.* 147:195.

Sonnhammer, E. L. and Durbin, R., 1993, A workbench for large-scale sequence homology analysis, unpublished manuscript.

Sulston, J., Du, Z., Thomas, K., Wilson, R., Hillier, L., Staden, R., Halloran, N., Green, P., Thierry-Mieg, J., Qiu, L., Dear, S., Coulson, A., Craxton, M., Durbin, R., Berks, M., Metzstein, M., Hawkins, T., Ainscough, R., and Waterston, R., 1992, The *C. elegans* genome sequencing project: a beginning, *Nature* 356:37.

Sulston, J., Schierenberg, E., White, J., and Thomson, J., 1983, The embryonic cell lineage of the nematode *Caenorhabditis elegans*, *Devel. Biol.* 100:64.

Wadsworth, W. G. and Hedgecock, E. M., 1992, Guidance of neuroblast migrations and axonal projections in *Caenorhabditis elegans*, *Curr. Opinion Neuro.* 2:36.

Wang, B. B., Muller-Immergluck, M. M., Austin, J., Robinson, N. T., Chisholm, A., and Kenyon, C., 1993, A homeotic gene cluster patterns the anteroposterior body axis of *C. elegans*, *Cell* 74:29.

Waterston, R., Martin, C., Craxton, M., Hunyh, C., Coulson, A., Hillier, L., Durbin, R., Green, P., Shownkeen, R., Halloran, N., Metzstein, M., Hawkins, T., Wilson, R., Berks, M., Du, Z., Thomas, K., Thierry-Mieg, J., and Sulston, J., 1992, A survey of expressed genes in *Caenorhabditis elegans*, *Nature Genet.* 1:114.

Watson, A., Smaldon, N., Lucke, R., and Hawkins, T., 1993, The *Caenorhabditis elegans* genome sequencing project: first steps in automation, *Nature* 362:569.

White, J., Southgate, E., Thomson, J., and Brenner, S., 1986, The structure of the nervous system of the nematode *Caenorhabditis elegans*, *Phil. Trans. R. Soc. Lond.* 314:1.

Wilson, C., Pearson, R. K., Bellen, H. J., O'Kane, C. J., Grossniklaus, U., and Gehring, W. J., 1989, P-element-mediated enhancer detection: an efficient method for isolating and characterizing developmentally regulated genes in *Drosophila*, *Genes Dev.* 3:1301.

Wilson, R., Ainscough, R., Anderson, K., Baynes, C., Berks, M., et al., 1993, The *C. elegans* genome project: nucleotide sequence of over two megabases from chromosome III, unpublished manuscript.

Wood, W. B., ed., 1988, *The Nematode* Caenorhabditis elegans, Cold Spring Harbor Laboratory, New York, NY.

Yuan, J., Finney, M., Tsung, N., and Horvitz, H. R., 1991, Tc4, a *Caenorhabditis elegans* transposable element with an unusual foldback structure, *Proc. Natl. Acad. Sci. USA* 88:3334.

APPENDIX

A. Putative homologues

The name of the predicted protein, its length, the length of the maximal scoring Smith/Waterman alignment to the database sequence, percentage residue identity over that alignment, and the name and description of the most similar database sequence in SWISSPROT and PIR are indicated.

Protein	length	align	id	hits	description
B0303.12	195	135	40.1%	RL11_ECOLI	50S RIBOSOMAL PROTEIN L11.
B0303.4	315	243	23.9%	PNMT_HUMAN	PHENYLETHANOLAMINE-N-METHYLTRANSFERASE ...
B0303.5	497	430	28.1%	THIK_YEAST	3-KETOACYL-COA THIOLASE PEROXISOMAL PRE...

B0303.8	576	365	18.1%	SLP1_YEAST	SLP1 PROTEIN (VACUOLAR PROTEIN SORTING ...
B0303.TC3	329	263	32.8%	YT31_CAEEL	HYPOTHETICAL 31.8 KD PROTEIN FROM TRANS...
B0464.1	531	501	62.4%	SYD2_HUMAN	ASPARTYL-TRNA SYNTHETASE ALPHA-2 SUBUNI...
B0464.5	1087	698	18.4%	S30783	PROTEIN KINASE HOMOLOG KNS1 - YEAST (SA...
B0523.1	363	218	32.5%	KROS_AVISU	ROS TYROSINE KINASE TRANSFORMING PROTEI...
B0523.5	848	748	27.7%	GELS_PIG	GELSOLIN PRECURSOR, PLASMA (ACTIN-DEPOL...
C02C2.3	458	350	20.3%	ACHG_RAT	ACETYLCHOLINE RECEPTOR PROTEIN, GAMMA C...
C02C2.4	568	472	24.4%	S27951	SODIUM/PHOSPHATE TRANSPORT PROTEIN, REN...
C02D5.1	332	321	31.2%	ACDL_RAT	ACYL-COA DEHYDROGENASE PRECURSOR, LONG-...
C02F5.3	573	364	56.7%	JC1349	DEVELOPMENTALLY REGULATED GTP-BINDING P...
C06E1.4	983	913	36.3%	GLRK_LYMST	GLUTAMATE RECEPTOR PRECURSOR.
C08C3.1	218	74	100.0%	HM11_CAEEL	HOMEOBOX PROTEIN CEH-11 (FRAGMENT).(egl-5)
C08C3.3	353	210	100.0%	HMMA_CAEEL	HOMEOBOX PROTEIN MAB-5 (FRAGMENT). (mab-5)
C14B9.1	110	92	40.2%	CRAB_MESAU	ALPHA CRYSTALLIN B CHAIN (ALPHA(B)-CRYS...
C14B9.2	664	601	48.1%	ER72_MOUSE	PROTEIN DISULFIDE ISOMERASE-RELATED PRO...
C14B9.7	161	160	62.5%	R5RT21	RIBOSOMAL PROTEIN L21 - RAT
C14B9.8	1257	1189	38.2%	KPBA_RABIT	PHOSPHORYLASE KINASE ALPHA CHAIN, SKELE...
C29E4.1	305	266	43.5%	CAC8_CAEEL	CUTICLE COLLAGEN 8.
C29E4.7	250	214	25.0%	S16268	AUXIN-INDUCED PROTEIN (CLONE PGNT35) - ...
C29E4.8	248	233	61.4%	KAD2_RAT	ADENYLATE KINASE ISOENZYME 2, MITOCHOND...
C30A5.3	378	260	30.2%	S30854	PHOSPHOPROTEIN PHOSPHATASE - YEAST (SAC...
C30A5.6	429	428	91.2%	UN86_CAEEL	TRANSCRIPTION FACTOR UNC-86.
C30A5.7	467	466	100.0%	UN86_CAEEL	TRANSCRIPTION FACTOR UNC-86.
C30C11.2	504	497	42.7%	DXA2_MOUSE	PROBABLE DIPHENOL OXIDASE A2 COMPONENT ...
C30C11.4	776	688	36.6%	S30788	HEAT SHOCK PROTEIN HOMOLOG MSI3 - YEAST...
C38C10.1	374	293	36.6%	NK1R_CAVPO	SUBSTANCE-P RECEPTOR (SPR) (NK-1 RECEPT...
C38C10.2	472	448	27.4%	S27951	SODIUM/PHOSPHATE TRANSPORT PROTEIN, REN...
C40H1.4	291	228	26.3%	YCS4_YEAST	HYPOTHETICAL 40.0 KD PROTEIN IN CRY1-RB...
C50C3.11	734	629	25.5%	CIC2_RABIT	DIHYDROPRYRIDINE-SENSITIVE L-TYPE... (unc-36)
F02A9.5	608	484	31.5%	PCCB_RAT	PROPIONYL-COA CARBOXYLASE BETA CHAIN PR...
F09G8.6	278	256	38.3%	A44984	COLLAGEN - NEMATODE (HAEMONCHUS CONTORTUS)
F22B7.7	335	312	22.2%	KCH_ECOLI	PUTATIVE POTASSIUM CHANNEL PROTEIN.
F44B9.8	447	346	38.3%	A45253	ACTIVATOR 1 37 KDA SUBUNIT, A1 37 KDA S...
F44B9.9	220	188	26.3%	S24264	PROTEIN PHOSPHATASE 1A - ARABIDOPSIS TH...
F44E2.1	1746	942	26.0%	POL2_DROME	RETROVIRUS-RELATED POL POLYPROTEIN (PRO...
F44E2.8	308	171	22.8%	YBIA_ECOLI	HYPOTHETICAL 18.7 KD PROTEIN IN RHLE-DI...
F54C8.1	298	286	45.9%	HCDH_PIG	3-HYDROXYACYL-COA DEHYDROGENASE (EC 1.1...
F54C8.5	207	189	35.4%	RAS_LENED	RAS-LIKE PROTEIN.
F54F2.1	1226	1194	27.3%	ITAV_HUMAN	VITRONECTIN RECEPTOR ALPHA SUBUNIT PREC...
F55H2.1	184	132	54.5%	SODC_BOVIN	SUPEROXIDE DISMUTASE (CU-ZN) (EC 1.15.1...
F55H2.2	257	251	51.0%	S30826	HYPOTHETICAL PROTEIN 11 - YEAST (SACCHA...
F55H2.5	266	243	33.7%	C561_BOVIN	CYTOCHROME B561.
F58A4.10	164	153	50.3%	UBC7_YEAST	UBIQUITIN-CONJUGATING ENZYME E2-18 KD (...
F58A4.4	410	403	39.3%	PRI1_MOUSE	DNA PRIMASE 49 KD SUBUNIT (EC 2.7.7.-) ...
F58A4.7	292	228	32.0%	A36394	TRANSCRIPTION FACTOR AP-4 - HUMAN (FRAG...
F58A4.8	444	436	43.8%	TBG_XENLA	TUBULIN GAMMA CHAIN. (tbg-1)
F58A4.9	144	129	27.9%	RPC9_YEAST	DNA-DIRECTED RNA POLYMERASES I AND III ...
F59B2.3	418	351	31.8%	NAGA_ECOLI	N-ACETYLGLUCOSAMINE-6-PHOSPHATE DEACETY...
F59B2.7	205	203	77.1%	RAB6_HUMAN	RAS-RELATED PROTEIN RAB-6.
GLP1A.cds	1295	1294	100.0%	GLP1_CAEEL	GLP-1 PROTEIN PRECURSOR.
K04H4.1	1744	1743	92.6%	CA14_CAEEL	COLLAGEN ALPHA 1(IV) CHAIN. (emb-9)
K06H7.4	377	371	40.5%	S24168	HYPOTHETICAL PROTEIN - HUMAN
K06H7.8	283	236	24.2%	HR25_YEAST	CASEIN KINASE I HOMOLOG HRR25 (EC 2.7.1...
K11H3.3	374	320	24.5%	A45763	UNCOUPLING PROTEIN, MITOCHONDRIAL - HUMAN
K12H4.4	180	177	48.6%	SPC2_CANFA	MICROSOMAL SIGNAL PEPTIDASE 23 KD SUBUN...
LIN12A.cd	1429	1428	100.0%	LI12_CAEEL	LIN-12 PROTEIN PRECURSOR.
R05D3.7	843	816	46.2%	KINH_LOLPE	KINESIN HEAVY CHAIN.
R08D7.5	173	163	31.7%	CATR_CHLRE	CALTRACTIN (20 KD CALCIUM-BINDING PROTE...
R08D7.6	841	510	36.8%	CNAG_BOVIN	CGMP-DEPENDENT 3',5'-CYCLIC PHOSPHODIES...
R107.2	285	281	25.4%	YJEF_ECOLI	HYPOTHETICAL PROTEIN IN AMIB 5'REGION (...
R107.7	208	207	100.0%	GTP_CAEEL	GLUTATHIONE S-TRANSFERASE P (EC 2.5.1.18).

R10E11.2	302	152	72.4%	VATL_MANSE	VACUOLAR ATP SYNTHASE 16 KD PROTEOLIPID...
R10E11.4	289	212	30.1%	A24148	N-ACETYLLACTOSAMINE SYNTHASE - BOVINE (...
T05G5.3	332	331	100.0%	S26572	P34 CDC2-LIKE PROTEIN (ncc-1)
T05G5.5	196	162	30.9%	S27735	HYPOTHETICAL PROTEIN A - THERMUS AQUATICUS
T05G5.6	288	273	60.2%	ECHM_RAT	ENOYL-COA HYDRATASE, MITOCHONDRIAL PREC...
T16H12.7	193	175	25.0%	MIPP_MOUSE	MIPP PROTEIN (MURINE IAP-PROMOTED PLACE...
T23G5.1	788	780	73.4%	A24050	RIBONUCLEOSIDE-DIPHOSPHATE REDUCTASE CH...
T23G5.5	514	494	44.7%	NTNO_HUMAN	SODIUM-DEPENDENT NORADRENALINE TRANSPOR...
T26G10.1	489	430	33.8%	DEAD_ECOLI	PUTATIVE ATP-DEPENDENT RNA HELICASE DEAD.
T26G10.3	91	89	24.0%	RS24_HUMAN	P16632 40S RIBOSOMAL PROTEIN S24 (S19).
ZC21.2	823	751	36.4%	JH0588	CALMODULIN-BINDING PROTEIN TRPL - FRUIT...
ZC84.2	772	562	51.4%	CGCC_HUMAN	CGMP-GATED CATION CHANNEL PROTEIN (CYCL...
ZC84.4	425	291	19.7%	SSRB_MOUSE	SOMATOSTATIN RECEPTOR TYPE 2B.
ZK1098.4	305	304	35.3%	GCN3_YEAST	TRANSCRIPTION ACTIVATOR GCN3.
ZK1236.1	581	578	38.2%	LEPA_ECOLI	LEPA PROTEIN.
ZK1236.3	1364	715	22.3%	B34751	HYPOTHETICAL PROTEIN - AFRICAN MALARIA ...
ZK353.6	491	366	30.5%	AMPA_RICPR	ASPARTATE AMINOPEPTIDASE (EC 3.4.11.7) ...
ZK370.5	401	346	30.1%	BCKD_RAT	[3-METHYL-2-OXOBUTANOATE DEHYDROGENASE ...
ZK507.6	409	231	36.3%	CG2A_PATVU	G2/MITOTIC-SPECIFIC CYCLIN A.
ZK512.2	578	499	22.4%	MS16_YEAST	ATP-DEPENDENT RNA HELICASE MSS116.
ZK512.4	76	75	50.7%	A34731	SIGNAL RECOGNITION PARTICLE 9K CHAIN - DOG
ZK512.6	466	399	25.4%	S27951	SODIUM/PHOSPHATE TRANSPORT PROTEIN, REN...
ZK632.1	810	658	30.3%	CD46_YEAST	CELL DIVISION CONTROL PROTEIN CDC46 (MI...
ZK632.4	411	391	33.3%	MANA_EMENI	MANNOSE-6-PHOSPHATE ISOMERASE (EC 5.3.1...
ZK632.6	619	612	40.4%	A37273	CALNEXIN PRECURSOR - DOG
ZK632.8	178	177	48.6%	S28875	GTP-BINDING PROTEIN - ARABIDOPSIS THALIANA
ZK637.1	456	391	20.7%	A43267	SYNAPTIC VESICLE PROTEIN 2, SV2=82.7 KD...
ZK637.10	499	465	33.8%	GSHR_ECOLI	GLUTATHIONE REDUCTASE (EC 1.6.4.2) (GR).
ZK637.13	159	158	44.7%	GLBH_TRICO	GLOBIN-LIKE HOST-PROTECTIVE ANTIGEN PRE...
ZK637.7	610	609	99.0%	LIN9_CAEEL	LIN-9 PROTEIN.
ZK637.8	935	908	55.1%	VPP1_RAT	CLATHRIN-COATED VESICLE/SYNAPTIC VESICL...
ZK643.3	482	418	25.3%	CALR_PIG	CALCITONIN RECEPTOR PRECURSOR (CT-R).
ZK652.10	420	317	25.5%	S27951	SODIUM/PHOSPHATE TRANSPORT PROTEIN, REN...
ZK652.4	123	122	68.0%	R5RT35	RIBOSOMAL PROTEIN L35 - RAT
ZK652.5	305	291	26.6%	HMES_DROME	EMPTY SPIRACLES HOMEOTIC PROTEIN. (ceh-23)
ZK652.6	580	390	23.4%	S29962	REF(2)PERECTA PROTEIN - FRUIT FLY (DROS...
ZK686.2	696	649	23.8%	S31248	PROBABLE RNA HELICASE, ATP-DEPENDENT - ...

B. Other significant similarities

Protein	length	align	id	hits	description
B0303.3	424	182	31.9%	S25770	P33 RSP-1 PROTEIN - MOUSE
B0303.7	359	156	27.1%	NCF2_HUMAN	NEUTROPHIL NADPH OXIDASE FACTOR (P67-PH...
C02C2.1	484	229	25.1%	A40957	MONOPHENOL MONOOXYGENASE PRECURSOR - HU...
C02F5.7	489	362	22.1%	GRR1_YEAST	GRR1 PROTEIN.
C02F5.9	564	227	41.0%	PRC5_HUMAN	PROTEASOME COMPONENT C5 (EC 3.4.99.46) ...
C05B5.5	585	259	25.2%	TENA_HUMAN	P24821 TENASCIN PRECURSOR (TN)...
C06E1.10	1152	407	39.5%	S22609	HYPOTHETICAL PROTEIN - FRUIT FLY (DROSO...
C08C3.2	274	128	26.9%	VA55_VACCC	PROTEIN A55.
C13G5.1	240	70	60.0%	HME6_APIME	HOMEOBOX PROTEIN E60 (FRAGMENT). (ceh-16)
C15H7.2	266	233	24.5%	KFPS_DROME	DFPS TYROSINE KINASE (EC 2.7.1.112).
C30A5.4	102	83	30.6%	SYB_DROME	SYNAPTOBREVIN.
C38C10.5	1112	279	22.8%	RGR1_YEAST	GLUCOSE REPRESSION REGULATORY PROTEIN R...
C40H1.1	372	307	28.3%	S24577	OVARIAN PROTEIN - FRUIT FLY (DROSOPHILA...
C50C3.2	1009	950	20.6%	SPCA_DROME	SPECTRIN ALPHA CHAIN.
C50C3.5	178	82	34.1%	CALM_ACHKL	CALMODULIN.
C50C3.7	398	241	23.0%	IT5P_HUMAN	75 KD INOSITOL-1,4,5-TRISPHOSPHATE 5-PH...
F22B7.5	943	344	37.9%	DNAJ_BACSU	DNAJ PROTEIN.
F42H10.3	209	88	36.0%	SRC8_CHICK	SRC SUBSTRATE P80/85 PROTEINS (CORTACTIN).
F42H10.4	221	59	39.0%	TSF3_HELAN	POLLEN SPECIFIC PROTEIN SF3.

F44B9.1	761	272	26.4%	ACPH_RAT	ACYLAMINO-ACID-RELEASING ENZYME (EC 3.4...
F44E2.3	244	205	25.5%	RU17_XENLA	U1 SMALL NUCLEAR RIBONUCLEOPROTEIN 70 KD.
F44E2.4	1609	209	22.8%	S03430	LOW DENSITY LIPOPROTEIN RECEPTOR PRECUR...
F44E2.6	152	105	41.0%	PILB_NEIGO	PILB PROTEIN.
F54C8.2	261	123	46.3%	JQ1343	HISTONE H3.3 - FRUIT FLY (DROSOPHILA ME...
F54C8.4	359	171	36.3%	A40781	ORF1 PROTEIN - AUTOGRAPHA CALIFORNICA N...
F54G8.2	827	231	51.0%	KDGL_DROME	PUTATIVE DIACYLGLYCEROL KINASE (EC 2.7....
F54G8.3	1139	144	32.9%	A41543	INTEGRIN ALPHA-6B CHAIN - HUMAN (FRAGMENT)
F54G8.4	932	354	19.9%	KRET_HUMAN	RET PROTO-ONCOGENE TYROSINE KINASE (EC ...
F54G8.5	413	399	19.1%	PATC_DROME	MEMBRANE PROTEIN PATCHED.
F58A4.1	258	170	30.1%	A43932	MUCIN - HUMAN (FRAGMENT)
F58A4.3	288	119	50.4%	H3_PEA	HISTONE H3.
F58A4.5	1222	594	27.4%	B34751	HYPOTHETICAL PROTEIN - AFRICAN MALARIA ...
F59B2.11	337	218	14.9%	AAC2_DICDI	AAC-RICH MRNA CLONE AAC11 PROTEIN (FRAG...
K02D10.1	786	212	28.0%	SN25_MOUSE	SYNAPTOSOMAL ASSOCIATED PROTEIN 25.
K06H7.1	547	278	55.2%	S22127	PROTEIN KINASE - FRUIT FLY (DROSOPHILA ...
K06H7.3	831	245	36.5%	IPPI_YEAST	ISOPENTENYL-DIPHOSPHATE DELTA-ISOMERASE...
K12H4.1	586	538	30.8%	PRO_DROME	PROTEIN PROSPERO.
K12H4.8	1822	323	23.3%	A31922	ATP-DEPENDENT RNA HELICASE HOMOLOG - FR...
M01A8.4	69	47	48.9%	BIK1_YEAST	NUCLEAR FUSION PROTEIN BIK1.
R05D3.1	2434	1139	41.5%	TOP2_SCHPO	DNA TOPOISOMERASE II (EC 5.99.1.3).
R10E11.1	2015	257	24.7%	FSH_DROME	FEMALE STERILE HOMEOTIC PROTEIN (FRAGIL...
R10E11.3	408	295	19.8%	S22158	TRANSFORMING PROTEIN (CLONE 213) - HUMAN
T02C1.1	160	88	29.5%	RA18_YEAST	DNA REPAIR PROTEIN RAD18.
T05G5.1	346	228	16.7%	S27770	HYPOTHETICAL PROTEIN 1 - AFRICAN MALARI...
T23G5.2	470	213	31.3%	SC14_YEAST	SEC14 CYTOSOLIC FACTOR.
ZC21.4	733	199	30.2%	S29956	BETA-CHIMAERIN - RAT
ZC262.6	466	158	51.5%	S24603	KINESIN HEAVY CHAIN - HUMAN
ZC84.1	2885	53	45.3%	ISHP_STOHE	KUNITZ-TYPE PROTEINASE INHIBITOR SHPI.
ZC97.2	296	278	23.7%	S31248	PROBABLE RNA HELICASE, ATP-DEPENDENT - ...
ZK112.7	3343	n.d.	n.d.	A41087	CADHERIN-RELATED TUMOR SUPPRESSOR PRECU...
ZK370.3	923	699	22.4%	TALI_MOUSE	TALIN.
ZK507.1	251	166	32.8%	HR25_YEAST	CASEIN KINASE I HOMOLOG HRR25 (EC 2.7.1...
ZK632.3	510	175	35.2%	S26727	HYPOTHETICAL PROTEIN 186 (RPOA2 3' REGI...
ZK637.11	316	183	33.5%	TWIN_DROME	CDC25-LIKE PROTEIN PHOSPHATASE TWINE (E...
ZK637.5	342	293	27.4%	A25937	ARSA PROTEIN - ESCHERICHIA COLI R-FACTO...
ZK652.9	568	255	38.8%	YIGO_ECOLI	HYPOTHETICAL 28.1 KD PROTEIN IN UDP-RFA...
ZK757.2	292	109	30.3%	TPCL_HUMAN	PROTEIN-TYROSINE PHOSPHATASE CL100 (EC ...

GENETIC ANALYSIS IN *CAENORHABDITIS ELEGANS*

Ann M. Rose,[1] Mark L. Edgley,[1] and David L. Baillie[2]

[1]Department of Medical Genetics
University of British Columbia
Vancouver, B. C. V6T 1W5 Canada

[2]Institute of Molecular Biology and Biochemistry
Simon Fraser University
Burnaby, B. C. V5A 1S6 Canada

INTRODUCTION

In order to contribute to the understanding of the organization and function of genes in the genome of *Caenorhabditis elegans*, we have undertaken a genetic approach. This type of approach relies upon the availability of mutant strains. Although there are many specific applications, in general genetic deduction depends upon the removal of a single component, and subsequent inference from phenotypic alterations as to the function of that component. The biology of *C. elegans* makes it very amenable to genetic manipulation (Brenner, 1974; reviewed in Wood, 1988). This nematode is a self-fertilizing hermaphrodite. The genetic material is organized into five autosomes and a sex chromosome. Males, which are XO, arise at a frequency of approximately one in 1,000 individuals by X-chromosome nondisjunction (Hodgkin et al., 1979). The complete cell lineage is known (Sulston and Horvitz, 1977; Sulston et al., 1983), and the structure of the 302-cell nervous system has been determined by serial section electron microscopy (White et al., 1986). Mutant phenotypes can be readily generated and easily maintained both in laboratory cultures and in frozen suspension. The *C. elegans* genetic map compiled by Edgley and Riddle (1993) includes 982 genes, 225 deficiencies, and 74 chromosomal duplications. Virtually the entire genome has been cloned in cosmid and yeast artificial chromosome vectors, and these clones have been ordered with respect to one another on a physical map of the genome (Coulson et al., 1988). Most especially the DNA of the genome is being sequenced by a joint U.S./U.K. effort headed by Drs. R. H. Waterston at Washington University in St. Louis, Missouri, and John Sulston at the MRC Laboratory of Molecular Biology in Cambridge, England (Sulston et al., 1992). Within the next decade, the sequence of every gene (within a genome of 10^8 base pairs) will be available to investigators by means of electronic databases. An integrated database program, ACEDB (for "A C. Elegans Data Base"), now links all the collected genetic and molecular information on *C. elegans* in a hypertext model (J. Thierry-Mieg and R. Durbin, pers. comm.). The current genetic map is compiled by J. Hodgkin at the MRC Laboratory of Molecular Biology in Cambridge, England and is circulated to the research community using the ACEDB database. Genetic strains are maintained by and available from the Caenorhabditis Genetics Center (CGC) run by R. K. Herman and T. Stiernagle, at the University of Minnesota, St. Paul, Minnesota.

LETHAL MUTANTS

Mutant phenotypes provide an approach to understanding what genes do, and how their functions relate to one another, since even in cases where sequence implies a function, the precise role in developmental processes is not always clear. In *C. elegans* there is a diversity of mutant phenotypes including morphological mutants that alter body shape, movement defective mutants, and mutations in genes affecting cell lineages, cell-cell interactions in development, axonal guidance, and neural connectivity (see Wood, 1988). A description of the more common mutant phenotypes in *C. elegans* is presented in Table 1. It can be seen that the largest phenotypic category are those that effect survival of the organism, not including the many genes in the other categories that also encode essential functions. There are a wide range of genes that are essential for development, vitality and fertility. We have broadly defined lethal mutants as any individuals having a heritable defect that does not allow it to grow, live or reproduce. Consequently genetic strains of lethals cannot be maintained in the homozygous state in the laboratory. The mutant genes include ones that encode essential cellular, tissue specific gene products, developmental pathway components, embryonic regulators, transcription factors, and fertility factors. In addition, many genes in *C. elegans*, which were originally identified by morphological, visible phenotypes, are now known to have lethal alleles, for example, *bli-4*, *him-1*, *rol-3*, *unc-60*, and others. Many other visible mutants, which although viable and fertile in the laboratory, would not survive in nature, such as paralyzed muscle mutants for example. The biological role of the gene products for a number of these genes has been deduced from genetic interactions, physiological, anatomical and biochemical studies, and from DNA sequencing. The literature describing these studies can be accessed using the ACEDB database or by Internet gopher connections.

Although biologically lethals are a very heterogeneous category, technically large numbers of lethal mutants can be worked with using similar genetic methodologies. Over the past years, we have developed a number of protocols for generating, maintaining and studying large numbers of lethal mutants. As stated above, homozygous genetic strains of lethals cannot be maintained, making it is necessary to keep them as heterozygous strains.

Table 1. The most abundant phenotypic classes.

Gene name	Phenotype	Normal gene function	Number of genes[1]
let-	lethality	essential function	375
unc-	uncoordinated	movement	120
egl-	egg laying defective	egg laying	50
lin-	lineage defective	cell division patterns	45
emb-	embryonic arrest	embryonic development	35
sup-	suppressor	interacting functions	34
dpy-	dumpy	body morphology	29
mel-	maternal effect	egg cytoplasm	26
daf-	dauer defective	dauer formation	24
mab-	male abnormal	cell division in male	21
spe-	sperm defect	male fertility	17
zyg-	zygotic arrest	zygotic development	14
che-	chemotaxis defect	sensation	14
him-	high incidence males	X chromosome disjunction	14
dyf-	dye filling	exclusion of molecules	12
osm-	osmotic	hypodermis	12
vab-	variable abnormal	morphology	10
ced-	cell death	cell maintenance	10

[1]From the 1993 genetic map

The approach we have taken is to genetically link them to visible (morphological) markers, for example, dumpy (Dpy). The advantage of this is that the lethal mutation can be followed using the absence of the visible marker. The lethal can be maintained indefinitely by selecting heterozygous individuals. These are easily identified if both homologues are marked with recessive markers. For example, individuals of the genotype, + *unc*/ *dpy* +, will look wild type but will segregate uncoordinated (Unc) and dumpy (Dpy) progeny as well as wild types. If a mutation is induced on the *dpy* chromosome, the heterozygotes will segregate Uncs, arrested Dpys (Let Dpy) and wild-type progeny. Since recombination will occur between the two homologous chromosomes, this genotype will rearrange. Thus, the lethal mutation may be lost. This undesirable event can be recognized by the appearance of viable fertile Dpys in the progeny. In order to prevent rearrangement of the genotype, genetic balancers are used. A genetic balancer refers to any situation which prevents or severely reduces recombination between homologues. In *C. elegans* balancers commonly are chromosomal rearrangements.

GENETIC BALANCERS

Three major classes of balancers have been used for generating, maintaining and mapping lethals. These are reciprocal translocations (Herman, 1978; Rosenbluth and Baillie, 1981), duplications (Meneely and Herman, 1979; Howell et al., 1987), and intrachromosomal rearrangements (Sigurdson et al., 1984; Zetka and Rose, 1992). Reciprocal translocations affect portions of two different chromosomes, for example the right half of chromosome III and the left half of chromosome V, in the case of *eT1* (Rosenbluth and Baillie, 1981). The regions affected are usually large, containing hundreds of genes. In individuals heterozygous for the translocation, recombination between the normal and translocated chromosome is absent, most likely due to an absence of pairing as has been discussed elsewhere (McKim et al., 1988). The absence of recombination makes these very effective balancers, and creates pseudo-linkage between the two affected regions, which can be useful for genetic manipulations. Many intrachromosomal balancers have not been well characterized, but those that have include specific deletions that suppress crossing-over (Rosenbluth et al., 1990) and an inversion (Zetka and Rose, 1992). These balancers are used in the heterozygous condition and usually do not prevent crossing-over, but greatly reduce it such that they can be used with care as effective balancers. Duplications of chromosomal regions are present in addition to the normal diploid genomic constitution. Most do not undergo genetic exchange (crossing over) with the normal chromosomes and can be useful for rescuing lethal mutations. Since the lethal is maintained in a homozygous condition (that is, both normal chromosomes carry the mutation), it is very easy to use these strains to make heterozygous males for crosses. However in certain cases, dosage effects of duplications (two mutant alleles with one wild type as compared to one mutant with one wild type in the heterozygous state) can be undesirable. We are currently involved in a collaborative project with D. L. Riddle, University of Missouri, Columbia USA to generate balancers for the entire genome. This "Genetic Tool Kit" project is generating and characterizing balancers of the type described above, and genetic deficiencies (deletions) to facilitate genetic analysis in *C. elegans*. Information about available balancers and how to use them is currently available electronically from edgley@genekit.medgen.ubc.ca.

ESSENTIAL GENES

In order to identify genes by classical mutational analysis, we use a mutagen such as ethylmethane sulfonate (EMS). EMS primarily causes single base pair alterations and thus is an appropriate choice for identifying mutations specific to a single gene. The frequency of recovering lethal mutants has been determined for different doses of EMS and doses in the range of 10mM to 25mM are recommended (Rosenbluth et al., 1983). Since mutations in many different genes give rise to lethal phenotypes, and since at the resolution of the dissecting microscope many lethal mutants look very similar, genetic complementation tests are necessary to sort the mutations according to their corresponding genes. Complementation tests are done using males carrying one lethal in crosses with hermaphrodites carrying a second lethal. In these crosses it is necessary to have the lethals genetically linked to a common marker. In the cross progeny there will be some individuals homozygous for a common marker, *dpy*, for example and heteroallelic (carrying both) lethal mutations. If viable fertile Dpys result from the cross, the two lethal mutations are interpreted as being in two separate genes. In this case one wild-type allele is present for each of the genes and provides the normal function for that gene. The mutations are said to complement (each other). If the Dpys are sterile, arrested in development, or die prematurely, the two lethal mutations are interpreted as being in the same gene. There is no wild type product produced by that gene from either chromosome. The two allelic mutations "fail to complement".

Using these procedures our laboratories have accumulated a large collection of mutations in essential genes for specific regions of the genome. Four regions of the genome have been intensively characterized with regard to lethal mutations, viz., the *unc-22* region of chromosome IV (Rogalski et al., 1982; Rogalski and Baillie, 1985; Clark et al., 1988), the left portion of chromosome V balanced by the translocation *eT1* (Rosenbluth et al., 1988; Johnsen and Baillie, 1991), the left portion of chromosome I balanced by *sDp2* (Rose and Baillie, 1980; Howell et al., 1987; Howell and Rose, 1990; McKim et al., 1992), and the middle to left portion of chromosome III balanced by *sDp3* (Stewart and Baillie, unpublished results).

CHROMOSOMAL REARRANGEMENTS

Chromosomal rearrangements are very useful for a range of genetic experiments. In order to generate rearrangements mutagens such as radiation (gamma, X-rays or ultraviolet light) or chemicals (e.g., formaldehyde) are used (Rosenbluth et al., 1985; Johnsen and Baillie, 1988; Stewart et al., 1991). These mutagens can cause breaks in the DNA, which may be repaired and isolated as lethal mutations. A frequent type of rearrangement recovered after treatment with these mutagens is a deletion (genetic deficiency of function). In order to test for the present of a DNA deletion, complementation tests are done using visible markers (Uncs for example) with genetic map position known to be in the region where the lethal maps. Exposure of the marker in the first generation (F1) of the cross (i.e. allelism) is an indication that a putative deletion of that gene may have been isolated. In order to confirm that the rearrangement being tested is a deletion, adjacent markers are tested. If the new mutation exposes two or more adjacent genetic markers, and in the absence of any conflicting information, it is interpreted to be a deletion. The analysis for duplications is similar except that the diagnostic category is suppression of a mutant phenotype (rescue by wild-type alleles carried on the duplication). We have used deficiencies (Clark et. al., 1988; Rosenbluth et. al., 1988; Johnsen and Baillie, 1991) and duplications (Howell et al., 1987; Howell and Rose, 1990; McKim and Rose, 1990; McKim et al., 1993) extensively for the mapping of essential genes.

Deficiencies are useful tools for fast and accurate mapping of mutations that convey either visible (non lethal morphological or behavioral phenotype) or lethal phenotypes. Deficiency breakpoints automatically left-right position genes with respect to each other. In addition, deficiencies provide the simplest and quickest method for determining if a given allele of an arbitrary locus is a hypermorph (less severe phenotype over a deficiency), hypomorph (more severe phenotype over a deficiency), or amorph (unchanged phenotype over a deficiency). This test has provided valuable molecular insights into the mechanism by which the gene in question brings about its phenotypic consequences. A special use of deficiencies has been to survey large areas of the genome for the existence of genes controlling specific features of development. Ahnn and Fire (1993) recently used deficiencies to look for genes that are required for the appropriate expression of two myosin isoforms. Deficiency homozygotes (arrested embryos) were stained with antibodies, and two deficiencies were found that failed to express these proteins, suggesting that genes required for proper muscle development are present in the region covered by these deficiencies. Lastly, combinations of deficiencies that overlap for small regions provide a means of assaying the information content of the overlap region. For example, if a heterozygote carrying two deficiencies is inviable but no known gene is deleted by both, then one can deduce the existence of at least one previously unidentified gene in the overlap region.

Large numbers of deficiencies have been generated in three regions, viz., the chromosome II cluster (Sigurdson et al., 1984), the *unc-22* region (Clark et al., 1988, Clark and Baillie, 1992) and the left half of chromosome V (Rosenbluth et al., 1988, Johnsen and Baillie, 1991). These deficiencies have provided breakpoints for the mapping of 165 genetic loci, restriction fragment length differences for correlating the genetic and physical map, and tools for developmental genetics. For example, deficiency breakpoints divide the two map unit interval in the *unc-22* region into 14 regions (zones). Clark and Baillie (1992) estimated that there are at least 50 essential genes in *sDf2*, of which 36 have been identified, resulting in an average of three to four genes per zone.

The study of chromosome rearrangements has provided insights into the structural organization and the meiotic functioning of chromosomes, such as the mechanisms of chromosome pairing, meiotic segregation and genetic recombination. The identification of localized homologue recognition regions (HRR) on the chromosomes of *C. elegans* came as a direct consequence of the characterization of rearrangements intended to be used to balance lethal mutations (Rose et al., 1984; McKim et al., 1988). Haplo-insufficient regions, one near *dpy-10* on chromosome II (Sigurdson et al., 1984) and one in zone 16 of chromosome V (Johnsen and Baillie 1991), have been identified. Although cytogenetic analysis of *C. elegans* chromosomes has demonstrated that the chromosomes behave holokinetically during mitosis (Albertson and Thomson, 1993), genetic analysis of the meiotic chromosomes has been consistent with the presence of only one segregator on each chromosome (Rosenbluth and Baillie, 1981). The analysis of a paracentric inversion of the right arm of chromosome I (Zetka and Rose, 1992) has shown that centromeric activity during meiosis is restricted to one site and the chromosomes are functionally monocentric. Like the majority of holocentric organisms such as *Parascaris* (Pimpinelli and Goday, 1989), *C. elegans* has localized centromere behaviour during meiosis.

CORRELATION OF MUTANT PHENOTYPE WITH DNA SEQUENCE

Essential genes, represented by lethal mutants, are initially positioned on the genetic map by recombination mapping. The frequency of crossing over between the lethal and linked visible marker is determined. A higher resolution map can be obtained using rearrangement breakpoints as described above. Ultimately one would like to have a one to one correspondence between mutant phenotype and DNA sequence. Towards this end a high resolution alignment of the genetic and physical maps is needed. Before the existence of a physical map of the genome, in order to obtain cloned DNA for essential genes we began genetically mapping DNA fragments (Rose et al., 1982). We took advantage of the fact that Tc1 transposable element insertions created restriction fragment length differences (RFLDs) between two strains of *C. elegans* (Emmons et al., 1983). By cloning and hybridizing randomly chosen fragments to the two strains we were able to produce the initial "genetic map" of cloned DNA fragments. Although the number of mapped fragments was not large, the generation of this early map was important for two reasons. It demonstrated that the DNA polymorphisms were distributed on several linkage groups and demonstrated the feasibility of producing a denser DNA polymorphism map covering the genome. In addition, it showed that probes for restriction fragments differences (RFLDs) between strains of nematodes could be readily obtained and used for strain identification in parasitic nematology and agriculture, as was suggested by Rose et al., (1982). Subsequent cloning and characterization of both single copy and repeat sequence probes made possible both strain and species identification in nematodes (Curran et al., 1985; Webster et al., 1990) and kelp (Fain et al., 1988). Additional studies showed that the transposable element Tc1 was present in other species and phyla demonstrating the potential for using this transposable element to generate probes for a wide range of organisms (Harris et al., 1988; Henikoff and

Plasterk, 1988). Sequence comparison between Tc1-like elements in different species revealed aspects of the functional organization of this mobile element, for example the existence of a putative intronic sequence in the coding region of Tc1 (Prasad et al., 1991).

One of the most powerful applications of the work on transposable elements has been their use for gene tagging (Greenwald, 1985; Moerman et al., 1986). Tc1 mutagenesis has also been used to develop direct approaches for obtaining mutant phenotypes for cloned and sequenced genes by R. Plasterk, Amsterdam, The Netherlands). Zwaal et al. (1993) constructed a library of clonal *C. elegans* lines containing transposable element (Tc1) inserts that can be screened by simple hybridization with PCR material to gridded filters. The PCR material is amplified using one primer containing Tc1 sequences and one primer containing sequences from the gene of interest. Thus, every interested lab can identify mutant addresses by hybridization, request the culture and be virtually guaranteed to have the insertion allele (almost) clonal.

Other approaches for aligning the two maps that have been taken include DNA polymorphism mapping (Rose et al., 1982; Baillie et al., 1985), physical mapping of rearrangement breakpoints by genomic blot hybridizations (Starr et al., 1989) PCR mapping of the left half of chromosome V (McKim et al., 1993a), cloning of known genetic loci (McKim et al., 1993a), and linear cosmid rescues (Clark and Baillie, 1992; McKay and Rose, unpublished results).

A very good method for correlating the genetic map with the physical map is to physically map the endpoints of deficiencies using PCR amplifications (McKim et al., 1993a). Primers are tested in a PCR reaction (Barstead and Waterston, 1991). If the primers are within the deleted region no PCR product will be produced. Deficiency homozygotes, which arrest as embryos, are picked. To identify the deficiency homozygotes, the progeny are synchronized by allowing the parents to lay eggs for a short period (e.g., 20 parents are allowed to lay eggs for 3 hours and removed). The next day the only unhatched eggs will be deficiency homozygotes. Approximately five homozygous mutant eggs or larvae are placed into 10 μl of lysis buffer (containing detergents and salts) and digested with Proteinase K. A 1-2 μl aliquot is used for PCR. Many DNA primers are available for sequenced genes (Coulson et al., 1988). In addition, more than 2,000 cDNA's have been sequenced (Waterston et al., 1992; McCombie et al., 1992) and placed on the physical map, providing a rich resource for designing PCR primers of known location. If at least one of the primers hybridizes within the deleted region then the PCR reaction will not amplify DNA from deficiency homozygotes. At the same time, wild types (which are heterozygous for the deficiency) can be picked and used as a PCR control. In these animals, the primers should amplify DNA. In order to ensure that both reactions are capable of amplification, a pair of primers from another chromosome should be added to both reactions as a positive control. The process is illustrated in Figure 1.

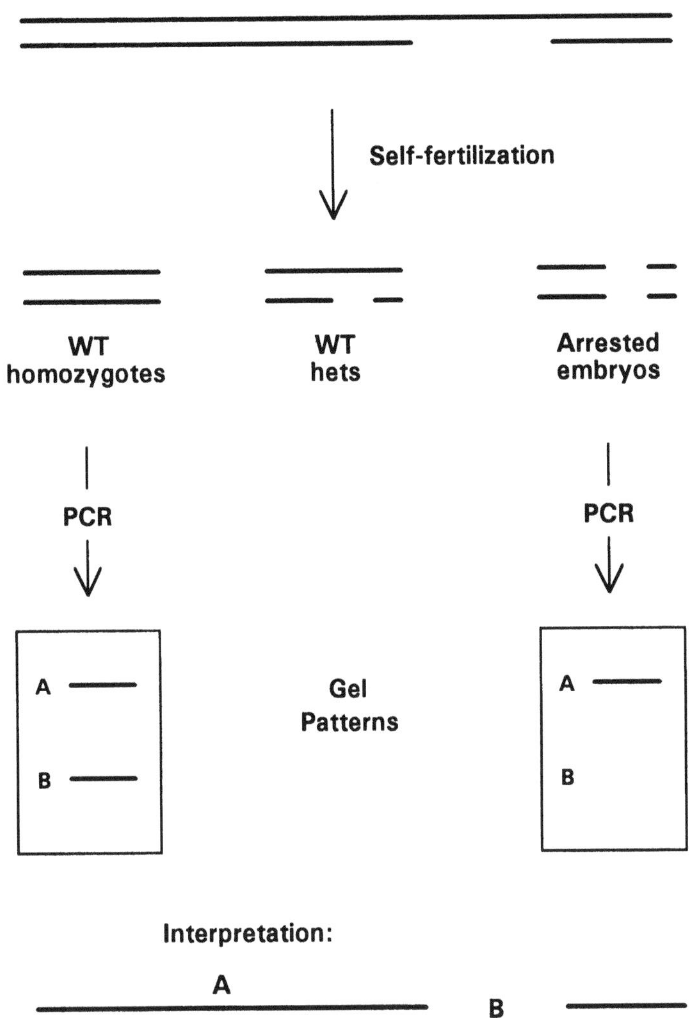

Figure 1. PCR mapping using segregants from a deficiency heterozygote.

A second highly informative approach towards aligning the maps makes use of the numerous genetic mutants that exist. Injected cosmid DNA is used to create transgenic individuals (Fire, 1986) which are tested for rescue of the mutant phenotype. Initially, direct rescue of a group of tightly linked lethal mutations was used by microinjection of cosmid DNA into hermaphrodites heterozygous for the allele to be tested. In this study, a region of approximately 200 kilobases (kbs) immediately to the left of *unc-22*(IV), six cosmids have been used to rescue six genes, *let-56* and *let-653* (Clark and Baillie, 1992), *let-92* (Jones and

Baillie, unpublished results), *par-5* (D. Shakes, pers. comm), *dpy-20* (D. Clark, unpublished results), and *let-60* (P. Sternberg, pers. comm.). Currently we use a plasmid, pRF4, containing a dominant morphological marker *rol-6* (from C. Mello, Seattle WA) and procedures described in Mello et al. (1991). Genetic strains are generated that contain one to three cosmids. The cosmids are introduced by injection of the DNA into the gonad and maintained in a heritable manner as extrachromosomal arrays. Positive transformations are indicated by the presence of Roller progeny and can be confirmed by PCR amplification of cosmid DNA using cosmid-specific primers. Genetic segregations indicate that Roller progeny carry extrachromosomal arrays containing several copies of the injected cosmid along with copies of the *rol-6* plasmid. Assembly of the extrachromosomal array in the syncytial gonad has been shown to be driven by homologous recombination between injected molecules (Mello et al., 1991). Some cosmids and pRF4 contain the sequences conferring ampicillin resistance. We have included a Kan^{R} cassette (Genblock^{TM}-Pharmacia) in a plasmid clone of pUC18 (Messing, 1983) in addition to the *rol-6* plasmid for co-injection with the Lorist vectors (cosmids beginning with K and T). Lorist vectors confer kanamycin resistance and have little sequence identity with pRF4. Strains segregating a high percentage of roller progeny have been established, used in genetic crosses, and frozen for future experiments.

The advantage of establishing strains that transmit the cosmids as extrachromosomal arrays is that these strains can be maintained and used repeatedly in genetic crosses designed to study aspects of the mutant phenotypes that they rescue. They are in effect duplication-bearing strains. In this case, however, the portion of the genome that is duplicated corresponds to a single (or few) cloned cosmid(s). In the two regions studied to date (Clark and Baillie, 1992; McKay and Rose, unpublished results), the analysis has produced a high resolution alignment of mutations in identified genes with a physical region containing characterized coding regions. In the regions investigated, a resolution of one cosmid rescue per lethal has been obtained. These results suggest that at our current lethal densities, somewhere between 1 in 10 to 1 in 3 genes are represented in the lethal mutant collections for those regions.

In addition to generating a high resolution correlation between the genetic and physical maps, this approach provides mutant strains corresponding to sequenced cosmids. In the region of chromosome III where the genome sequencing project has begun, EMS-induced lethals have been recovered and are being rescued using DNA from the sequencing project (provided by R. Wilson and R. Waterston, St. Louis Missouri, and A. Coulson and J. Sulston, Cambridge England). A summary of our approach for the study of essential genes is shown in Figure 2.

THE ACEDB DATABASE

An important aspect of genetic analysis is the ready availability of existing results. The

ACEDB information storage and retrieval system enables dynamic links to connect the various items of data in the database, so that one can move from any piece of information (for example, a gene name) to any related entry in any of the other data classes. Thus, by choosing a gene name from a list, the user opens a genetic or physical map window that displays that section of the appropriate map with the gene name highlighted in color. Double-clicking the gene name with a mouse button opens a summary window containing a phenotypic descriptions of animals mutant for the gene, a list of all genetic mapping crosses involving the gene, a list of all strains in the CGC collection carrying mutations in it, and a list of articles concerning it. Within the last two or three years, ACEDB has been modified

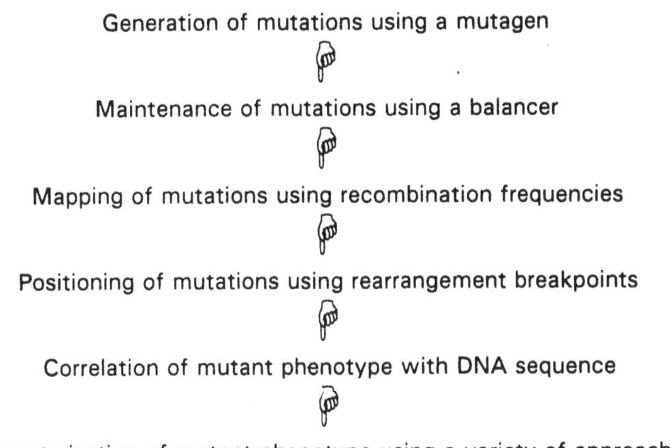

Figure 2. An outline of the steps in the analysis of lethal mutations.

to assist in the compilation of the genetic map. It determines ordering through multi-factor and deficiency/duplication complementation experiments, calculates two-factor genetic distances, and allows the map preparer to determine relatively quickly the optimum arrangement of genes for a given set of experiments. This development may dramatically increase the rate at which the genetic and physical maps evolve, since the time lag between submission of data and appearance of the map should be substantially reduced.

ACEDB is in the public domain, and regular updates of both the program and the data are made available over the Internet through a number of anonymous ftp sites (including the NCBI server that also handles GenBank, PIR and SwissProt databases). Thus, any laboratory with access to UNIX systems can easily acquire ACEDB for local use. The system is quite powerful, with no constraints on the number of data items it can handle. It is an excellent vehicle for making accessible to the *C. elegans* community the large amount of data collected on this organism. The database program at its core is generic and has been applied to many different organisms. Versions of ACEDB derived databases are being used by researchers who work with human (single chromosome), *Drosophila*, *Arabidopsis*, and yeast genomes.

BIOLOGICAL FUNCTION

Genetic analysis of lethal mutations has contributed to our understanding of the biological function of several genes such as, *lin-3* (cell lineage), *lag-2* (lineage and germline proliferation), *rol-3* (roller) and *unc-60* (uncoordinated). The *lin-3* gene encodes a TGF-alpha-like protein that is the putative ligand for the *let-23* (lethal) receptor (Aroian et al., 1990). Some alleles result in a vulvaless phenotype, but the existence of lethal alleles shows that *lin-3* is also required earlier in development (Hill and Sternberg, 1992). *lag-2* has the same phenotype as the double mutant *lin-12 glp-1* (germline proliferation), that is early larval lethality, with the absence of an excretory pore and anus (Johnsen and Baillie, 1991; Lambie and Kimble, pers. comm; Tax and Thomas, pers. comm.). This gene produces a product that is similar to the *Drosophila* serrate gene and may interact with the *lin-12* and/or *glp-1* gene products *in vivo*. *rol-3* has two alleles that produce a left-handed rolling phenotype, but in addition it has 12 lethal alleles that demonstrate an essential developmental function for this gene. The analysis of two gene-specific suppressors of this gene has led to the proposal that they are involved in the regulation of expression of both *rol-3* and genes involved in male tail development (Barbazuk et al., 1994). Another example is *unc-60*, which has more than 15 alleles that give rise to a paralyzed phenotype. The paralysis results from a failure to properly polymerize and position the thin filaments of muscle. The isolation of a lethal allele of this locus shows that the gene also plays an essential role in development. Molecular analysis has demonstrated that by means of alternate RNA splicing, the gene encodes two cofilin-like proteins, one or both of which are required for completion of embryonic development (McKim et al., 1993a).

One of the future challenges for genetic research in *C. elegans* is characterizing the function of the genes represented by the large collection of existing mutants. Mosaic analysis can be used to determine cell autonomous and lineage specific functions. This approach has been applied to lethal mutants and indicated that many genes are expressing in specific cells (Bucher and Greenwald, 1991). These results are encouraging and suggest that a genetic approach such as mosaic analysis could be applied generally to sort many of the lethal mutants according to lineage specific functions. A second methodology that may be promising for functional characterization of lethal mutants is suppressor analysis. Using linked drug resistance Marra and Baillie (unpublished results) selected dominant suppressors of lethal mutants. These and other approaches are being developed to expedite the functional analysis of the essential genes.

It is not known how many of the genes in the *C. elegans* genome will mutate to lethality. In the regions where we have identified large numbers of essential genes, the percentage appears to be between 10 to 30%. What do the rest of the genes do? A small number mutate to morphologically or biochemically detectable phenotypes. Some have been shown to have no detectable phenotype under the conditions investigated, although of course the expression of certain genes will be affected by the culture environment. Other genes, which would be expected not to produce visible mutant phenotypes, are those that are present in multiple copies, such as rRNA or some collagen genes. As well, genes that are functionally redundant, where one gene product although biochemically different from the modified gene can act as its functional replacement, will not have easily identifiable mutant phenotypes.

Most of the genes in *C. elegans* are expected to be homologous to genes in other organisms, and thus, *C. elegans* genetic resources will be of value for the understanding of gene function in many other species.

Acknowledgements

Research in the laboratories of DLB and AMR was supported by grants from the National Science and Engineering Research Council (NSERC) and the Medical Science Council (MRC) of Canada and the National Institutes of Health (NIH) USA. M.L.E. was supported by Department of Health and Human Sciences grant No. RR07127. This manuscript was prepared while on sabbatical in the laboratories of Drs. Danielle and Jean Thierry-Mieg, Centre National De La Researche Scientifique (CNRS), Montpellier France.

REFERENCES

Ahnn, J. and Fire, A., 1993, A screen for genetic loci required for body-wall muscle development during embryogenesis in C. elegans, *Genetics* in press.

Albertson, D.G. and Thomson, J.N., 1993, Segregation of holocentric chromosomes at meiosis in the nematode, *Caenorhabditis elegans*, *Chromosome Research* 1:15.

Aroian, R.V., Koga, M, Mendel, J.E., Ohshima, Y. and Sternberg, P.W., 1990, The *let-23* gene necessary for *Caenorhabditis elegans* vulval induction encodes a tyrosine kinase of the EGF receptor subfamily, *Nature* 348:693.

Baillie, D.L., Beckenbach, K.A. and Rose, A.M., 1985, Cloning within the *unc-43* to *unc-31* interval (LGIV) of the *C. elegans* genome by Tc1 transposon tagging, *Can. J. Genet. & Cytol.* 27: 457.

Barbazuk, W.B., Johnsen, R.C. and Baillie, D.L., 1994, srl-Suppressors of **R**oller **L**ethal: the generation and genetic analysis of suppressors of lethal mutations in the *Caenorhabditis elegans rol-3*(V) gene, *Genetics* 136:in press.

Barstead, R.J. and Waterston, R.H., 1991, Vinculin is essential for muscle function in the nematode, *J. Cell Biol.* 114:715.

Bucher, E.A. and Greenwald, I., 1991, A genetic mosaic screen of essential genes in *C. elegans*, *Genetics* 128:281.

Brenner, S., 1974, The genetics of *Caenorhabditis elegans*, *Genetics* 77:71.

Clark, D.V. and Baillie, D.L., 1992, Genetic analysis and complementation by germ-line transformation of lethal mutations in the *unc-22 IV* region of *Caenorhabditis elegans*, *Mol. Gen. Genet.* 232:97.

Clark, D.V., Rogalski, T.M., Donati, L.M. and Baillie, D.L., 1988, The *unc-22(IV)* region of *Caenorhabditis elegans*: Genetic analysis of lethal mutations, *Genetics* 119:345.

Coulson, A., Waterston, R.H., Kiff, J., Sulston, J.E. and Kohara, Y., 1988, Genome linking with yeast artificial chromosomes, *Nature* 335:184.

Curran, J., Baillie, D.L. and Webster, J.M., 1985, Use of genomic DNA restriction fragment length differences to identify nematode species, *Parasitology* 90:137.

Edgley, M.L. and Riddle, D.L., 1993, The nematode *Caenorhabditis elegans*, in "Genetic Maps Locus maps of complex genomes," S.J. O'Brien, ed, Cold Spring Harbor Laboratory, Cold Spring Harbor, N.Y.

Emmons, S.W., Yesner, L., Ruan, K.-S. and Katzenberg, D., 1983 Evidence for a transposon in *C. elegans*, *Cell* 32:55.

Fain, R.S., Baillie, .D.L and Druehl, L.D, 1988, Repeat and single copy sequences are differentially conserved in the evolution of kelp chloroplast, *J. Phycol.* 24:292.

Fire, A., 1986 DNA transformation in *C. elegans*, *EMBO Journal* 5:2673.

Greenwald, I., 1985, *lin-12*, a nematode homeotic gene, is homologous to a set of mammalian proteins that includes epidermal growth factor, *Cell* 43:583.

Harris, L.J., Baillie, D.L. and Rose, A.M., 1988, Sequence identity between an inverted repeat family of transposable elements in *Drosophila* and *Caenorhabditis*, *Nucleic Acid Res.* 16:5991.

Henikoff, S. and Plasterk, R.H.A., 1988, Related transposons in *C. elegans* and *D. melanogaster*, *Nucleic Acid Res.* 16:6234.

Herman, R.K., 1978, Crossover suppressors and balanced recessive lethals in *C. elegans*, *Genetics* 88:49.

Hodgkin, J.A., Horvitz, H.R. and Brenner, S., 1979, Nondisjunction mutants of the nematode *C. elegans*, *Genetics* 91:67.

Howell, A.M., Gilmour, S.G., Mancebo, R.A. and Rose, A.M., 1987, Genetic analysis of a large autosomal region in *C. elegans* by the use of a free duplication, *Genetical Research* 49:207.

Howell, A.M. and Rose, A.M., 1990, Essential genes in the *hDf6* region of chromosome I in *Caenorhabditis elegans*, *Genetics* 126:583.

Hill, R.J. and Sternberg, P.W., 1992, The gene *lin-3* encodes an inductive signal for vulval development in *C. elegans*, *Nature* 358:470.

Johnsen, R.C. and Baillie, D.L., 1988, Formaldehyde mutagenesis of the *eT1* balanced region in *C. elegans*: Dose-response curve and the analysis of mutational events, *Mutation Research* 201:137.

Johnsen, R.C. and Baillie, D.L. 1991, Genetic analysis of a major segment [LGV(left)] of the genome of *Caenorhabditis elegans*, *Genetics* 129:735.

McCombie, W.R., Adams, M.D, Kelley, J.M., FitzGerald, M.G., Utterback, T.R., Khan, M., Dubnick, M., Kerlavage, A.R., Venter, J.C. and Fields, C., 1992, *Caenorhabditis elegans* expressed sequence tags identify gene families and potential disease gene homologues, *Nature genetics* 1:124.

McKim, K.S., Howell, A.M. and Rose, A.M., 1988, The effects of translocations on recombination frequency in *Caenorhabditis elegans*, *Genetics* 120:987.

McKim, K.S. and Rose, A.M., 1990, Chromosome I duplications in *Caenorhabditis elegans*, *Genetics* 124:115.

McKim, K.S., Starr, T. and Rose, A.M., 1992, Genetic and molecular analysis of the *dpy-14* region in *Caenorhabditis elegans*, *Mol. Gen. Genet.* 233:241.

McKim, K.S., Matheson, C., Marra, M.A., Wakarchuk, M.F. and Baillie, D.L., 1993a, The *C. elegans unc-60* gene encodes proteins homologous to a family of actin binding proteins, *Mol. Gen. Genet.* in press.

McKim, K.S., Peters, K. and Rose, A.M., 1993b, Two types of sites required for chromosome pairing in *Caenorhabditis elegans*, *Genetics* 134:749.

Mello, C.C., Kramer, J.M., Stinchcomb, D. and Ambros, V., 1991, Efficient gene transfer in *C. elegans*: extrachromosomal maintenance and integration of transforming sequences, *EMBO Journal* 10:3959.

Meneely, P.M. and Herman, R.K., 1979, Lethals, steriles and deficiencies in a region of the X chromosome of *C. elegans*, *Genetics* 92:99.

Messing, J., 1983, New M13 vectors for cloning, *Methods Enzymol.*, 101:20.

Moerman, D.G. Benian, G.M. and Waterston, R.H., 1986, Molecular cloning of the muscle gene *unc-22* in *C. elegans* by Tc1 transposon tagging, *Proc. Natl. Acad. Sci USA* 83:2579.

Pimpinelli, S. and Goday, C., 1989, Unusual kinetochores and chromatin diminution in *Parascaris*, *TIGS* 57:310.

Prasad, S., Harris, L.J. Baillie, D.L. and Rose, A.M., 1991, Evolutionarily conserved regions in *Caenorhabditis* transposable elements deduced by sequence comparison, *Genome* 34:6.

Rogalski, T.M., Moerman, D.G. and Baillie, D.L., 1982, Essential genes and deficiencies in the *unc-22 IV* region of *C. elegans*, *Genetics* 102:725.

Rogalski, T.M. and Baillie, D.L., 1985, Genetic organization of the *unc-22 IV* gene and the adjacent region in *C. elegans*, *Mol. Gen. Genet.* 201:409.

Rose, A.M. and Baillie, D.L., 1980, Genetic organization of the region around *unc-15(I)*, a gene affecting paramyosin in *C. elegans*, *Genetics* 96:639.

Rose, A.M., Baillie, D.L., Candido, E.P.M., Beckenbach, K.A. and Nelson, D., 1982, The linkage mapping of cloned restriction fragment length differences in *Caenorhabditis elegans*, *Mol. Gen. Genet.* 188:286.

Rose, A.M., Baillie, D.L. and Curran, J., 1984, Meiotic pairing behavior of two free duplications of linkage group I in *C. elegans*, *Mol. Gen. Genet.* 195:52.

Rosenbluth, R.E. and Baillie, D.L., 1981, The genetic analysis of a reciprocal translocation, *eT1(III;V)*, in *C. elegans*, *Genetics* 99:415.

Rosenbluth, R.E., Cuddeford, C. and Baillie, D.L., 1983, Mutagenesis in *Caenorhabditis elegans*: I. A Rapid eukaryotic mutagen test system using the reciprocal translocation, *eT1*(III;V). *Mutation Research* 110:39.

Rosenbluth, .E., Cuddeford, C. and Baillie, D.L., 1985, Mutagenesis in *Caenorhabditis elegans*. II. A spectrum of mutational events induced with 1500 R of gamma-radiation, *Genetics* 109:493.

Rosenbluth, R.E., Rogalski, T.M., Johnsen, R.C., Addison, L.M. and Baillie, D.L., 1988, Genomic organization in *Caenorhabditis elegans*: deficiency mapping on linkage group V(left), *Genetical Research* 52:105.

Rosenbluth, R.E., Johnsen, R.C. and Baillie, D.L., 1990, Pairing for recombination in LG V of *Caenorhabditis elegans*: A model based on recombination in deficiency heterozygotes, *Genetics* 124:615.

Sigurdson, D.C., Spanier, G.J. and Herman, R.K., 1984, *C. elegans* deficiency mapping, *Genetics* 108:331.

Starr, T., Howell, A.M., McDowall, J., Peters, K. and Rose, A.M., 1989, Isolation and Mapping of DNA probes within the linkage group I gene cluster of *Caenorhabditis elegans*, *Genome* 32:365.

Stewart, H.I., Rosenbluth, R.E. and Baillie, D.L., 1991, Most ultraviolet irradiation induced mutations in the nematode *Caenorhabditis elegans* are chromosomal rearrangements, *Mutation Research* 249:37.

Sulston, J.E. and Horvitz ,H.R., 1977, Post-embryonic cell lineages of the nematode *Caenorhabditis elegans*, *Dev. Biol.* 56:110.

Sulston, J.E., Schierenberg, E., White, J.G. and Thomson, J.N., 1983, The embryonic cell lineage of the nematode *Caenorhabditis elegans*, *Dev. Biol.* 100:64.

Sulston, J. Du, Z., Thomas, K., Wilson, R., Hillier, L., Staden, R., Halloran, N., Green, P., Thierry-Mieg, J., Qiu, L., Dear, S., Coulson, A., Craxton, M., Durbin, R., Berks, M., Metzstein, M., Hawkins, T., Ainscough, R. and Waterston, R., 1992, The *C. elegans* genome sequencing project: a beginning, *Nature* 356:37.

Waterston, R., Martin, C., Craxton, M., Huynh, C., Coulson, A., Hillier, L., Durbin, R., Green, P., Shownkeen, R., Halloran, N., Metzstein, M., Hawkins, T., Wilson, R., Berks, M., Du, Z., Thomas, K., Thierry-Mieg, J. and Sulston, J., 1992, A survey of expressed genes in *Caenorhabditis elegans*, *Nature genetics* 1:114.

Webster, J.M., Baillie, D.L., Beckenbach, K. and Rutherford, T.A., 1990, DNA probes for differentiating isolates of the pinewood nematode species complex, *Revue Nématol.* 3:255.

White, J.G., Southgate, E., Thomson, J.N. and Brenner, S., 1986, The structure of the nervous system of *Caenorhabditis elegans*, *Philos. Trans. R. Soc. Lond. B Biol. Sci.* 341:1.

Wood, W.B., ed., 1988, "The nematode *Caenorhabditis elegans*," Cold Spring Harbor Laboratory, Cold Spring Harbor, N.Y.

Zetka, M.-C. and Rose, A. M., 1992, The meiotic behavior of an inversion in *Caenorhabditis elegans*, *Genetics* 131:321.

Zwaal, R.R., Broeks, A., van Meurs, J., Groenen, J.T.M. and Plasterk, R.H.A., 1993, Target-selected gene inactivation in *Caenorhabditis elegans* by using a frozen transposon insertion mutant bank, *Proc. Natl. Acad. Sci USA* 90:7431.

TRANSPOSABLE ELEMENTS IN NEMATODES

Pierre Abad

I.N.R.A.
Laboratoire de Biologie des Invertébrés
B.P. 2078
06606 Antibes Cedex, France

INTRODUCTION

Transposable elements are present in the genomes of most, if not all, organisms. Because of their ability to insert into and excise from the chromosomes of their hosts transposons are a significant source of spontaneous mutations in organisms. Therefore they can be used as a tool for cloning genes that have been identified by mutations and for which no gene products are known. Transposable elements may be divided into two classes according to their mechanisms of transposition. Class I elements transpose by reverse transcription of an RNA intermediate (a DNA-RNA-DNA mechanism), while class II elements transpose directly from DNA to DNA (Fig. 1).

Transposable elements can be used in reverse genetics as molecular tools for many aspects:

- gene tagging: the first step is to find a spontaneous mutant for the gene of interest in a transposing strain. The method for identifying the relevant transposon is to outcross the mutant (from the transposing strain) to strains without transposon or without transposition of transposable element in order to have homozygous mutant. The transposon fragment that invariably cosegregates with the mutation of interest could be identified by the hybridization of the mutant genomic DNA with the transposon. Isolation of the flanking sequence of the transposon can be used subsequently for probing the wild type gene. Mutant rescue following transformation with the isolated wild type gene provides a good evidence that the correct DNA sequence was cloned.
- gene mutagenesis: transposons inactivate genes when they are inserted in exons or in regulatory sequences. Therefore the function of a gene can be studied by isolation of an animal in which this gene has been altered in targeted fashion. PCR experiments have been used to investigate such events by using a specific oligonucleotide for the transposon and one specific for the gene of interest. In a mutant animal, PCR products are only obtained if a transposon is inserted in the area of interest.
- germ-line transformation: this property has been demonstrated for only some transposable elements. The P transposable element from *Drosophila melanogaster* represents the best example for the germ-line transformation since the use of a P-transposable element as vector to introduce specific DNA segments into the *Drosophila* germline revolutionized the study of gene regulation and function in *Drosophila*. This element integrates as a single copy and the transduced genes are expressed in an appropriate temporal and spatial manner during *Drosophila* development (for review, Engels, 1989).

In nematodes most information about transposons is known in *Caenorhabditis elegans*, because of the strong interest in this species as an experimental organism for developmental

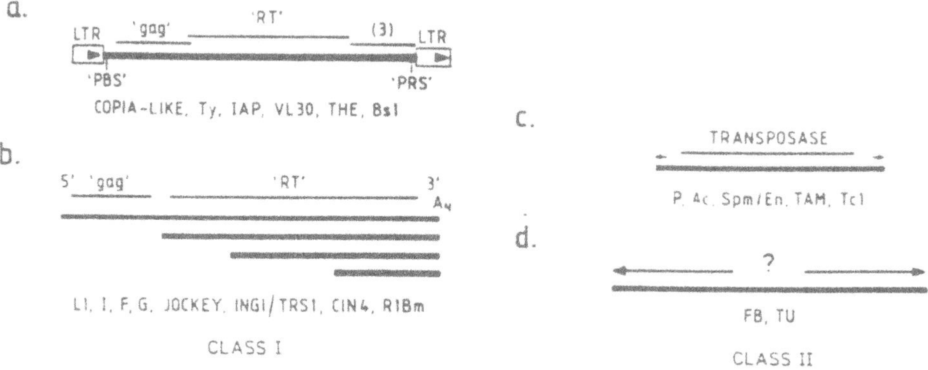

Figure 1. The structure of four different types of eukaryotic transposable elements. Examples of each type of transposable element are given. (a,b) class I : transposable elements believed to transpose via RNA intermediates. (c,d) class II : transposable elements believed to transpose directly from DNA to DNA. (from Finnegan, 1989).

genetic studies. This article focuses principally on Tc1 (Tc= transposon of *C. elegans*), the most clearly understood element in nematodes (for review, Moerman and Waterston, 1989; Anderson et al., 1992). However, other transposable elements in *C. elegans* and in other species such as *C. briggsae*, *Ascaris lumbricoides* and *Panagrellus redivivus* have been described. This article also deals with these elements and also the different strategies for cloning transposable elements in species other than the ones described above.

TC1

Identification of Tc1

Tc1 was first characterized as a consequence of polymorphism in restriction endonuclease patterns observed on a Southern blot between the two strains of *C. elegans*, Bristol N2 (English strain) and Bergerac BO (French strain). Unique sequence probes identified restriction fragments that were 1.6-kb larger in BO than in N2 (Emmons et al., 1979). This polymorphism was due to the insertion of a dispersed sequence present in N2 in about 30 copies and in BO in over 300 copies per haploid genome (Emmons et al., 1983; Liao et al., 1983).

Tc1 is 1, 610 bp long with 54 bp perfect terminal inverted repeats. It also contains two long open reading frames (ORF) on the same DNA strand (Fig.2). This ORF has a coding potential for a 273 amino acid protein and is flanked by putative regulatory signals.

An unusual feature of the Tc1 family is that, in contrast to P elements of *Drosophila* (O'Hare and Rubin, 1983) and Ac- Ds elements of maize (Fedoroff et al., 1983) with which Tc1 shares significant structural features, most Tc1 family members are 1.6-kb in length even in the high copy number strains.

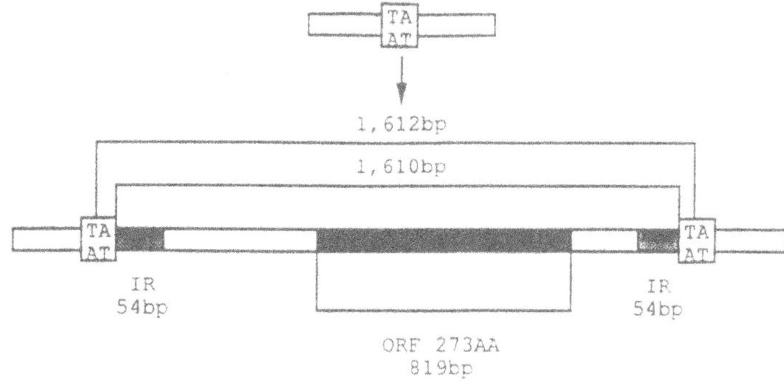

Figure 2 . The structure of Tc1 and its insertion target site based on nucleotide sequence of Tc1, from Rosenzweig et al., 1983. The stippled boxes represent the terminal inverted repeats (IR) and the filled box represent the longest open reading frame (adapted from Herman and Shaw, 1987).

This is clearly shown using *Eco*RV which cuts Tc1 in its terminal inverted repeats. The genomic digests with *Eco*RV reveal mainly a single 1.6-kb Tc1-hybridizing band (Emmons et al., 1985). However, sequence micro-heterogeneity has been revealed within the Tc1 family with *Eco*RI, *Hin*dIII and *Tha*I (Eide and Anderson, 1985a; Rose et al., 1985). The sequence of *st* 137::Tc1, an element recently inserted in the gene *unc-22* of the BO strain, has been established. Compared to the canonical element pCe (Be) T1, it shows six nucleotide alterations in the large ORF. One of them is an amber triplet that results in a shorter version of the gene (Plasterk, 1987). Tc1 elements from N2 also show microheterogeneity. Out of the 17 Tc1 elements cloned from the 30 Tc1 elements in the N2 strain, 2 have modified restriction sites, 2 have different internal deletions of about 700 bp, 1 an 89-bp terminal deletion, and 1 a 54-bp insertion (Harris and Rose, 1989).

Somatic Excision

Evidence for mobility of Tc1 can be shown on a Southern blot by hybridizing genomic DNA with unique sequences flanking several Tc1 elements. Such experiments show that at Tc1 insertion sites DNA from populations of worms is a mixture of sequences, some containing the Tc1 element and some lacking it. The empty sites are always 1.6 kb smaller and therefore suggest that excision is often precise or nearly precise (Eide and Anderson, 1985a; Emmons et al., 1986; Moerman et al., 1986). In contrast to the behavior of P and I elements in *Drosophila* during hybrid dysgenesis where excision and transposition are germ-line specific, excision of Tc1 is higher in somatic cells than in the germ line (Brégliano and Kidwell, 1983).

Tc1 elements excise in somatic cells from genomic sites in both N2 and BO strains at high frequency. This was shown by analyzing different developmental stages. The fraction of empty sites is low in embryos, increases during larval development, and then decreases as more germ line cells are added in the adult (Emmons and Yesner, 1984). Furthermore, it was demonstrated that excision continued in worms stopped in their development as first stage larvae by starvation in buffer (Emmons et al., 1985; Emmons et al., 1986). Thus Tc1 excision does not require DNA replication in contrast to Ac-Ds elements in maize, where transposition is correlated to DNA replication (Fedoroff, 1983; Greenblatt and Brink, 1962).

In addition, somatic excision was demonstrated directly by observing mosaic animals among populations of Tc1-induced mutants of *unc-54* and *unc-22* (Eide and Anderson, 1985a; Moerman et al., 1988).

Germ Line Transposition

Germ line transposition of Tc1 has been detected by analysis of spontaneous mutations at specific genetic loci. Among all the tested *C. elegans* strains only the BO strain had mutator activity. Transposition rate was measured by Tc1 insertions into *unc-54* or *unc-22* genes in a Bergerac genetic background (Moerman and Waterston, 1984). These two spontaneous mutations can be studied because a sensitive and specific screen is available for detecting mutant worms. In *unc-22* , heterozygous or homozygous mutant worms can be recognized because they twitch or vibrate in a 1% nicotine alkaloid solution whereas wild-type worms are paralyzed (Moerman and Baillie, 1979). Mutations in *unc-54* result in paralysed animals unable to lay eggs because their muscles are defective.

The frequency of spontaneous mutation at *unc-22* locus was found to be 10^{-4} (per gamete). This was at least 100 times greater than the frequency observed in several other strains at the same locus. Similarly, in the BO strain, two thirds of spontaneous *unc-54* are due to insertion of a Tc1 element and the frequency of spontaneous *unc-54* mutants is approximately 10^{-6} (Eide and Anderson, 1985a,b).

These mutations are unstable and revert to wild type. These unstable spontaneous mutations result from excision of this element (Moerman et al., 1986). In germinal tissue, excision of Tc1 occurs but is less frequent than in somatic cells. Therefore, Tc1 excision is under tissue-specific regulation . Germ line reversion of *unc-22*:: Tc1 alleles occurs at frequency of about 2×10^{-3} (Moerman et al., 1986), wheareas reversion of *unc-54*:: Tc1 alleles occurs at a frequency from 10^{-5} to less than 10^{-7}, depending upon their location within the gene

(Eide and Anderson, 1985a,b; Eide and Anderson, 1988). However, for other genes such as *unc-52* and *unc-15*, spontaneous alleles have reversion frequencies of 10^{-2} (Moerman, unpublished results), which approach those estimated for somatic reversion rates using Southern blot analysis.

Site Specificity For Tc1 Insertion

Most transposable elements, including procaryotic and eucaryotic elements, show a preference for particular target sequences (Halling and Klecker, 1982; Ikenaga and Saigo, 1982; O'Hare and Rubin, 1983; Berg et al., 1984; Freund and Meselson, 1984; Inouye et al., 1984; Zerbib et al., 1985). The site of each insertion was determined by comparing sequences of the wild-type genes with those of the insertional junctions (Eide and Anderson, 1988; Mori et al., 1988a). This indicated that Tc1 also displays a striking target site preference with a 9-bp consensus sequence (Fig. 3). The only absolute requirements for Tc1 insertion are the -1 T and the +1 A.

Each of the Tc1 insertions is flanked by TA dinucleotide and only one copy of the TA dinuclotide is found at the corresponding site in the wild-type sequence. These TA dinucleotides might represent a 2-base-pair target site duplication which is formed during insertion, like those found for most of transposable elements (Grindley, 1978; O'Hare and Rubin, 1983; Bucheton et al., 1984).

Position Relative to Insertion Site

Number of Occurrences Among 16 DNA Strands		-8	-7	-6	-5	-4	-3	-2	-1	+1	+2	+3	+4	+5	+6	+7	+8
	G	6	3	3	$\underline{8}$	1	$\underline{4}$	$\underline{4}$	0	0	2	$\underline{7}$	2	3	6	3	3
	A	5	5	7	3	$\underline{12}$	1	$\underline{8}$	0	$\underline{16}$	1	1	3	7	5	6	5
	T	1	1	3	2	3	$\underline{9}$	2	$\underline{16}$	0	$\underline{7}$	1	$\underline{10}$	4	2	4	4
	C	4	7	3	3	0	2	2	0	0	$\underline{6}$	$\underline{7}$	1	2	3	2	3
Consensus:					G	A	G/T	A/G	T	A	T/C	G/C	T				

Figure 3 . Consensus site for Tc1 integration. The matrix derived from the number of occurrences among 16 DNA strands. The sequence of Tc1 insertion sites is shown from -8 to +8, flanking the TA dinucleotide which is probably duplicated upon insertion. Underlined values indicate significant bases (adapted from Moerman and Waterson, 1989).

However, since TA is itself an inverted repeat, these nucleotides might be part of Tc1. In this case, no target site duplication is formed upon insertion, and Tc1 might be 1, 612 bp long and contain T at one end and A at the other. Therefore, Tc1 insertion between T and A would result in a flush cleavage of the target site. This strategy is unusual but not without precedent for an element with an insertion sequence-like structure (Halling and Klecker, 1982).

Although Tc1 exhibits a preference for particular target sequences, the primary sequence alone does not determine the frequency of integration. Insertion of Tc1 is also strongly gene-specific. Tc1 insertion into the BO *unc-22* gene is about 100-fold more frequent than insertion in the BO *unc-54* gene (Eide and Anderson, 1985a; Collins et al., 1987). This difference can not be explained by a postulated larger number of target sites in *unc-22* which can account for only sevenfold more occurrences of Tc1 insertion site consensus sequence than does the *unc-54* gene (Mori et al., 1988a). Therefore, Tc1 clearly prefers some DNA sequences, but factors in addition to primary DNA sequence must influence Tc1 target site specificity.

Excision Footprints

Germ-line and somatic excisions of the Tc1 were analyzed by sequencing cloned sites from anonymous sites, *unc-54* and *unc-22* genes after Tc1 excision.

Information on 20 *unc-54* alleles indicates that Tc1 excision is usually imprecise in both germ line and somatic cells. Imprecise excision generates new *unc-54* alleles which show amino acid substitutions, amino acid insertions, and, in some cases, probably altered mRNA splicing (Eide and Anderson, 1988).

Among revertants from a single *unc-22*:: Tc1 insertion, phenotypically wild type and partial revertants were obtained. Sequence analysis of empty Tc1 sites of full revertants and partial revertants indicates that only one site presents the ancestral wild-type sequence. The others exhibit sequence duplications, subtitutions or large deletions (Kiff et al., 1988). The fact that most of the Tc1 events are imprecise is not surprising since imprecise excision is a common feature of eucaryotic transposable elements. Such events have been described for yeast cells (Roeder and Fink, 1983), *Drosophila melanogaster* (Searles et al., 1986; Tsubota and Schedl, 1986), maize (Shure et al., 1983, Sutton et al., 1984), mice (Copeland et al., 1983) and *Antirrhinum majus* (Bonas et al., 1984; Coen et al., 1986).

Surprising transposons in animals and plants leave behind some of their terminal sequences when they excise. Two explanations have been proposed, in the first one transposons themselves are excised in an imprecise fashion, as it was generally believed for transposons in plants (Schwartz-Sommer et al., 1985; Coen et al., 1989). If the transposon excises imprecisely, leaving behind its ends, then the excised transposon would have to be repaired to make it integration-competent again.

In the second explanation, transposons may be excised in precise fashion and the repair may be done in an imprecise way. The work on the P tranposable element of *D. melanogaster* points in that direction. Engels et al. (1990) found that a P insertion mutant of the *white* gene which was known to require precise loss of the transposon for reversion, reverted at an elevated frequency when the homologous chromosome had the wild-type sequence at that location. They suggest that the double-strand DNA break that is left after P transposon excision is repaired using the homologous chromosome as a template, and that the frequent presence of internally deleted P elements results from interrupted repair.

In order to test this model for Tc1excision, Plasterk (1991) investigated wheter the frequency of loss of Tc1 elements was affected by the sequence of the homologous chromosome (Fig.4).

```
wt:        caattttgggatagtcgttgaacgt
st192:     caattttgggat A C A G T G C-------G C A C T G T atgtcgtt

Revertants from heterozygote:
Rev.Het1   caattttgggat                               atgtcgtt
Rev.Het2   caattttgggat                               atgtcgtt
Rev.Het3   caattttgggat                               atgtcgtt
Rev.Het4   caattttgggat                               atgtcgtt
Rev.Het5   caattttgggat                               atgtcgtt
Rev.Het6   caattttgggat                               atgtcgtt

Revertants from homozygote:
Rev.Hom1   caattttgggat A C A                   T G T atgtcgtt
Rev.Hom2   caattttgggat A C A A                   G T atgtcgtt
Rev.Hom3   caattttgggat A                                 cgtt
Rev.Hom4   caattttgggat A caattttggga           T G T atgtcgtt
Rev.Hom5   caattttgggat A C G                   T G T atgtcgtt
Rev.Hom6   caattttgggat A C                           atgtcgtt
Rev.Hom7   caattttgggat                           G T atgtcgtt
```

Figure 4 . Sequence of the area containing Tc1 insertion *st 192* in revertants. Tc1 sequence is in upper cases and flanking sequences are in lower cases. The two nucleotides which immediately flank the integrated Tc1 element are underlined. As shown all revertants heterozygous for Tc1 insertion have a sequence which is identical to that of the gene in the homologous chromosomes. All revertants homozygous for Tc1 insertion have a footprint. Most contain a few pairs from both ends of Tc1 but some have deletion or duplication (from Plasterk, 1991).

The *unc-22* gene was chosen for three reasons. Firstly, because several sequenced Tc1 alleles which can revert in an imprecise fashion are available. Secondly, characterized EMS-induced alleles are available and thirdly the reversion of this gene is easily observed (Moerman et al., 1988; Benian et al., 1989). Plasterk (1991) found that in diploid animals having a homozygous Tc1 insertion the frequency is around 10^{-4}, and a Tc1 footprint is found. When the corresponding sequence on the homologous chromosome is wild-type, the reversion frequency is 100 times higher, and the reverted sequence is precise. Therefore, Tc1 footprints result from incomplete gene conversion from the homologous chromosome, and not from imprecise excision. This indicates that Tc1 excision leaves a double-strand DNA break, which can be repaired using the homologous chromosome or the sister chromatid as a template. In heterozygotes, repair can lead to reversion and in homozygotes, Tc1 is copied into the empty site, and only rare interrupted repair leads to reversion (Fig. 5) (Plasterk, 1991).

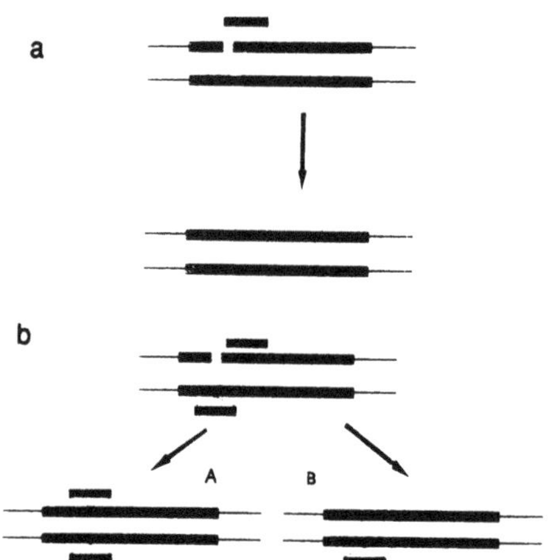

Figure 5 . Homologue-dependent repair of the site of Tc1 excision. (a) In an animal heterozygous for Tc1 insertion, an excision is followed by repair from the homologue that leads to copying the wild-type sequence. (b) In animal homozygous for Tc1 insertion, repair from the homologue template either results in the new copy on Tc1 into the old site (arrow A) or partial restoration of the gene (arrow B) (from Plasterk, 1991).

Futhermore, Plasterk and Groenen (1992) have shown that the double strand DNA break resulting from Tc1 excision can be repaired by the extrachromosomal DNA as template. Thus the flanking sequences of the break can be replaced by sequences from the transgene. Therefore, transgene-instructed break repair provides a method for the targeted introduction of precise alterations into the *C. elegans* genome.

Extrachromosomal Copies Of Tc1

Extrachromosomal copies of Tc1 have been identified in BO, at a level between 0.1 to 1.0 copy per cell (Rose and Snutch, 1984; Ruan and Emmons, 1984). The predominant form is a 1.6 kb linear molecule with ends corresponding to the ends of the integrated Tc1. Linear and circular forms exist and are very similar or identical in sequence to the integrated form since digestions with some restriction enzymes produced intact Tc1 elements.

It has been suggested that the extrachromosomal copies of Tc1 are the products of high-frequency somatic excision (Rose and Snutch, 1984; Ruan and Emmons, 1984). However, the concentration of extrachrosomal copies does not increase in starved BO larvae (Rose et al., 1985), whereas the fraction of empty sites does increase. Therefore, if the extrachromosomal copies are products of somatic excision, they should be unstable or perhaps they reinsert (Ruan and Emmons, 1984).

Regulation Of Tc1 Transposition

Tc1 copy number. Tc1 copy number is clearly not solely responsible for the Tc1 transposition frequency. Transposition, measured by spontaneous *unc-54* and *unc-22* mutations, is greater in BO (high-copy number strain) than in N2 (low-copy number strain). However, DH424, FR1, and BL1 are also high-copy number strains for Tc1 (Emmons et al. 1983; Liao et al., 1983) and show little if any transposition for Tc1.

Mori et al. (1988b) undertook the genetic analysis of a spontaneous *unc-22*:: Tc1 mutation, iniatially isolated in BO, stabilized in an N2 background and then crossed again to BO. From this cross, several isogenic mixed N2/BO lines containing the *unc-22*::Tc1 mutation were established. Measurement of the Tc1 copy number and reversion frequency at the *unc-22* indicate that the Tc1 copy number is not the major variable regulating Tc1 transposition since the reversion frequencies are not correlated with the Tc1 copy number. Conversely, Tc1-transposing strains with only about 45-60 copies of Tc1 have been isolated (Mori et al., 1988b).

Genetic background. The correlation of germ line excision and insertion has been done by analysing unstable *unc-22*:: Tc1 alleles. Moerman and Waterston (1984) demonstrated that both the induction frequency of *unc-22*:: Tc1 mutations and their stability are sensitive to genetic background. An *unc-22* region from N2 becomes mutable when put in a BO background, and a BO *unc-22* region is stabilized in an N2 background. Strains in which the *unc-22* alleles become unstable also exhibit elevated forward mutation rates at the *unc-22* locus. These observations suggest that excision and transposition of some Tc1 elements are regulated by "trans" acting factors. Further studies have confirmed that Tc1 insertion and excision activities are correlated (Waterston et al., 1986; Collins et al., 1987; Mori et al., 1988b). Thus germ line excision and insertion may be regulated in the same manner and, as suggested earlier, may reflect different stages of the transposition process.

Mutator activity. The mutator activity necessary for germline Tc1 transposition in the BO strain is under intensive investigation. A previous study suggested that multiple genetic components are involved. This was demonstrated by systematic chromosome replacement using marked N2 chromosomes, showing that the mutator activity of the BO strain was not restricted to a single region (Moerman and Waterston, 1984). Thus the mutator activity in the BO strain could be ascribed to a few chromosomal sites or numerous sites in the BO genome. To further analyze this mutator activity, Mori et al. (1988b) undertook genetic analysis by constructing hybrid strains between BO and N2 strains. One of the hybrid strains exhibits a single locus of mutator activity, named *mut-4*, which maps on chromosome I. Two additional mutators, *mut-5* (on chromosome II) and *mut-6* (on chromosome IV), arise spontaneously in *mut-4* -harboring strains. This spontaneous appearance of mutator activity at new sites suggests that the mutator itself transposes. These genetic properties of the germline mutator activity parallel those of autonomous elements of other systems such as the *Ac/Ds* system of maize (Fedoroff, 1983) and the P factor and *mariner* element of *Drosophila* (Engels, 1983; Medhora et al., 1988). In all these systems there are autonomous and nonautonomous elements. However, the mutator for Tc1 transposition, if it is indeed an autonomous Tc1 element, remains unknown. In this connection, the microheterogeneity observed on some Tc1 elements may underlie the differences between inactive and autonomous Tc1 elements.

EMS-mutator strain. Tc1 activity is regulated in a tissue-specific manner. This was demonstrated by ethyl methanesulfonate (EMS)-mutants in which Tc1 transposition and excision are elevated in the germ line but unaffected in the soma (Collins et al., 1987). From 1,500 EMS-mutagenized N2/BO hybrid animals, eight strains were isolated for their elevated reversion frequencies of an *unc-54*:: Tc1 mutant (4 to 100 fold higher than the parental strain). Two of the eight EMS-induced mutations were mapped and one of them, the *mut-2* locus located on chromosome I, appeared very interesting since it elevated Tc1 transposition frequency 20 fold above the activity reported for BO (Moerman and Waterston, 1984; Eide and Anderson, 1985a).

The relationship between these EMS-induced mutators, (e.g., *mut-2*) and the naturally-occuring mutator activities in the BO strain, (e.g., *mut-4*) is unclear. The mutator activities could be transposons that promote transposition in "trans" of Tc1 elements, but also host regulatory genes, whose products may interact with transposons. An apparent distinction between these two types of mutants has been revealed by the fact that *mut-4* has been shown to activate only Tc1 in more than 25 mutations examined, whereas *mut-2* has been shown to cause transposition of both Tc1 and other transposable elements such as Tc3 and Tc4.

The strain carrying *mut-2*, called TR679, and some of its derivatives have been used for Tc1 transposon tagging. This strategy of gene cloning was first undertaken in BO strain, where the frequencies of spontaneous Tc1 induced mutants for *unc-54*, *lin-12* and *unc-22* genes are about 5 x 10^{-7}, 5x 10^{-5} and 10^{-4} respectively. For each of these loci, selective methods for identifying rare mutant individuals were used and led to the isolation of *lin-12* and *unc-22* genes (Greenwald, 1985; Moerman et al., 1986). However, for loci for which mutant selection is not possible, the frequency of Tc1 insertion may be too low to make BO useful for Tc1 transposon tagging. In these cases, mutator strains such as TR679 have been successfully used since the germline excision and transposition are elevated up to 100 fold. This allows the tagging of numerous genes of interest, e.g., *lin-14* gene (Ruvkun et al., 1989).

Futhermore, the high frequency of Tc1 insertion can be used in order to analyze genes that have been altered in targeted fashion. This is carried out on frozen mutant banks by PCR experiments using one primer that corresponds to the ends of the transposons and another that correspond to the gene of interest. PCR products are only obtained from animals in which transposons are inserted in the area of the analyzed gene (Zwaal et al., 1993).

Hybrid dysgenesis. In the course of mapping new mutator loci, an unusual finding was made which consisted of a high number of *unc-22* revertants in hybrids derived from certain crosses (Mori et al., 1990). This elevation of the excision rate of Tc1 at the *unc-22* locus might reflect unusual features in the regulation of Tc1 transposition, just as hybrid dysgenesis reveals aspects of the regulation of transposable element systems in *Drosophila.*

Hybrid dysgenesis is a syndrome of spontaneous traits, including sterility, male recombination, mutation, reversion, and chromosomal rearrangement, induced in hybrids derived from certain crosses (Bregliano and Kidwell, 1983). Three independent systems of hybrid dysgenesis are known in *Drosophila.*: P-M, I-R and E-H. In each of these systems active transposable elements have been characterized: the P factor (Rubin et al., 1981; Bingham et al., 1982) in the P-M system, the I factor (Bucheton et al., 1984; Fawcett et al., 1986; Abad et al., 1989) in the I-R system and the Hobo element (Blackman and Gelbart, 1989) in the H-E system. The syndrome of hybrid dysgenesis is caused by the elevated rate of transposition of these active transposable elements in each of these systems.The transpositional increasing in the hybrids results from alterations of mechanisms which normally regulate the activity of transposable elements. In the P-M as well as in the I-R system, a repressor activity of P and I factors exists in females of the P and I strains. Introduction of P or I factors into M or R oocytes which lack the repressor results in germline activation of P and I transpositions.

In the case of germline Tc1 activity, Mori et al. (1990) found that the reversion of *unc-22*:: Tc1 alleles is elevated in a range of 50 to 100 fold by certain crosses. In a *mut-6* IV strain, the *unc-22*:: Tc1 reversion is 10^{-3} whereas it is less than 10^{-6} in a non-mutator strain. However, in the *unc-22*:: Tc1 progeny of a cross between *mut-6* hermaphrodites and non-mutator males, the reversion observed is 10^{-1}. The reciprocal cross does not induce this enhancement of reversion. Similar results were obtained using a *mut-5* II strain. Therefore, this reversion enhancement appears to depend on a maternal component inherited from the mutator strain. However, since this phenomenon is observed in hybrids derived from hermaphrodites in which Tc1 is transpositionally active in the germline, the male can be from either a mutator or a non-mutator strain. Thus this phenomenon is quite different from either P-M and I-R hybrid dysgenesis in *D. melanogaster* in which males with an active transposable element must be crossed with females with no element or inactive elements to induce the phenomenon (Bregliano and Kidwell, 1983; Engels, 1989). There are nonetheless some similarities between enhancement of Tc1 excision and hybrid dysgenesis in *Drosophila.* The enhancement of Tc1 excision is initiated by intercrosses, as in dysgenesis in *Drosophila.* Also, both enhanced Tc1 excision rates and dysgenesis involve a maternal component. The nature of the crosses used to observe Tc1 excision demonstrates that *mut-5* and *mut-6* act dominantly.

Thus the mutators act as gain-of-function mutations and since the mutator is also transposable (Mori et al, 1988b), it is quite likely that the mutator itself provides an active transposase. In addition, the observations reported by the unusual excision rates of Tc1 from the *unc-22* locus imply the existence of a repressor of Tc1 activity in self-propagating mutator strains. Perhaps, Tc1 transpositional activity in these strains might then be a balance between the transposase and repressor, a balance which is disturbed by the interstrain crosses (Mori et al., 1990).

Mechanism Of Tc1 Transposition

Tc1 has one large ORF of 273 triplets which might encode the transposase of Tc1. The first experiments were conducted by using antibodies against peptides corresponding to the predicted putative transposase sequence. Although these antibodies are very specific in detecting transposase from recombinant sources, they do not reproducibly recognize a specific protein in nematode extracts (Schukkink and Plasterk, 1990; Abad et al., 1991a). Therefore, one can conclude that the amounts of transposase protein are very low in the nematode.

In order to produce high inducible expression of transposase and to test its DNA binding properties, its production was carried out in expression vectors. Two strategies were developed. In Plasterk's laboratory, they chose to get expression of transposase in *Escherichia coli* (Schukkink and Plasterk, 1990). The gene of the putative transposase was cloned behind the tac-promoter. However, expression in a bacterial system leads generally to insoluble protein with absence of post-translational modifications which could be required for normal function of the protein. Therefore, Abad et al. (1993) have chosen to overproduce the putative transposase in the baculovirus system, *Autographa californica* nuclear polyhedrosis virus (AcMNPV) (Doerfler, 1986; Lucknow and Summers, 1988). This virus propagates in culture insect cells and has strong temporally regulated promoters. The coding sequence of interest is inserted by *in vivo* recombination and proteins expressed in insect cells may be correctly folded with appropriate post-translational modifications (Bustos et al., 1988).

In these two stategies of overproduction of Tc1 transposase, it has been assumed that the 273 triplet ORF encodes for the transposase.

The transposase produced in *E.coli* is insoluble and a denaturation step followed by a slow renaturation is needed for its purification (Schukkink and Plasterk, 1990). The renatured protein is only soluble in 1M NaCl, but quickly precipitates in lower salt concentrations. Therefore the DNA binding of the purified transposase can not be tested in gel retardation assays. To assay transposase DNA binding activity, South-Western blot experiments were performed. It was found that Tc1 transposase has a strong affinity for DNA (both single stranded and double stranded) (Fig. 6). A fusion protein of Tc1 transposase with β-galactosidase also exhibits DNA binding and deletion derivatives of this fusion protein were tested for DNA binding. Although a deletion of 108 C-terminal amino acids does not affect DNA binding, transposase DNA binding is abolished by a deletion of 39 amino acids at the N-terminal region. This shows that the DNA binding domain of transposase is near the N-terminal region (Schukkink and Plasterk, 1990).

The transposase overproduced in baculovirus also exhibits poor solubility and solubilization must be done in 6M guanidinium chloride. However, instead of renaturation by several steps of decreasing NaCl concentrations, Abad et al. (1993) chosed renaturation by dilution, which allows testing the DNA binding of the protein by gel-retardation assays against different Tc1 sequences. Thus, it was shown that the transposase overexpressed in the baculovirus system binds *in vitro* DNA fragments covering, respectively, 279 bp and 210bp from the 5' and 3' terminus ends of the Tc1 element (the putative targets for transposase DNA binding). A non specific affinity for DNA was also demonstrated (Fig.7). The relationship between the binding of the Tc1 ends and non specific binding was assayed by filtration on nitrocellulose. The ability of non-relevant DNA to bind the transposase is 6 times (as compared with the $K_{0.5}$ values) or 20 times (as compared with the Kd values) lower than the relevant DNA (Fig. 8). The non specific DNA binding of Tc1 transposase could be the result of imperfectly folded protein leading to partial activity of transposase (Schukkink and Plasterk, 1990; Abad et al., 1993). However, high non specific DNA binding is also exhibited by other transposases such as those of the bacteriophage Mu, Tn 10 and the *Drosophila* P element

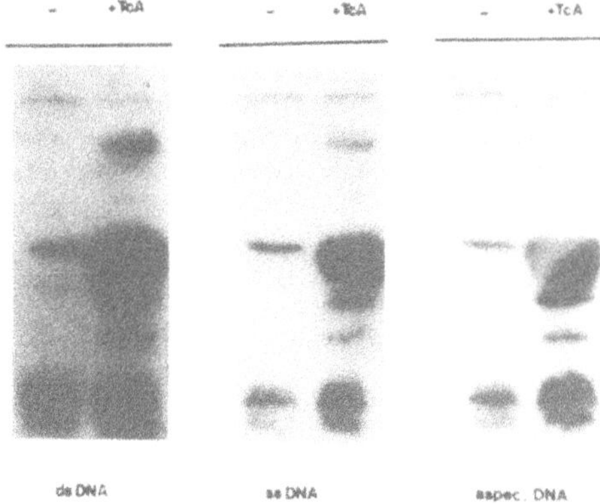

Figure 6 . DNA binding properties of TcA. South-western blot of lysates of the wild-type strain indicated (-) and TcA recombinant strain indicated (TcA). They were probed with the entirely Tc1 sequence (pIM55 plasmid) either in single strand DNA (ssDNA) or in double strand DNA (dsDNA). Aspecific DNA (Aspec-DNA) is salmon sperm (from Schukkink and Plasterk, 1990).

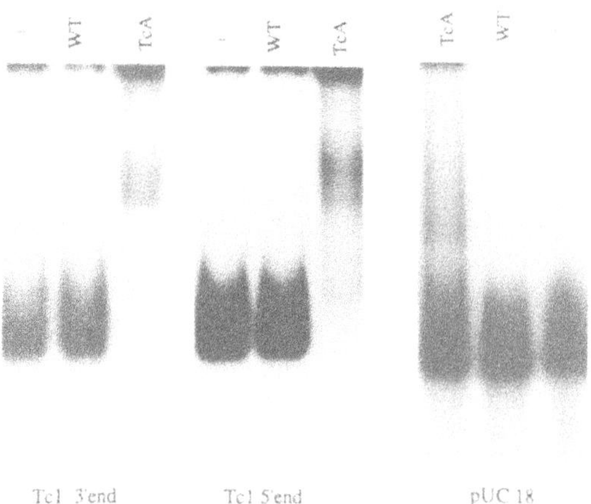

Figure 7 . (A) Gel-retardation assays with the 2 fragments covering 279 bp and 210 bp from 5' and the 3' terminus end of the Tc1 element, respectively. (B) Gel-retardation assays with the 216 bp *NdeI/Hin*dIII fragment from pUC18 (from Abad et al., 1993).

(Kaufman et al., 1989). This non specific DNA binding activity may also be involved in determining sites of transposable element insertion (Kaufman et al., 1989). The ratio value of approximatively 10 to 1 obtained for the equilibrium dissociation constants of Tc1 transposase binding to specific versus non specific DNA has been already observed for other transposases like those of P element (Kaufman et al., 1989) or prokaryotic transposons (Chandler, pers. comm.). The direct binding of transposase to the Tc1 ends indicates the existence of two families of binding sites for Tc1 ends. The first family of sites has a higher affinity (with the dissociation constant Kd, of 0.26 nM) compared to the second family, which exhibits decreased affinity (with the Kd value of 5.2 nM).

The existence of two families of binding sites for Tc1 DNA could be explained by two hypotheses. In the first one, these two families could result from different activities of two main

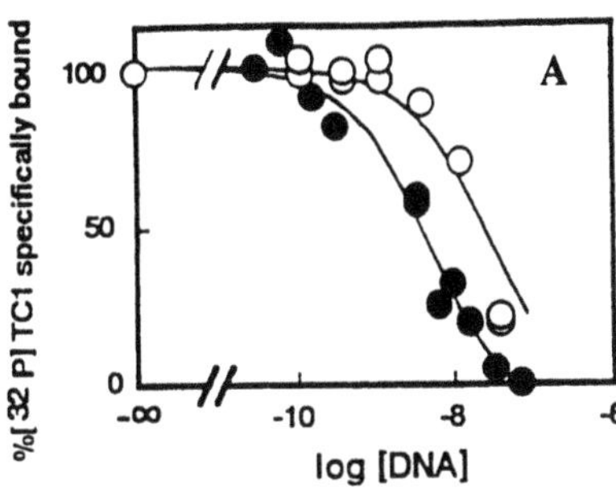

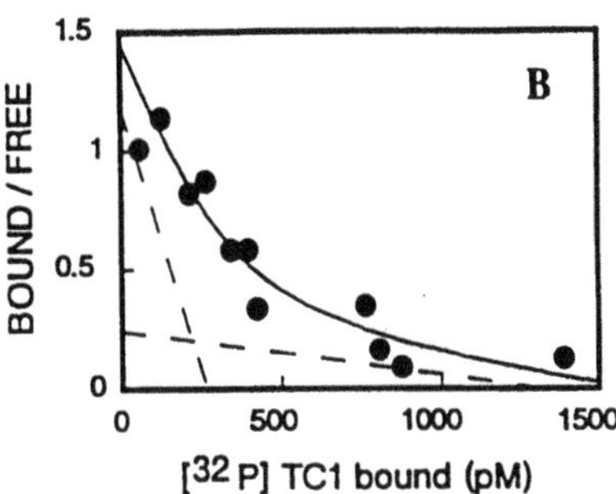

Figure 8 . (A) Protection of [^{32}P] Tc1 binding by unlabeled homologous (279 bp from 5' terminus end of the Tc1 element) and non-homologous (sonicated calf thymus sized at ≈ 300 bp) competitor DNA fragments. Increasing concentrations of the fragment of the Tc1 element (•) or sonicated calf thymus (o) were incubated with renatured TcA Then [^{32}P] labeled 5' end of the Tc1 element was added and the radioactivity bound was measured after an incubation of one hour. (B) Scatchard plot of the data concerning the specific direct binding of [^{32}P] labeled 5' end fragment of Tc1 element to the renatured TcA protein. TcA protein was incubated with increasing concentrations of [^{32}P] labeled 5' end fragment of Tc1 element (from Abad et al., 1993).

classes of transposase obtained after the denaturation/renaturation procedure which led to a kind of active transposase with high affinity and a kind of incorrectly folded protein with partial activity and low affinity for DNA. In the second hypothesis, the existence of two families of binding sites for Tc1 DNA could be correlated with results of protection experiments, where Tc1 transposase would be able to recognize specific motives of Tc1 but also less conserved sequences which might be found in non-relevant DNA. In this case, the first class of binding sites with high affinity could correspond to target sequences for the specific DNA binding activity of transposase and the second class with lower affinity could correspond to non-relevant sequences recognized by the non specific DNA binding activity of transposase. These two classes of DNA binding activity could play a role in the Tc1 transposition. The specific DNA binding activity of transposase may be involved in the specific interaction with Tc1 termini whereas the nonspecific DNA binding activity may be involved in determining sites of Tc1 insertion in the genome during the transposition process.

However, the non specific DNA activity exhibited by these overproduced putative transposases could be due to the fact that the true Tc1 transposase is encoded by a larger ORF than the generally assumed 273 triplet ORF (Plasterk, pers. comm.). In attempts to clarify the ORF encoding for the Tc1 transposase, the most 5' ATG start codon of Tc1 was fused to the *hsp-16* promoter. After establishing transgenic worms with this construct, the analysis of the heat shock induced RNA was done. A 41 nt intron was removed from the primary transcripts and the Tc1 putative transposase could therefore encode a 343 amino acid polypetide (Vos et al., 1993). The 343 amino acid putative transposase has been expressed in *E. coli* and a soluble protein was partially purified. Gel retardation assays have shown that this protein contains two DNA binding domains. A domain in the 69 N-terminal amino acids that binds a region of 5-25 bp from the transposon ends and a general DNA binding domain in the center of the transposase sequence. The specific DNA binding domain seems to be masked in the full size transposase, since only it was observed in truncated derivatives (Vos et al., 1993).

TC2

The Tc2 transposable element was first identified in an experiment designed to detect copies of the Tc1 transposon that might function as an autonomous component of a two-element system. When genomic DNA from any *C. elegans* strain is digested with *Eco*RV and probed in a Southern hybridization experiment with a Tc1 sequence, an intense band of 1.6kb is detected, representing multiple individual copies of Tc1. Therefore, it is possible to identify variants of Tc1 elements as restriction fragments with different mobilities. In the three Tc1-high-copy number strains, in addition to the 1.6 kb *Eco*RV fragment, an *Eco*RV fragment of

4.8 kb was observed. This fragment was further studied as a candidate for a different form of Tc1. In fact this 4.8 kb polymorphic Tc1 fragment contained another member of a repetitive element family, called Tc2 (Levitt and Emmons, 1989).

Tc2 elements are polymorphic in structure. The sequence diversity of the Tc2 element family may be due to a preponderance of rearranged, inactive copies of the elements, as is found for the *Ac/Ds* elements in maize (Doring and Starlinger, 1984) and *P* elements in *Drosophila* (Rubin et al., 1982). A transposed copy of Tc2 has been sequenced (Ruvolo et al., 1992). It is 2074 bp in length and has perfect inverted terminal repeats of 24 bp.The structure of the transposon suggests that it may have the capacity to code for a transposase protein. Three open reading frames on one strand show non-random codon usage and may represent exons. The first coding region is preceded by a potential CAAT box and TATA box. In addition to its inverted terminal repeats, Tc2 has subterminal degenerated direct repeats which are arranged in an irregular overlapping pattern. This pattern is the same as for *Spm* elements and seems to be an essential structure for transposition (Masson et al., 1987). In this repetitive region, the 12 bp motif of the *Spm* repeats is similar at seven positions to a 10 bp motif in Tc2 (Ruvolo et al., 1992). Tc2 element is not a member of the Tc1-like group of transposons, since it is 500 bp has shorter and more than one large open reading frame. The structure of Tc2 is more similar to the *Ac* and *Spm* elements of maize and P elements of *Drosophila*. Concerning the insertion sites, except for the TA dinucleotide, the sites of Tc1 and Tc2 are not similar to each other.

Tc2 is present at 20-25 copies in all of the high-copy-number Tc1 strains whereas all of the low-copy-number Tc1 strains contain 4 or 5 copies of Tc2. This suggests that the two elements may be regulated by the same genomic factor or one element may control the other. However, the last hypothesis is less plausible since these two elements are not structurally related. Conversely, in the *mut-4* strain which is known to activate Tc1, Tc2 is also activated. The fact that *mut-4* is itself mobile, makes this an especially attractive possibility, since the controlling elements in maize and *Drosophila* are known to be autonomous transposons. However, in those cases the controlling element usually has detectable sequence similarity with the elements it regulates. But *mut-4* and Tc2 do not crosshybridize and no single member of the polymorphic Tc2 family is linked with mutator activity (Levitt and Emmons, 1989).

Tc2 is highly active in the progeny of certain interstrain crosses. This was observed in Bristol/Bergerac (BO) recombinant strains and is frequent enough to be detected by blot hybridization. Four out the 17 recombinant strains contain Tc2-homologous restriction fragments which are not present in either parent. Since these observations involve a small number of events, it appears that Tc2 is significantly active in the recombinant strains.

TC3

Tc3 transposable elements were first identified in the mutator strain TR679 by analysing spontaneous mutations affecting the *unc-22* gene. The isolation and the characterization of three mutant alleles of *unc-22* gene by restriction maps of the inserted region indicate that the inserted elements are indistinguishable from each other, but are very different from *unc-22* mutant alleles induced by Tc1 (Collins et al., 1989).The cloning of the element responsible for *unc-22* (*r750*::Tc3) indicates that a complete Tc3 element (2.5kb) is inserted within a 2.6 kb restriction fragment of the wild type *unc-22* gene.

Tc3 is 2335 bp long with inverted repeats of 462 bp at its ends and exists in about 15 copies in the genomes of most *C.elegans* strains examined. Sequence analysis of Tc3 elements suggests that the transposase is encoded by a two-exon open reading frame of 987 nucleotides, separated by a 48 bp intron which is removed in mRNA encoding for Tc3 transposase (Tc3A) (Anderson, unpublished results; Van Luenen et al., 1993).

Tc3 transposition and excision in both somatic and germline cells are not detected in N2 and in all different wild-type varieties of *C. elegans*. However, in the strain TR679, Tc3 transposition and excision occur at high frequencies (Collins et al., 1989). In this strain, Tc3-induced mutations are unstable, and revertants result from precise or nearly precise excision of Tc3.

Although Tc1 and Tc3 do not hybridize to each other, they are clearly related. The terminal eight nucleotides of Tc3 are identical to eight of the nine terminal nucleotides of Tc1. Like Tc1, Tc3 inserts into a TA dinucleotide of the target site, and TA sequences flank the inserted Tc3 element. The similarities between Tc1 and Tc3 termini and between their sites of insertion suggest that these two elements are both activated in TR679 because *mut-2* affects a component that is common to the regulation of transposition of both transposons. One element of this common regulator could be the transposase encoded by these elements. However, the recent work of Van Luenen et al. (1993) has shown that Tc3A does not interact with Tc1 elements.

Since Tc1 and Tc3 are both activated in TR679, it is clear that activities of these two elements are not always correlated. The copy number and genomic positions of Tc3 are relatively constant in different wild-type varieties of *C. elegans* (Collins et al., 1989), whereas Tc1 copy number varies over 10 fold in the same tested strains (Emmons et al., 1983). Therefore the frequencies of Tc3 transcription and excision are low or absent in these genetic backgrounds and none of the 200 spontaneous mutations that have been isolated is due to Tc3 insertion (Eide and Anderson, 1985a,b; Greenwald, 1985; Moerman et al;, 1986; Collins et al., 1987; Eide and Anderson, 1988); these spontaneous mutations are due to Tc1 insertion. Furthermore, Tc1 exhibits a very high frequency of somatic excision, but such excisions cannot be detected for Tc3 (Emmons and Yesner, 1984; Collins et al., 1989). Since Tc3 is activated only in *mut-2* (*r459*), transposition and excision of Tc1 are activated at high levels in each of the eight different mutator strains (Collins et al., 1989).

Tc1 and Tc3 share some identical nucleotides in their termini. Sequence analysis of other transposable elements indicates that many of them with short terminal inverted repeats share common sequences at their termini. The comparison of inverted repeat termini of 9 different transposons isolated from 6 different species showed that each of these elements begins with the sequence CA at or near one terminus and ends with the sequence TG at or near the other terminus (Fig. 9) (Collins et al., 1989; Brezinsky et al., 1990). For some of these elements, similarities extend even farther into the element (Doring and Starlinger, 1986; Streck et al., 1986). The similarity of their termini may reflect common ancestry, cross-species transmission, or convergent evolution. Therefore their mechanism of transposition and excision may be similar as was recently shown for P and Tc1 elements (see above Engels et al., 1990; Plasterk, 1991).

Another common feature of transposons with inverted repeat termini is the heterologous size of individual elements. This is the case for P elements, *Ac/Ds* and *Spm/En* where defectives elements with large deletions or internal rearrangements are widespread (Finnegan, 1989). In contrast, most if not all copies of the Tc1 family and Tc3 family are very similar in structure (Emmons et al., 1983; Collins et al., 1989). One possible explanation is the occurrence of a strong active mechanism for maintaining the homogeneity of multigene families in *C. elegans*. Another explanation is that transposons in the genome of *C. elegans* result from recent events and therefore have not had sufficient time to accumulate deleted copies.

Element	Species	5'-Inverted repeat-3'
Tc1	*C. elegans*	**CAG-TG**CTG
Tc3	*C. elegans*	**CAG-TG**TGG
P	*D. melanogaster*	**CA--TG**ATG
Hobo	*D. melanogaster*	**CAG**AGAACT
Mariner	*D. mauritiana*	C**CAGGTG**TAC
Spm/En	*Z. mays*	**CACTACAA**G
Tam1	*A. majus*	**CACTACAA**C
Tam2	*A. majus*	**CACTACAA**C
Tgm1	*G. max*	**CACTA**TTAG

Figure 9 . Comparison of inverted repeat termini of nine transposable elements. The terminal inverted repeats of these elements are aligned to show their similarities in bold (from Collins et al., 1989).

Tc3 elements are immobile in the N2 genome. However, in transgenic lines expressing the *hsp-16/* Tc3A construct (the ORF of Tc3mRNA fused to a inducible strong heat-shock promoter of *C. elegans*), frequent excision and transposition of endogenous Tc3 elements were observed (Van Luenen et al., 1993). This shows that the Bristol N2 genome contains Tc3 elements that could be actived for transposition, but are immobile in a wild type N2 background because Tc3A is absent. However, N2 contains the same amount of Tc3A mRNA and the same ratio of unspliced to spliced mRNA as in the mutator strain TR679. One possible explanation is that in spite of the presence of Tc3A mRNA no transposase protein is produced in N2 genome. Alternatively, the transposase protein may be produced at the wrong time or the wrong place (Van luenen et al., 1993). Futhermore, the expression of Tc3A results in excision of Tc3 elements. In this extrachromosomal fraction, the Tc3 element is mainly as a linear fragment. These extrachromosomal copies may be intermediates in Tc3 transposition since this result is consistent with a model in which Tc3 transposase makes double strand breaks at the ends of Tc3 elements and then the excised Tc3 would be inserted into other genomic target sequence (Van Luenen et al., 1993).

Using recombinant Tc3A, Van luenen et al.(1993) show that Tc3A binds specifically to the terminal part of Tc3 but does not act on Tc1 element. In addition, in the *hsp-16*/ Tc3A transgenic lines, no extrachromosomal Tc1 elements or elevated Tc1 transposition are detected. Therefore, the *mut-2* in TR679 that both increased Tc1 and Tc3 transposition, seems to be a mutated host factor (Van Luenen et al., 1993).

TC4 AND TC5

The first member of the Tc4 family is Tc4-*n1416*. It is a fold-back element composed of two almost perfect 774bp terminal inverted repeats joined by a 57-bp unique internal sequence (Yuan et al., 1991). No long open reading frame was found within Tc4-*n1416*.

Tc4 elements have been shown to transpose in the mutator strain which carried a *mut-2* gene and Tc4-*n1416* was identified as an insertional mutation in *ced-4* gene (Yuan et al., 1991). Other mutation analysis indicates the presence of Tc4 elements. In Southern blots hybridized with a Tc4-*n1416* probe, additional copies of Tc4 elements in TR 679 which are not present in the parental strains, N2 and BO, have been identified. In these strains, these copies are estimated to contain about 20 copies of Tc4 per haploid genome (Yuan et al., 1991).

In their first analysis, Yuan et al. (1991) indicated that the Tc4 family is heterogeneous. Since other Tc4 sequenced elements such as Tc4-*n1416*, Tc4-*n1351* and Tc4-*rh1030* were proved to be 1.6 kb long, the Tc4-*n1030* is 3483 bp long (Li and Shaw, 1993). The major difference between Tc4-*n1416* and Tc4-*n1030* is a novel 2343 bp sequence in Tc4-*n1030* which replaces a 477 bp segment near the middle of the right inverted repeat of Tc4-*n1416* (Fig. 10), whereas the left inverted repeats of the two elements are nearly identical. Therefore, the fold-back structure of the Tc4-*n1030* element is drastically different from that of Tc4-*n1416*.

The novel sequence of Tc4-*n1030*, named Tc4v, is present at about 5 copies per haploid genome and appear to be fairly homogeneous in structure with few differences among these elements. Tc4v subfamily comprises approximately one-fifth and one-sixth of the Tc4 elements found in strains N2 and TR 679, respectively (Li and Shaw, 1993). Sequence analysis of three cDNA clones suggests that a Tc4v element contains at least five exons that could encode a novel basic protein of 537 amino acid residues with an appropriate polyadenylation signal and poly(A)tail. The coding region is located entirely within the novel Tc4v sequence. Therefore, Tc4v elements may encode a Tc4-specific transposase. The predicted novel polypeptide produced by Tc4v shows no similarity to any other protein present in the sequence databank. However, it may have similarity to the potential Tc5 polypeptide (Collins, unpublished results).

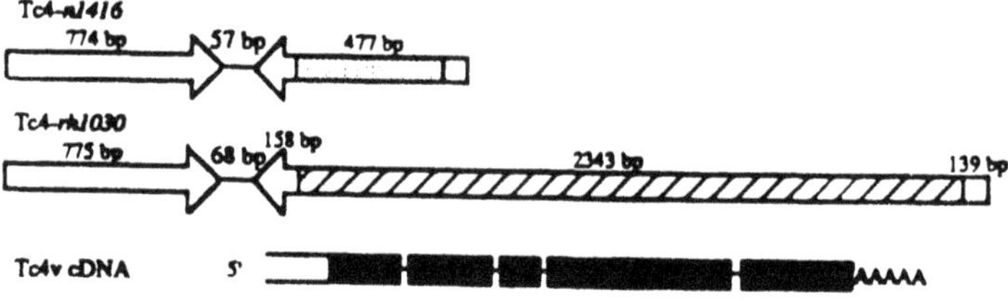

Figure 10 . Major features of Tc4-*n 1416* and Tc4-*rh 1030*. The dotted box represents the 477-bp segment in Tc4-*n 1416* which is replaced by a 2343-bp novel sequence in Tc4-*rh 1030*, represented by a striped box. At the bottom is represented the stucture of the Tc4 cDNA derived from the three cDNA clones. Filled boxes are ORFs. Open boxes represent the untranslated region (from Li and Shaw, 1993).

Tc4 transposition appears to be activated by the *mut-2* mutation carried by the strain TR 679, as Tc4v transcripts are more abundant in the mutator strain. A 1.6 kb long Tc4v RNA was detected in the mutator strain whereas it was absent from the N2 strain. Thus a correlation was found between activation of Tc4 transposition and enhanced Tc4v transcript levels. Therefore, one can imagine that the Tc4 transposon family may function as a two-element system, where deleted elements, e.g. Tc4-*n1416*, were trans-mobilized by autonomous components, e.g., Tc4-*n1030*.

The Tc5 transposon family is represented by elements of 3.2 kb which contain a long ORF possibly encoding for a transposase (Collins, unpublished results). From the DNA sequence, a mRNA composed of five exons which could encode a 495 amino acid protein is predicted (Collins, unplublished results). This protein shares significant amino acid identity (33%) with the putative Tc4v transposase (Li and Shaw, 1993). Like Tc1, Tc2, Tc3 and Tc4, Tc5 excises at high frequency from the germ line in *mut-2* strain. Only four to six copies of the element are present in non-*mut-2* strains. Like Tc3, Tc5 does not exhibit somatic excision detectable by Southern blot (Collins, unpublished results). The expression and the function of the Tc5 putative transposase are under investigation in N2, *mut-2*, and transgenic strains containing the Tc5 ORF fused to a heterologous promoter. In Northern blot, Tc5 RNA seems not to be present in N2 or *mut-2* strains, whereas 2.2 and 2.8 kb RNAs are detected in transgenic lines (Collins, unpublished results).

MARINER

The first mariner element was cloned from studies concerning the genome sequencing project (Sulston et al., 1993). Another copy was cloned in trying to clone *unc-80* gene, a gene which alters the sensitivity of the volatile anesthetics (Sedensky et al., 1993).They also found a mariner cDNA of 1.3 kb long. Unlike the insect mariner elements, the worm sequence lacks the inverted repeats on either side of the mariner sequence. However, like insect homologues, both flanking sequences are AT rich. There are 20 copies of mariner in the N2 strain and this number is roughly the same in Bergerac strains. However, three out of the four sequences vary among individual Bergerac strains. At the moment, we do not know whether mariner itself is mobile in BO strains. There is only one copy of mariner in *C. briggsae*, *C. remanii*, and *C. vulgaris* (Sedensky et al., 1993).

TRANSPOSABLE ELEMENTS IN OTHER NEMATODE SPECIES

Tc1-like sequences have been characterized in *C. briggsae*. Genomic hybridization of three strains of *C. briggsae* indicates that some transposable elements of this family, named Tcb (also known as barney element) are mobile (Harris et al., 1988). In the Indian strain G16, Tcb is represented by 30 bands that cross hybridize with Tc1 at low stringency (Fodor, unpublished results). Unlike the Tc1 element, this transposable element family seems to be quite heterogeneous. A composite barney element constructed from 2 deletion mutants, is 1, 616 bp long. It includes 80 bp imperfect ITR's and ends with the TA dinucleotide possibly representing a duplicated target site (Prasad et al., 1991).

Upon comparison with Tc1, the sequence identity of barney element is found to be restricted to the ORF. A barney element , Tcb1, has also a large ORF that starts and stops at the same position on the sequence as the large Tc1 ORF (Rosenzweig et al., 1983; Harris et al., 1988). These two ORF sequences are 71% identical at the DNA level and are 75% identical at the amino acid level (Harris et al., 1988).

In addition, similarities observed between ORF of Tc1 and Tcb1 elements are found to be conserved in ORFs of HB1 transposable elements from *D. melanogaster* and Uhu transposable elements from *D. heteroneura* (Harris et al., 1988, Brezinsky et al., 1990). Furthermore, Tc1, Tcb1, HB1 and Uhu show sequence identity in the inverted repeat termini (Fig. 11). The presence of related transposable elements in such distantly related phyla may reflect their presence in common ancestral genomes, horizontal transmission or convergent evolution (Brezinsky et al., 1990).

LEFT ITR

Nematodes	TC1	taCAGTGCTGGCCAAAAAGATATC
	TC3	taCAGTGTGGGAAAGTTCTATA
	BARNEY	taCAGTACTGGCCATAAAGAATGC
Drosophila	HB1	TAGCAGTGC
	HB2	TAGCAGTGC
	HB3	TAGCAGTGC
	HB4	TAGCAGTGC
	UHU	tataCAGTGTCTTACAGCTCAACTGG

Figure 11 . Comparison of nematodes Tc ITR's with ITR's from other transpososons in this class of elements (adapted from Brezinsky et al., 1990).

A transposable element, designated as PAT element, has been isolated from *Panagrellus redivivus*, a free-living nematode similar to *C. elegans* (Link et al., 1987). A mutated *Panagrellus unc-22* gene was cloned from a spontaneous "twitcher" mutant in the C15 strain. The mutated gene contained a 4.8kb insert of repetitive DNA. This repetitive element differs in copy number (10-50 copies) and location between the different *P. redivivus* strains. The PAT elements family has no detectable homology with Tc1 or any other sequence in *C. elegans*. Since the PAT element has direct terminal repeats of at least 170 bp, structurally it resembles a retrotransposon-like element (Link et al., 1987). A 900 bp transcript is detected when PAT sequences are used to probe a blot of C15 total RNA. Neither this transcript nor DNA sequences homologous to PAT are found in *C. elegans*. These experiments in *P. redivivus* suggest that the *unc-22* gene may be a general "transposon trap", and that some other transposons could be recovered by isolating spontaneous twitcher mutants.

Tas is the first retrovirus-like element found in any nematode species (Aeby et al., 1986). It was isolated by analysing clones at the junction of the satellite and non-satellite fraction from the germ line of the chromatin-eliminating nematode, *Ascaris lumbricoides*. It was named 'Tas' which stands for transposon-like element of *Ascaris*. This element is 7.5 kb long with two LTRs of 256 bp, each being bound by TG and CA. The first and last 12 nucleotides of these LTRs are inverted repeats. Thus Tas has a structure typical for retroviruses and related transposable elements such as *copia* in *Drosophila* (Xiong and Eickbush, 1990). The target site duplication of 5 bp upon insertion is a common feature of transposable elements (Aeby et al., 1986). The distribution of Tas element in the *Ascaris* genome shows the presence of 50 copies dispersed over 20 different chromosomal sites. Tas elements can be subdivided into two major classes according to their restriction patterns. The most abundant form, Tas-1, accounts for two thirds of Tas copies. Tas-2, the second form, is similar in length but contains two additional *Eco*RI sites (Aeby et al., 1986). These two subfamilies are different in structure but also in behavior since chromatin diminution only removes some Tas-1 elements but all Tas-2 copies.

STRATEGIES FOR CLONING TRANSPOSONS IN PLANT-PARASITIC NEMATODES

With a view to cloning transposons in plant-parasitic nematodes, four main general strategies can be applied.

The first one is to analyze repetitive sequence families of the genome. Very often these repetitive sequences have structural features common to transposable elements. Since transposable elements are mobile, their positions in the genome can be observed to change during relatively short periods of time, such as those separating strains of a single species. Therefore, analysis of polymorphisms between different strains could indicate the presence of transposable elements. The Tc1 element was initially identified as a mobile repetitive sequence using this strategy.

The second one is to probe sequences corresponding to cloned transposable elements on genomic DNA from plant-parasitic nematodes. This experiment was previously carried out by Abad et al. (1991b) in order to examine the distribution and the conservation of homologues to Tc transposons in different nematode species belonging to the Secernentea class. On the basis of hybridization patterns, most of the Tc transposons display a patchy distribution restricted to the Rhabditidae family with generally poor conservation. Conversely, Tc3 is widely distributed among nematode species. Tc3 homologues seem to be present in the majority of the Rhabditidae but also in two genera within the Panagrolaimidae family, and in the plant-parasitic nematode *Bursaphelenchus*, which belongs to the Aphelenchida order. However, due to the weak hybridization signal obtained in low stringency conditions, it seems risky to undertake the cloning of Tc- homologous elements in plant-parasitic nematodes by the strategy of genomic library hybridization with the previously cloned transposable elements of *C. elegans*.

The third approach is to look for homologous sequences by PCR experiments using primers from conserved domains of transposable elements. For example, Tc1-like elements seem to be widely dispersed since Tc1 homologues are found in *Drosophila* and in fish (Brezinsky et al., 1990; Heierhorst et al., 1992). These elements share significant sequence identity with some short conserved domains. In the case of retrotransposons, reverse transcriptase domains are also very conserved at the protein level between different members of this class of transposons which are distributed throughtout the animal and plant kingdoms (Xiong and Eickbush, 1990). However, the occurrence of homologous transposons in a species does not necessarily imply the presence of mobile copies of these transposons and further work is need to use transposons for gene tagging experiments.

The last strategy is to look for phenotypic mutants because it has been assumed that most natural mutations are due to transposon insertions in genes. Therefore, the first step is to clone the homologous target gene in plant-parasitic nematode species by using probes from *C. elegans* genes. At the moment, it is the only way, since the genetics of plant-parasitic nematodes is very poor. The comparison of the wild-type gene and the mutated gene allows the isolation of transposable elements. This strategy has been succesfully applied for the isolation of PAT element in *P. redivivus*, in which spontaneous twitcher mutants resulted from transposon insertion into the *unc-22* gene. However, this strategy is restricted by the sequence homology of the target gene between *C. elegans* and plant-parasitic nematodes.

REFERENCES

Abad, P., Cerutti, M., and Devauchelle, G., 1991a, DNA binding activity of the putative transposase of Tc1 element (TcA) overexpressed in baculovirus system, *8th International C. elegans Meeting, Madison, WI, USA*.

Abad, P., Cerutti, M., Pauron, D., Quiles, C., Palin, B., Devauchelle, G., and Dalmasso, A., 1993, Expression and biochemical characterization of the DNA binding activity of TcA, the putative transposase of *Caenorhabditis elegans* transposable element Tc1, *Biochem. and Biophy. Res. Comm.* 192 : 1445.

Abad, P., Quilès, C., Tarès, S., Piotte, C., Castagnone-Sereno, P., Abadon, M., and Dalmasso, A., 1991b, Sequences homologous to Tc(s) transposable elements of *Caenorhabditis elegans* are widely distribued in the phylum of nematoda, *J. Mol. Evol.* 33 : 251.

Abad, P., Vaury, C., Pélisson, A., Chaboissier, M.C., Busseau, I., and Bucheton, A., 1989, A long interspersed repetitive element - the I factor of *Drosophila teissieri* - is able to transpose in different *Drosophila* species, *Proc. Natl. Acad. Sci.* USA 86 : 8887.

Aeby, P., Spicher, A., De Chastonay, Y., Muller, F., and Tobler, H., 1986, Structure and genomic organization of proretrovirus-like elements partially eliminated from the somatic genome of *Ascaris lumbricoides*, *EMBO J.* 5 : 3353.

Anderson, P., Emmons, S.W., and Moerman, D.G., 1992, Discovery of Tc1 in the nematode genome *Caenorhabditis elegans*, *in*: "The dynamic genome," N. Federoff, ed., Cold Spring Harbor Laboratory Press, New York.

Benian, G.M., Kiff, J.E., Neckelmann, N., Moerman, D.G., and Waterston, R.H., 1989, Sequence of an unusually large protein implicated in regulation of myosin activity in *C. elegans*, *Nature* 342 : 45.

Berg, D.E., Lodge, J., Sasakawa, C., Nag, D.K., Phadnis, S.H., Weston-Hafer, K., and Carle, G.F., 1984, Transposon Tn5 : specific sequence recognition and conservative transposition, *Cold Spring Harbor Symp. Quant. Biol.* 49 : 215.

Bingham,P.M., Kidwell, M.G., and Rubin, G.M., 1982, The molecular basis of P-M hybrid dysgenesis: the role of the P element, a P strain-specific transposon family, *Cell* 29: 995.
Blackman, R.K., and Gelbart, W.M., 1989, The transposable element *hobo* of *Drosophila melanogaster*, *in* "Mobile DNA," D.E. Berg and M.M. Howe, eds., American Society for Microbiology, Washington DC.
Bonas, U., Sommer, H., and Saedler, H., 1984, The 17kb Tam1 element of *Antirrhinum majus* induces a 3bp duplication upon integration into the chalcone synthase gene, *EMBO J.*, 3 : 1015.
Bregliano, J.C., and Kidwell, M.G., 1983, Hybrid dysgenesis determinants, *in* "Mobile genetic elements," J. Shapiro, ed., Academic Press, New-York.
Brezinsky, L., Wang, G.V.L., Humphreys, T., and Hunt, J., 1990, The transposable element Uhu from Hawaïan *Drosophila* member of the widely dispersed class of Tc1-like transposons, *Nucleic Acids Res.* 18: 2053.
Bucheton, A., Paro, R., Sang, H.M., Pelisson, A., and Finnegan, D.J., 1984, The molecular basis of I-R hybrid dysgenesis in *Drosophila melanogaster* : Identification, cloning and properties of I factor, *Cell.* 338 : 153.
Bustos, M.M., Luckow, V.A., Griffing, L.R. Summers, M.D., and Hall, T.C., 1988, Expression glycosylation and secretion of phaseolin in a baculovirus system, *Plant Mol. Biol.* 10 : 475.
Coen, E.S., Carpenter, R., and Martin, C., 1986, Transposable elements generate novel spatial patterns of gene expression in *Antirrhinum majus*, *Cell.* 47 : 285.
Coen, E.S., Robbins, T.P. Almeida, J., Hudson, A., and Carpenter, R., 1989, Consequences and mechanisms of transposition in *Antirrhinum majus*, *in* "Mobile DNA," D.E.Berg and M.H. Howe, eds., American Society for Microbiology, Washington DC.
Collins, J., Forbes, E., and Anderson, P., 1989, The Tc3 family of transposable genetic elements in *Caenorhabditis elegans*, *Genetics* 120 : 621.
Collins, J., Saari, B., and Anderson, P., 1987, Activation of a transposable element in the germ line but not the soma of *Caenorhabditis elegans*, *Nature* 328 : 726.
Copeland, N.G., Jenkins, N.A., and Lee, B.K., 1983, Association of the lethal yellow (A^y) coat color mutation with an ecotropic murine leukemia virus genome, *Proc. Natl. Acad. Sci.* USA 80 : 247.
Doerfler, W., 1986, Expression of the *Autographa californica* nuclear polyhedrosis virus genome in insect cells: Homologous viral and heterologous vertebrate genes - The baculovirus system, *Curr. Top. Microbiol. Immunol.* 131 : 51.
Doring, H.P., and Starlinger, P., 1984, Barbara Mc-Clintock's controlling elements: Now at the DNA level, *Cell* 39: 253.
Doring, H.P., and Starlinger, P., 1986, Molecular genetics of transposable elements in plant, *Annu. Rev. Genet.* 20: 175.
Eide, D., and Anderson P., 1985a, Transposition of Tc1 in the nematode *C.elegans*, *Proc. Natl. Acad. Sci.* USA 82 : 1756.
Eide, D., and Anderson P., 1985b, The gene structures of spontaneous mutations affecting a *Caenorhabditis elegans* myosin heavy chain gene, *Genetics* 109 : 67.
Eide, D., and Anderson, P., 1988, Insertion and excision of the *Caenorhabditis elegans* transposable element Tc1, *Mol.Cell.Biol.* 8 : 737.
Emmons, S.W., Klass, M.R., and Hirsh, D., 1979, Analysis of the constancy of DNA sequences during the development and evolution of the nematode *Caenorbabditis elegans*, *Proc. Natl. Acad. Sci.* USA 76 : 1333.
Emmons, S. W., Roberts, S., and Ruan, K.S., 1986, Evidence in a nematode for regulation of transposon excision by tissue-specific factors, *Mol. Gen. Genet.* 202 : 410.
Emmons, S. W., Ruan, K.S., Levitt, A., and Yesner, L., 1985, Regulation of Tc1 transposable elements in *Caenorbabditis elegans*, *Cold Spring Harbor Symp. Quant. Biol.* 50 : 313.
Emmons, S.W., and Yesner, L., 1984, High-frequency excision of transposable element Tc1 in the nematode *Caenorhabditis elegans* is limited to somatic cells, *Cell* 36 : 599.
Emmons, S.W., Yesner, L., Ruan, K.S., and Katzenberg, D., 1983, Evidence for a transposon in *Caenorhabditis elegans*, *Cell* 32 : 55.
Engels, W.R., 1983, The P Family of transposable elements in *Drosophila*, *Annu. Rev. Genet.* 17 : 315.
Engels, W.R., 1989, P elements in *Drosophila melanogaster*, *in* "Mobile DNA," D.E. Berg and M.H. Howe ,eds., American Society for Microbiology, Washington DC.
Engels, W.R., Johnson-Schlitz, D.M., Eggleston, W.B., and Sved, J., 1990, High-frequency P lossin *Drosophila* is homolog dependent, *Cell* 62 : 515.
Fawcett, D.H., Lister, C.K., Kellett, E., and Finnegan, D.J., 1986, Transposable elements controlling I-R hybrid dysgenesis in *D. melanogaster* are similar to mammalian LINEs, *Cell* 47 : 1007.
Fedoroff, N., 1983, Controlling elements in maize, *in* "Mobile Genetic Elements," J.A. Shapiro, ed., Academic Press, Inc., N.Y.
Fedoroff, N., Wessler, S., and Shure, M., 1983, Isolation of the transposable maize controlling element *Ac* and *Ds*, *Cell* 35 : 243.
Finnegan, D.J., 1989, Eukaryotic transposable elements and genome evolution, *Trends Genet.* 5 : 103.

Freund, R., and Meselson, M., 1984, Long terminal repeat nucleotide sequence and specific insertion of the gypsy transposon, *Proc. Natl. Acad. Sci.* USA 81 : 4462.

Greenblat, I.M., and Brink, R.A., 1962, Twin mutations in medium variegated pericarp maize, *Genetics* 47 : 489.

Greenwald, I., 1985, *lin-12*, a nematode homeotic gene, is homologous to a set of mammalian proteins that includes growth factor, *Cell* 43 : 583.

Grindley, N. D. F., 1978, IS1 insertion generates duplication of a nine base pair sequence at its target site, *Cell* 13 : 419.

Halling, S., and Kleckner, N., 1982, A symmetrical six-base-pair target site sequence determines Tn10 insertion specificity, *Cell.* 28 : 155.

Harris, L.J., Baillie, D.L. and Rose, A.M., 1988, Sequence identity between an inverted repeat family of transposable elements in *Drosophila* and *Caenorhabditis*, *Nucleic Acids Res.* 16 : 5991.

Harris, L.J., and Rose, A.M., 1989, Structural analysis of Tc1 in *Caenornabditis elegans* var. Bristol (strainN2), *Plasmid* 22 : 10.

Heierhorst, J., Lederis, K., and Richter, D., 1992, Presence of a member of the Tc1-like transposon family from nematodes and *Drosophila* within the vasotocin gene of a primitive vertebrate, the pacific hagfish *Eptatretus stouti*, *Proc. Natl. Acad. Sci.* USA 89 : 6798.

Herman, R., and Shaw, J., 1987, The transposable genetic element Tc1 in the nematode *Caenorhabditis elegans*, *Trends Genet.* 3 : 222.

Ikenaga, H., and Saigo, K., 1982, Insertion of a movable genetic element, 297, into the TATA box for the H3 histone gene of *Drosophila melanogster*, *Proc. Natl. Acad. Sci.* USA 79 : 4143.

Inouye, S., Yuki, S., and Saigo, K., 1984, Sequence-specific insertion of the *Drosophila* transposable genetic element 17.6, *Nature* 310 : 32.

Kaufman, P.D., Doll, R.F., and Rio, D.C., 1989, *Drosophila* P element transposase recognizes internal P element DNA sequences, *Cell.* 59 : 359.

Kiff, J.E., Moerman,D.G., Schriefer, L.A., and Waterston, R.H., 1988, Transposon-induced deletions in *unc-22* of *Caenorhabditis elegans* associated with almost normal gene activity, *Nature* 331 : 631.

Levitt, A., and Emmons, S.W., 1989, The Tc2 transposon in *Caenorhabditis elegans*, *Proc. Natl. Acad. Sci.* USA 86 : 3232.

Li, W., and Shaw, J.E., 1993, A variant Tc4 transposable element in the nematode *C. elegans* could encode a novel protein, *Nucleic Acids Res.*, 21, 59.

Liao, L.W., Rosenzweig B., and Hirsh, D., 1983, Analysis of a transposable element from *Panagrellus redevivus*, *Proc.. Natl. Acad. Sci. U.S.A* 84 : 5325.

Link, C., Graft-Witsel, J., and Wood, W.B., 1987, Isolation and characterization of a nematode transposable element from *Panagrellus redivivus*, *Proc. Natl. Acad. Sci. U.S.A.* 84 : 5325.

Luckow, V.A., and Summers, M.D., 1988, Trends in the development of baculovirus expression vectors, *BioTechnology* 6: 47.

Masson, P., Surosky, R., Kingsbury, J., and Fedoroff, N.V., 1987, Genetic and molecular analysis of the *Spm* dependent a-m2 alleles of the maize a locus, *Genetics* 177 : 117.

Medhora, M.M., MacPeek, A.H., and Hartl, D.L., 1988, Excision of the *Drosophila* transposable element *mariner* : identification and characterization of the *Mos* factor, *EMBO J.* 7 : 2185.

Moerman, D.G., and Baillie, D.L., 1979, Genetic organization in *Caenorhabditis elegans* : fine structure analysis of the *unc-22* gene, *Genetics* 91 : 95.

Moerman, D.G., Beninan, G.M., Barstead, R.J., Schiefer, L., and Waterston, R.H., 1988, Identification and intracellular localization of the *unc-22* gene product of *Caenorhabditis elegans*, *Genes Dev.* 2 : 93.

Moerman, D.G., Beninan, G.M., and Waterston, R.H., 1986, Molecular cloning of the muscle gene *unc-22* in *Caenorhabditis elegans* by Tcl transposon tagging, *Proc. Natl. Acad. Sci. USA* 83 : 2579.

Moerman, D.G., and Waterston, R.H., 1984, Spontaneous unstable *unc-22 IV* mutations in *C. elegans* var. Bergerac, *Genetics* 108 : 859.

Moerman, D.G., and Waterston, R.H., 1989, Mobile elements in *Caenorhabditis elegans* and other nematodes, *in* "Mobile DNA," D.E. Berg and M.H. Howe ,eds., American Society for Microbiology, Washington DC.

Mori, I., Benian, G.M., Moerman, D.G., and Waterston, R.H., 1988a, The transposon Tcl of *C. elegans* recognizes specific target sequences for integration, *Proc. Natl. Acad. Sci.* USA 85 : 861

Mori, I., Moerman, D.G., and Waterston, R.H., 1988b, Analysis of a mutator activity necessary for germline transposition and excision of Tc1 transposable elements in *Caenorhabditis elegans*, *Genetics* 120 : 397.

Mori, I., Moerman, D.G., and Waterston, R.H., 1990, Interstrain crosses enhance excision of Tc1 transposable elements in *Caenorhabditis elegans*, *Mol. Gen. Genet.* 220 : 251.

O'Hare, K., and Rubin, G.M., 1983, Structure of P transposable elements and their sites of insertion and excision in the *Drosophila melanogaster* genome, *Cell* 34 : 25.

Plasterk, R.H.A., 1987, Differences between Tcl elements from the *C. elegans* strain Bergerac, *Nucleic Acids Res.* 15 : 10050.

Plasterk, R.H.A., 1991, The origin of footprints of the Tc1 transposon of *Caenorhabditis elegans*, *EMBO J.*, 10 : 1919.
Plasterk, R.H.A., and Groenen, J.T.M., 1992, Targeted alterations of the *Caenorhabditis elegans* genome by transgene instructed DNA double strand break repair following Tc1 excision, *EMBO J.* 11: 287.
Prasad, S.S., Harris, L.J., Baillie, D.L., and Rose, A.M., 1991, Evolutionary conserved regions in Caenorhabditis transposable elements deduced by sequence comparison,*Genome* 34 : 6.
Roeder, G.S. and Fink, G.R., 1983, Transposable elements in yeast, *in* "Mobile Genetic Elements" J.A. Shapiro, ed., Academic Press, Inc. New-York.
Rose, A.M., Harris, L.J., Mawji, N.R., and Morris, W.R., 1985, Tcl (Hin) : a form of the transposable element Tcl in *Caenorhabditis elegans*, *Can. J. Biochem. Cell. Biol.* 63 : 752.
Rose, A.M., and Snutch, T.P., 1984, Isolation of the closed circular form of the transposable element Tcl of *Caenorhabditis elegans*, *Nature* 311 : 485.
Rosenzweig, B., Liao, L., and Hirsh, D., 1983, Sequence of the *C. elegans* transposable element Tcl, *Nucleic Acids Res.* 11: 4201.
Ruan, K.S., and Emmons, S.W., 1984, Extrachromosomal copies of transposon Tcl in the nematode *Caenorhabditis elegans*, *Proc. Natl. Acad. Sci.* USA 81 : 4018.
Rubin, G.M., Kidwell, M.G. and Bingham, P.M., 1981, The molecular basis of P-M hybrid dysgenesis : The nature of induced mutations, *Cell* 29 : 987.
Ruvkun G., Ambros, V., Coulson, A., Waterston, R., Sulston, J., and Horvitz, H.R., 1989, Molecular genetics of the *Caenorhabditis elegans* heterochronic gene lin-14, *Genetics* 121 : 501.
Ruvolo, V., Hill, J.E., and Levitt, A., 1992, The Tc2 Transposon of *Caenorhabditis elegans* has the structure of a self-regulated element, *DNA and Cell. Biol.* 11: 111.
Schukkink, R.F., and Plasterk, R.H.A., 1990, TcA, the putative transposase of the *C. elegans* Tc1 transposon, has an N-terminal DNA binding domain, *Nucleic Acids Res.* 18: 895.
Schwartz-Sommer, Z., Gierl, A., Cuypers, H., Peterson, P.A., and Saedler, H., 1985, Plant transposable elements generate the DNA sequence diversity needed in evolution, *EMBO J.* 4 : 591.
Searles, L., Greenleaf, A., Kemp, W., and Voelker, R., 1986, Sites of P element insertion and structures of P element deletions in the 5' region of *Drosophila melanogaster RpII215*, *Mol. Cell. Biol.* 6 : 3312.
Sedensky, M., Hudson, S., Everson, B., and Morgan, P., 1993, Mariners grand finale, *9th International C. elegans Meeting*, Madison, WI, USA.
Shure, M., Wessler, S., and Fedoroff, N., 1983, Molecular identification and isolation of the *Waxy* locus in maize, *Cell* 35 : 235.
Streck, R.D., MacGaffey, J.E. and Beckendorf, S.K., 1986, The structure of *hobo* transposable elements and their insertion sites, *EMBO J.* 5 : 3615.
Sulston, J., Ainscough, R., Berks, M., Coulson, A., Craxton, M., Dear, S., Du, Z., Durbin, R.K., Gleeson, T., Green, P., Halloran, N., Hawkins, T., Hillier, L., Metzstein, M., Qiu, L., Staden, R., Thierry-Mieg, J., Thomas, K., Wilson, R., and Waterston, R., 1992, The *C. elegans* sequencing project : A begining, *Nature 356* : 37.
Sutton, W., Gerlach, W., Schwartz, D., and Peacok, W., 1984, Molecular analysis of *Ds* controlling element mutations at the *Ash1* locus of maize, *Science* 223 : 1265.
Tsubota, S., and Schedl, P., 1986, Hybrid dysgenesis-induced revertans of insertions at that 5' end of the *rudimentary* gene in *Drosophila melanogaster* : transposon-induced control mutations, *Genetics* 114 : 165.
Van Luenen, H.G.A.M., Colloms, S.D., and Plasterk, R.H.A., 1993, Mobilization of quiet, endogenous Tc3 transposons of *Caenorhabditis elegans* by forced expression of Tc3 transposase, *EMBO J.* 12 : 2513.
Vos, J.C., Van Luenen, H.G.A.M., and Plasterk, R.H.A., 1993, Characterization of *Caenorhabditis elegans* Tc1 transposon in vivo and in vitro, *Genes Dev.* 7 : 1244.
Waterson, R.H., Moerman, D.G., Benian, G.M., Barstead, R.J., Mori I., and Francis, R., 1986, Muscle genes and proteins in *Caenorhabditis elegans*, *Mol. Cell. Biol.* 29 : 605.
Xiong, Y., and Eickbush, T.H., 1990, Origin and evolution of retroelements based upon their reverse transcriptase sequences, *EMBO J.* 9 : 3353.
Yuan, J., Finney, M., Tsung, N., and Horvitz, R., 1991, Tc4, a *Caenorhabditis elegans* transposable element with an unusual fold-back structure, *Proc. Natl. Acad. Sci.* USA 88 : 3334.
Zerbib, D., Gamas, P., Chandler, M., Prentki, P., Bass, S., and Galas, D., 1985, Specificity of insertion of IS1, *J. Mol. Biol.*, 185 : 517.
Zwall, R.R., Broeks, A., Van Meurs, J., Groenen, J.T.M., and Plasterk, R.H.A., 1993, Target-selected gene inactivation in *Caenorhabditis elegans* by using a frozen transposon insertion mutant bank, *Proc. Natl. Acad. Sci.* USA 90 : 7431.

DISCUSSION

Niebel asked the perspectives of the use of heterologous transposon systems, such as Tc1, in plant parasitic nematodes.

Abad replied that there are two limitations for the use of Tc1 in plant parasitic nematodes. First the need to succeed in plant parasitic nematode transformations and second the need to test the excision and the transposition of Tc1 in a range of different nematodes species. If one considers the P element of *Drosophila melanogaster*, it has been shown that the P element does'nt excise outside species of Drosophilidae. Therefore it seems that the transposition of invertebrate transposable elements is restricted to relatively closely related taxa. Until now, we have no idea about transposition of Tc1 outside the *C.elegans* species.

LINKAGE MAPPING IN POTATO CYST NEMATODES

Jeroen N.A.M. Rouppe van der Voort, Jan Roosien,
Peter M. van Zandvoort, Rolf T. Folkertsma, Ellen L.J.G.
van Enckevort, Richard Janssen,[1] Fred J. Gommers and Jaap Bakker

Department of Nematology, Wageningen Agricultural University
P.O. Box 8123, 6700 ES Wageningen, The Netherlands

INTRODUCTION

A Mendelian proof for a gene-for-gene relationship between virulence in *Globodera rostochiensis* and the H_1 resistance gene from *Solanum tuberosum* spp. *andigena* CPC 1673 was obtained by Janssen et al., (1991). It was shown that virulence to the H_1 gene is recessively inherited at a single locus. As expected from the epigenetic nature of sex determination, this locus is not sex-linked. The resistance conferred by the H_1 gene is only effective against avirulent juveniles developing into females. It was shown that avirulent juveniles are still able to develop into males on the resistant cultivar (Janssen et al., 1992).

As a strategy to localise the avirulence gene, a backcross scheme was designed to find DNA markers that are tightly-linked to the avirulence locus. Screening of the progeny is done using random amplified polymorphic DNA (RAPD) markers. The RAPD technique allows linkage mapping in organisms for which only minute amounts of genomic DNA are available (Williams et al., 1990; Tulsieram et al., 1992). In this report the backcross scheme and the usefulness of the RAPD technique for linkage mapping in potato cyst nematodes are evaluated.

ORIGIN OF AVIRULENT AND VIRULENT LINES

Pure avirulent or virulent populations of *G. rostochiensis* are very unlikely to be found in the field. Since pure lines are required for a genetic approach to study virulence in potato cyst nematodes, a nematode-breeding program was carried out. A system was used that allowed controlled crosses between virgin females and males. About five years breeding and selection resulted in two lines homogeneous for virulent and avirulent alleles to the H_1 resistance gene (Janssen et al., 1990). Some of the outlines of this program are given below.

---

[1]Present address: DLO-Centre for Plant Breeding and Reproduction Research (CPRO-DLO) P.O.Box 16, 6700 AA Wageningen, The Netherlands.

For the production of a virulent line, virgin females from a relatively virulent field population Ro_5-Harmerz, were grown on roots of the resistant potato cultivar Saturna, carrying the H_1 gene, and crossed with a single male (AA, Aa or aa). The cysts of these controlled crosses (F_1) were multiplied on a resistant potato cultivar. From this full sib mating, juveniles were subsequently inoculated on cultivar Saturna to obtain an F_2, F_3, and F_4 generation. An avirulent line was selected following a similar procedure except that the susceptible cultivar Eigenheimer was used. This line originated from a field population Ro_1-Mierenbos.

Crosses between the avirulent line Ro_1-19 (genotype AA) and the virulent line Ro_5-22 (genotype aa) showed that virulence to the H_1 resistance gene is autosomally inherited at a single locus (Janssen et al., 1990). Resistance of the H_1 gene from *S. tuberosum* spp. *andigena* CPC 1673 is only effective against avirulent females (genotype AA or Aa). Only homozygous recessive females (aa) are able to induce a proper syncytium. The H_1 gene is ineffective against males, irrespective their genotype (AA, Aa, aa).

BACKCROSS SCHEME

To obtain progeny lines for segregation analysis we developed a backcross scheme (Figure 1). In the P generation, 26 females of line Ro_5-22 were crossed with males of line Ro_1-19. In the F_1, crossing-over events between parental and maternal chromosomes can occur. From this F_1, virgin females were back crossed with males of line Ro_5-22 resulting in B_1-individuals. Virulence was expected to segregate in a 1:1 ratio since heterozygous females (Aa) were backcrossed with virulent males (aa). This was tested by inoculating one quarter of the juveniles of the B_1 on the root tips of the susceptible cultivar Eigenheimer. The remaining part of the juveniles was inoculated on the resistant cultivar Saturna. The percentages of juveniles developing into females on the resistant cultivar were used as a measure for the number of virulent individuals in the B_1. An additional proof that virulence was inherited at a single locus was given since virulence segregated in a 1:1 ratio.

The cultivar Saturna carrying the H_1 gene was used to select 486 homozygous recessive females which were independently back crossed with males of line Ro_5-22. Since the H_1 gene is not effective against males, this second backcross is essential. Selfing of B_1-individuals is not possible because avirulent males (AA or Aa) are also able to develop. The B_2-individuals were multiplied independently by inoculating potato plants with juveniles from single cysts. To prevent aberrant distortions of allele frequencies in the progeny, only those B_2-lines were maintained that produced more than 20 cysts after the first multiplication. Selfing B_2-lines for three to four generations gave rise to 125 useful progeny lines.

SEGREGATION OF VIRULENCE AND RAPD MARKERS.

To enable the detection of sufficient polymorphisms between the parents, the lines were selected from field populations having distinct gene pools. The populations Ro_1-Mierenbos (Wageningen, The Netherlands) and Ro_5-Harmerz (Germany) differ widely in their capability to overcome various genes for resistance present in potato. The protein composition of females of the original field populations was studied with two dimensional gel electrophoresis. It was demonstrated that approximately 5% of the protein spots were distinct (Bakker and Bouwman-Smits, 1988a). This order of intraspecific variation at the protein level has been reported for various organisms such as fruit flies, rodents and man (Ayala, 1983). With regard to the interspecific variation between *G. rostochiensis* and *G. pallida* it is noted that these species are differentiated by 70% of their polypeptides (Bakker and Bouwman-Smits, 1988b).

Due to the two back crosses with the virulent line Ro_5-22, only RAPD markers specific for the avirulent line Ro_1-19 are informative. This is illustrated in Figure 1, in which the virulent parent line (aamm) and the avirulent parent line (AAMM) are fixed for alternate

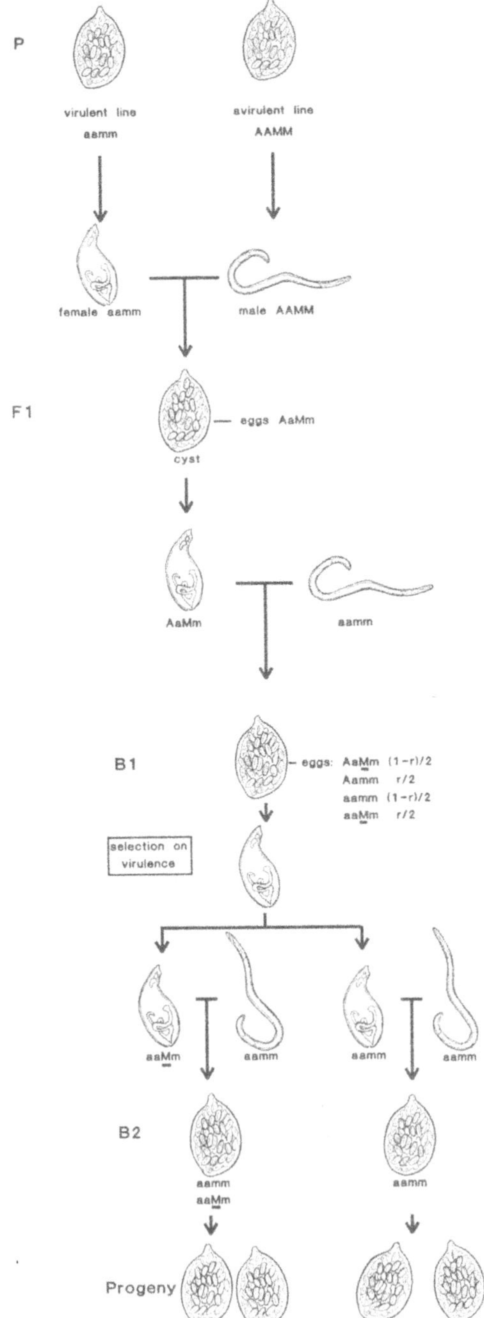

Figure 1. Backcross scheme for identification of markers linked to virulence in *G. rostochiensis* to the H_1-resistance gene. Locus M is defined by a single fragment specific for the avirulent line Ro_1-19 and the recombination frequency is given by r. If marker M is unlinked to the virulence locus than $r = 0.5$. For further explanation see text.

alleles of locus M. Marker M is defined by a single fragment specific for line Ro_1-19 which behaves as a dominant marker and the recombination frequency between A and M is given by r. In the case marker M is unlinked to the virulence locus (r = 0.5), the following distribution of genotypes will be obtained in the B_1: 25% AaMm, 25% Aamm, 25% aamm and 25% aaMm. Only juveniles homozygous for their virulence alleles will develop into females. Thus if marker M is not linked, 50% of the progeny lines will contain marker M. If marker M is linked to the virulence locus, the presence of M in the progeny results from a crossing-over between marker M and the virulence locus in the F_1. The number of progeny lines containing marker M will decrease proportional to the distance of M to the virulence locus. In that case,

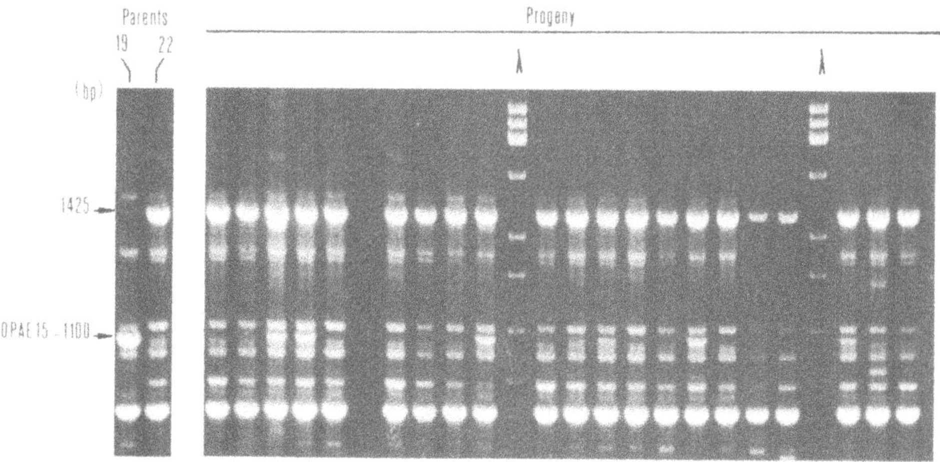

Figure 2. Segregation of RAPD marker $OPAE15_{1100}$ in progeny lines from the backcross population. The parent lines are designated by "19" (Ro_1-19) and "22" (Ro_5-22). A DNA fragment specific for the virulent line is indicated by an arrow at 1,425 bp. Lane λ represents *Eco*RI/*Bam*HI/*Hind*III digested lambda DNA as a standard for molecular weights.

segregation of fragment M is an estimate of the recombination frequency r between this marker and the virulence locus in the F_1. From this scheme it is also evident that RAPD markers specific for line Ro_5-22 can not be used for linkage studies. Due to the second backcross, they will be present in all the progeny lines.

Not all RAPD markers specific for line Ro_1-19 can be used to study the segregation patterns in the progeny lines. Weak bands can not be used. Only abundant DNA fragments were selected for studying the segregation patterns. This was tested in artificial mixtures of DNA template. Only those Ro_1-19 markers were informative which could be amplified from a mixture of genomic DNA that contained up to 75% line Ro_5-22 DNA. The importance of these experiments is that the introgressed Ro_1-19 markers can be identified in the progeny lines. As shown in Figure 1, the Ro_1-19 markers are present in an allele frequency of 25% in the progeny lines.

IDENTIFICATION OF RAPD MARKERS

The reaction conditions for PCR on potato cyst nematodes are described by Roosien et al., 1993. Experiments were done to determine the effect of the template concentration, G + C content of the primer and the sensitivity of the reaction in artificial mixtures of genomic template DNA. In general, decamer primers of random sequence (Williams et al., 1990) amplified an average of eight fragments per primer from potato cyst nematodes. The reproducibility of RAPD markers was evaluated by repeating each primer template combination for at least four times.

Figure 2 shows the results of the amplification using primer OPAE15 with genomic DNA of the parent lines and 21 progeny lines. Marker $OPAE15_{1100}$ is identified in 6 of the presented 21 progeny lines. The fragment of 1,425 bp specific for line Ro_5-22 is not suitable for linkage studies. Due to the second backcross all the progeny lines contain this fragment.

CONCLUSIONS

This research project aims at the mapping of an avirulence gene in *G. rostochiensis*. The production of a backcross progeny which segregates around a major virulence gene brings cloning of this gene into focus. The backcross scheme requires that the following considerations must be taken into account. Firstly, the zygosity of the parents for any locus besides the virulence locus is unknown. Thus the genotypes of these individuals must be deduced from the analysis of the offspring. Secondly, identification of DNA markers by the RAPD technique is performed on a mixture of genotypes.

The segregation patterns observed in the progeny lines indicate that the markers specific for the avirulent line may be used for the construction of a genetic map. Formerly, restriction fragment length polymorphisms (RFLPs) were used as DNA markers. High amounts of genomic DNA are needed for RFLP analysis which hampers the use of this technique for the genetic characterisation of potato cyst nematodes. In contrast, the RAPD technique amplifies fragments from genomic DNA to such an extent that polymorphisms can be seen directly on the gel without the use of radioisotopes. Sometimes DNA fragments are amplified that are absent in the parents but present in some of the progeny lines. This is one of the caprices of the RAPD technique. On the other hand, these extra bands can be ignored, they do not interfere with the interpretation of the segregation patterns.

The Mendelian segregation patterns indicate that this backcross scheme can be used for linkage studies in a gene-for-gene system which involves a plant parasitic nematode. Up to now, we found 30 segregating RAPD markers among an offspring of 52 progeny lines. Surprisingly, only one linkage group of 22 markers could be constructed. Moreover, these markers are also linked to the virulence locus. These preliminary data however, need to be confirmed in an extended number of progeny lines. The linked molecular markers can be used to construct a genetic and physical map. Ultimately, they will be used as starting points for further characterisation of this locus.

ACKNOWLEDGEMENTS

The research is supported by a grant of the Programme Committee on Agricultural Biotechnology of the Netherlands (PcLB) and the EC-grant AIR3 CT 92.0062.

REFERENCES

Ayala, F.J., 1983, Enzymes as taxonomic characters, *in*: "Protein Polymorphism: Adaptive and Taxonomic Significance," G.S. Oxford and D. Rollinson, eds., Academic Press, London and New York.

Bakker, J. and Bouwman-Smits, L., 1988a, Genetic variation in polypetide maps of two *Globodera rostochiensis* pathotypes, *Phytopathology* 78:894.

Bakker, J. and Bouwman-Smits, L., 1988b, Contrasting rates of protein and morphological evolution in cyst nematode species, *Phytopathology* 78:900.

Janssen, R., Bakker, J. and Gommers, F.J., 1990, Selection of virulent and avirulent lines of *Globodera rostochiensis* for the H_1 resistance gene in *Solanum tuberosum* ssp.*andigena* CPC 1673, *Revue Nématol.* 13:265.

Janssen, R., Bakker, J. and Gommers, F.J., 1991, Mendelian proof for a gene-for-gene relationship of *Globodera rostochiensis* and the H_1-gene in *Solanum tuberosum* ssp.*andigena* CPC 1673, *Revue Nématol.* 14:213.
Janssen, R., Bakker, J. and Gommers, F.J., 1992, The effect of heterozygosity for (a)virulence genes on the development of potato cyst nematodes, *Fundam.appl.Nematol.* 15:463.
Jong, A.J. de, Bakker, J., Roos, M. and Gommers, F.J., 1989, Repetitive DNA and hybridization patterns demonstrate extensive variability between the sibling species *Globodera rostochiensis* and *G.pallida*, *Parasitology* 99:133.
Roosien, J., Van Zandvoort, P.M., Folkertsma, R.T., Rouppe van der Voort, J.N.A.M., Goverse, A., Gommers, F.J. and Bakker, J., 1993, Single juveniles of the potato cyst nematodes *Globodera rostochiensis* and *G. pallida* differentiated by randomly amplified polymorphic DNA, *Parasitology* 107:567.
Tulsieram, L.K., Glaubitz, J.C., Kiss, G. and Carlson, J.E., 1992, Single tree genetic linkage mapping in conifers using haploid DNA from megagametophytes, *Bio/Technology* 10:686.
Stam, P., 1993, Construction of integrated genetic linkage maps by means of a new computer package: Joinmap, *Plant Journal* 3:739.
Williams, J.G.K., Kubelik, A.R., Livak, K.J., Rafalski, J.A. and Tingey, S.V., 1990, DNA polymorphisms amplified by arbitrary primers are useful as genetic markers, *Nucl.Acids Res.*18:6531.

DISCUSSION

BAILLIE, BIRD and WILLIAMSON expressed concern that the apparent linkage of the RAPD markers to the virulence locus might be an artefact. WILLIAMSON suggested that during the crosses performed to construct the mapping strain, there may have been selection for the virulence-carrying chromosome, but that the other chromosomes might have been averaged out, thereby masking polymorphisms on the other loci. BAILLIE noted that when his group first performed RFLP mapping using hybrids of the Bristol and Bergerac strains of *C. elegans*, all the polymorphisms appeared to map to one linkage group, as the result of selection, and that this is now known that the polymorphisms are spread throughout all the linkage groups. BAKKER replied that they also are very surprised by finding so many markers linked to the virulence locus. The linkage map is difficult to explain by artefacts such as biased selection for markers and reduced recombination events between the chromosomes of the avirulent line Ro_1-19 and virulent line Ro_5-22. Firstly, the linkage map is confident by linkage analysis with the Joinmap program (Stam, 1993). Data were analysed with a logarithm of odds (LOD) score of 6.0 for linkage threshold. The average mean square of our linkage group is 1.594 which implies a statistical reliable map. Secondly, not all our markers are linked. However, the number of the remaining markers (eight) is too small to embody separate linkage groups. Thirdly, virulence tests in the B_1 gave no indication for selection on phenotypic traits other than virulence. The absolute numbers as well as the segregation ratio for virulence was as expected. Fourthly, the reproducibility of our analysis was guaranteed since the primers were tested on different batches of isolated DNA and scored independently by two different persons. So far, it seems that most of the polymorphisms between the two lines are linked to a phenotypic difference, i.e. virulence to the H_1 resistance gene. The observation that most of the markers might be linked to the avirulence gene suggests that the chromosome carrying the avirulence gene is more variable than the other chromosomes.

BIRD proposed that the basis for the 70% difference in protein patterns between the two *Globodera* species is not only genetically determined when the protein differences are partly due to phosphorylation or glycosylation. BAKKER stated that the large genetic distance is not determined by phosphorylation or glycosylation events. Both at the protein (2-DGE) and the DNA level (RAPDs, RFLPs), the two sibling species *G. rostochiensis* and *G. pallida* show extensive genetic differentiation. RAPDs for example showed that only 9 out of 250 fragments were shared between the two species (Folkertsma, unpublished). Similar data were obtained with RFLP analysis (de Jong et al., 1989). These data indicated that those species have diverged millions of years ago. The fact that they are morphological nearly indistinguishable implies that they have experienced hardly any morphological evolution in this time period. Slow morphological divergence is probably not rare throughout the phylum Nematoda since the same phenomenon has been observed for the species *Caenorhabditis elegans* and *C. briggsae*, which are morphological also nearly identical.

GRUNDLER mentioned that according to the concept of epigenetic sex determination, under favourable conditions all penetrated juveniles develop into females, while under adverse conditions a high proportion of males develop. These effects can occur in susceptible and resistant plants due to a positioned effect on the feeding sites. He asked if these findings could be brought into accordance with the proposed gene-for-gene relationship. BAKKER replied that during their experiments they didn't find any evidence that sex determination is genetically linked to virulence (Janssen et al., 1991).

GENETIC ANALYSIS OF THE SOYBEAN - *HETERODERA GLYCINES* INTERACTION

Charles H. Opperman[1,2], Ke Dong[1], and Stella Chang[1]

[1]Department of Plant Pathology
[2]Department of Genetics
North Carolina State University
Raleigh, NC 27695-7616

INTRODUCTION

An understanding of the genetics of nematode-plant interaction is essential to development of both classically and biotechnologically derived resistant host cultivars. Unfortunately, the study of this interaction has been very one-sided, focusing primarily on the genetics of plant resistance and almost not at all on the genetics of nematode parasitism. The small size and obligately parasitic life habit of phytophagous nematodes has hindered genetic analysis of nematode parasitism. In addition, many of the most important sedentary endoparasitic forms exhibit modified reproductive strategies (e.g., mitotic or meiotic parthenogenesis in *Meloidogyne* spp.) that preclude classical genetic approaches to analysis. The cyst nematodes, however, are primarily amphimictic and are amenable to genetic analyses.

Nematode-host interactions are complex and poorly understood. The relationship between cyst nematodes and their hosts appears to have co-evolved, and as a result there are numerous genes for host resistance that are complemented by nematode parasitism genes (Triantaphyllou, 1987). Different alleles of these genes may interact in various combinations to give a range of host-nematode interactions. Because there may be numerous genes for resistance in a given host species, the interpretation of these interactions is complicated. A complete lack of knowledge regarding the functions of either resistance or parasitism genes further confuses this picture.

There are a number of cyst nematode species for which genetic variation in virulence has been observed. One of the most thoroughly understood systems is the interaction between potato cyst nematode (*Globodera rostochiensis* and *G. pallida*) (PCN) and potato. In this case, a gene-for-gene relationship appears to be in operation (Janssen et al., 1991). In this system, nematode genes for virulence are recessive. Potatoes carrying the dominant H_1 gene are resistant to certain pathotypes of PCN, but those nematodes carrying recessive virulence genes can reproduce. Pure virulent and avirulent lines of PCN have been selected, and crosses using these lines have revealed that parasitism is inherited at a single locus in a recessive manner (Janssen et al., 1990; 1991). Reciprocal crosses suggested that there was no evidence of sex-linked inheritance of parasitism. The expected segregation patterns of 3:1 avirulent to virulent combined with the dominant nature of the H_1 resistance gene suggests that this interaction functions in a classical gene-for-gene type of mechanism (Janssen et al., 1991). This topic is reviewed more thoroughly elsewhere in this volume.

Ten pathotypes of the cereal cyst nematode (*Heterodera avenae*) have been reported, based on the response of resistant barley cultivars (Andersen and Andersen, 1982). Extensive variations in pathogenicity have been reported within these pathotypes, however, and it is clear that there is more genetic variation within this nematode species than is

Advances in Molecular Plant Nematology
Edited by F. Lamberti *et al.*, Plenum Press, New York, 1994

represented by the current schemes. Additionally, the use of an alternative host plant, such as oat or wheat, further complicates the picture (Cook and York, 1982). *Heterodera avenae* populations from Europe and Australia exhibit extensive variation in their relative abilities to parasitize resistant cultivars, indicating the presence of multiple genes conferring parasitic ability on a given host genotype (Triantaphyllou, 1987). Limited data from crossing experiments indicates that at least two dominant and one or more recessive genes are involved in *H. avenae* pathogenicity on cereal crops (Andersen, 1965; Person and Rivoal, 1979).

The interaction of soybean cyst nematode (*H. glycines*) (SCN) with soybean (*Glycine max*) has been extensively studied in the United States. This system has been chosen as a model to dissect the genetics of nematode parasitism due to both the extensive characterization of soybean resistance genes and the tractability of classical genetic manipulation of the nematode. The remainder of this review will focus on the interaction of SCN with its host and the progress that has been made in unraveling the complex genetic systems controlling parasitic behavior.

SOYBEAN CYST NEMATODE BIOLOGY AND MANAGEMENT

Soybean cyst nematode is one of the most devastating pathogens of soybean on a worldwide basis, accounting for most of the approximately $2.5 billion in annual crop loss due to nematodes (Sasser and Freckman, 1987; Noel, 1992). In the United States, SCN accounted for a 5.8% yield loss during 1985, costing growers over $200 million in revenue and control expenditures (Sasser and Freckman, 1987). As management options become more limited and the nematode continues to spread throughout the soybean growing areas of the United States, these figures are likely to increase dramatically (Schmitt and Noel, 1984; Noel, 1992). Currently, very little is known of the molecular basis of either susceptible or resistant responses of soybean to this nematode, and even less is understood regarding the nematode parasitic biology. It is critical to have a detailed understanding of this interaction in order to develop durable and reliable management tactics to combat soybean cyst nematode. The rapid advances in molecular biology during the past few years have made it possible to study this interaction at a level of detail not attainable previously.

Soybean cyst nematode is a sedentary endoparasite with an extremely complex and intimate relationship to the host plant. The infective second-stage juvenile (J2) in the soil is arrested developmentally after hatching. After the J2 locates a soybean root, it penetrates the root and migrates to the developing vascular cylinder tissue (Schmitt and Noel, 1984; Wyss and Zunke, 1986). The nematode aligns itself parallel to the cylinder and injects glandular secretions into a single undifferentiated root cell adjacent to its head (Hussey, 1989). This cell undergoes a remarkable process, becoming enlarged with highly invaginated and partially degraded cell walls, and extremely active metabolically (Jones and Dropkin, 1975; Endo, 1992; Jones, 1981). Other cells surrounding the initial syncytial cell are incorporated by partial cell wall degradation, resulting in a multinucleate feeding site. Syncytia function as super transfer cells and provide nourishment to the developing nematode during its entire life cycle (Wyss and Zunke, 1992). The presence of the nematode is essential to the maintenance of the syncytia (Jones, 1981; Endo, 1992). After 20-30 days, the adult female nematode is fertilized by the male and then produces 200-600 eggs, which remain primarily in her hardened body (i.e., a cyst) (Triantaphyllou and Hirschmann, 1962; Young, 1992). There may be several generations per growing season, and the nematode egg is able to survive in the cyst for a number of years under very harsh environmental conditions (Alston and Schmitt, 1988; Young, 1992). Clearly, this is a highly evolved parasitic relationship.

In the past several years, management options for SCN have gradually eroded. Many effective nematicides either have been banned or restricted due to environmental and toxicity concerns (Rodriguez-Kábana, 1992). In addition, resistant soybean cultivars have become less effective as increased genetic variability in SCN populations has been observed. For example, in North Carolina, over 60% of the soybean fields infested with SCN have races to which there is no host resistance available (K. R. Barker, pers. comm.). Although traditional plant breeding efforts continue, attempts to develop agronomically acceptable cultivars with broad resistance are hampered by a lack of knowledge regarding the genetics of SCN virulence. Understanding the relationship between host response and nematode parasitism has been constrained by the oligogenic nature of soybean resistance combined with the extreme genetic variability encountered in *H. glycines* populations. It is essential to understand genetic variability in the nematode if host resistance for *H. glycines* management is to remain durable .

GENETIC RESISTANCE TO SOYBEAN CYST NEMATODE

Within two years of the discovery of SCN in the United States, screening of existing soybean germplasm resulted in the identification of several lines carrying resistance to the known field isolate of SCN (now known to be race 1) (Ross and Brim, 1957). These lines, Ilsoy, Peking, PI 90763, and PI 84751, were used in breeding programs to develop resistant cultivars of soybean. The first resistant cultivar released was Pickett, which resulted from a cross involving Peking (Brim and Ross, 1966). As variability in SCN parasitic ability was characterized, further germplasm screening identified lines resistant to some of the new nematode races, including PI 88788, PI 89772, PI 87631-1, Cloud, Columbia, Peking, PI 84751, and PI 90763 (Epps and Hartwig, 1972; Caviness, 1992). Many new cultivars carrying resistance incorporated from some of these lines, particularly PI88788, were subsequently released. New sources of resistance to various SCN races have been identified since that time, including PI 209332, PI 437654, and others (Anand and Brar, 1983; Anand and Gallo, 1984). Recently, the cultivar Hartwig has been released and carries resistance to all characterized races of SCN (Anand, 1992). However, this cultivar is not adapted to all soybean growing regions.

Resistance to SCN is oligogenic and inheritance patterns may be complex. Resistance to SCN has been demonstrated to be controlled by at least four recessive and one or more dominant major genes (Ross and Brim, 1957; Caldwell et al., 1960; Matson and Williams, 1965; Thomas et al., 1975; Rao-Arelli et al., 1992). Each of these genes may have multiple allelic states. Resistance to race 5 carried by PI 437654 is controlled by 2 dominant and 1 recessive genes, and at least some of these genes are also present in Peking and PI 90763 (Anand et al., 1988). Analysis of resistance to race 3 in Peking and PI 90763 indicates that 1 dominant and 2 recessive genes confer resistance (Rao-Arelli et al., 1992). Other experiments have shown that some resistance genes may be shared among the various lines, and that some genes may be linked or have multiple alleles involved (Hartwig, 1985; Hancock et al., 1987; Rao-Arelli and Anand, 1988; Anand and Rao-Arelli, 1989). The complex genetic nature of soybean resistance to SCN combined with the variability observed in nematode populations has impeded progress in the development of improved varieties.

SOYBEAN CYST NEMATODE GENETICS

Karyotype

The chromosomes of cyst nematodes are larger and more distinct than those of *Meloidogyne* spp. (Triantaphyllou, 1982). Unlike the root-knot nematodes, most *Heterodera* spp. that have been characterized are amphimictic and carry a haploid chromosome number of 9. No morphologically distinct sex chromosomes have been observed in cyst nematodes by microscopic examinations. There is one meiotic parthenogenetic species known, *H. betulae*, which also carries an increased chromosome number of n=12 or 13 (Triantaphyllou, 1970). Two polyploid forms appear to reproduce by mitotic parthenogenesis. *Heterodera sacchari* is a triploid (3n=27) that is thought to be an example of a divergent evolutionary line (Netscher, 1969). There are two cytological forms of *H. trifolii*. The most common form is a triploid (3n=26-28), but a tetraploid (4n=33-35) has also been reported (Triantaphyllou and Hirschmann, 1978). Previous studies have shown that SCN is a diploid nematode with nine chromosomes (Triantaphyllou and Hirschmann, 1962). In addition, a tetraploid form with 18 bivalents has been isolated (Triantaphyllou and Riggs, 1979). The tetraploid carries approximately 1.5 times the amount of DNA per nucleus than the diploid (Goldstein and Triantaphyllou, 1979; Triantaphyllou and Riggs, 1979). The juveniles and adults have significantly larger body size than those of the diploid form as well, which is typical of tetraploid forms (Triantaphyllou and Riggs, 1979). Crosses between the diploid and tetraploid forms have yielded viable aneuploid (n=14) hybrids (Goldstein and Triantaphyllou, 1979).

Soybean Cyst Nematode Races

Genetic variability in *H. glycines* was detected almost as soon as host resistance was identified (Ross, 1962; Miller, 1970). Initially, four races of *H. glycines* were designated based on reproduction on four resistant host differentials: Pickett, Peking, PI 88788, and PI 90763 (Golden et al., 1970). A fifth race was soon added to this system, but field populations did not always fit into one of these races (Riggs et al., 1981). As a result, 16 races of *H. glycines* are now recognized in the fully expanded scheme (Riggs and Schmitt, 1988) (Table 1). Nematode reproduction on one of the host differentials is considered positive if it exceeds 10% of that observed on the fully susceptible cultivar, Lee. In essence, races of *H. glycines* are field populations that possess a number of genotypes (Triantaphyllou, 1975; Leudders, 1983). Selection pressure by cropping resistant cultivars is likely to alter the frequency of alleles for parasitism, and therefore the race designation. The race concept as it is applied to *H. glycines* is not based on genotype, but rather on the predominant phenotype encountered at a particular time (Niblack, 1992). This has made it extremely difficult to compare populations of nematodes for diagnostic purposes and has obscured the understanding of host-parasite interaction. The genetics of the soybean cyst nematode-soybean interaction do not conform to a conventional pattern of gene-for-gene interaction (i.e., dominant resistance genes/recessive parasitism genes). Nevertheless, it has been possible to make certain observations regarding SCN genetics.

Table 1. Full expansion of the race classification for *Heterodera glycines* (Riggs and Schmitt, 1988).

	Reaction on differential			
Race	Pickett	Peking	PI 88788	PI 90763
1	-	-	+	-
2	+	+	+	-
3	-	-	-	-
4	+	+	+	+
5	+	-	+	-
6	+	-	-	-
7	-	-	+	+
8	-	-	-	+
9	+	+	-	-
10	+	-	-	+
11	-	+	+	-
12	-	+	-	+
13	-	+	-	-
14	+	+	-	+
15	+	-	+	+
16	-	+	+	+

+ = Number of females and cysts recovered was >10% of 'Lee'.
- = Number less than 10%.

Early genetic studies on SCN variability were mainly by directional selection experiments on the various host differentials. In these studies, selection on a resistant host resulted in a gradual increase in the ability of the nematode population to reproduce on that host (Triantaphyllou, 1975; Young, 1982; McCann et al., 1982). However, selection on one resistant differential had no effect on the nematode population's ability to parasitize a different host. From these studies, it was supposed that multiple genes in the nematode were involved in parasitism of resistant cultivars, and that these genes could be separated into three relatively independent groups (Triantaphyllou, 1975; Young, 1982; McCann et al., 1982). These groups corresponded to the ability to parasitize PI 88788, PI 90763, and Pickett, respectively. Soybean lines carrying genes derived from one of these differentials responded to the selected nematode populations in the same manner as the original differential, although some lines actually carry genes derived from several sources. These lines can often be attacked with equal intensity by the selected nematode populations (Young, 1984). Later tests combining a primary directional selection on a resistant cultivar with a secondary selection on a cultivar with different resistance genes have further complicated the issue. In these experiments, secondary selection resulted in increased parasitic ability on the secondary host, but a loss of parasitism on the primary host (Leudders and Dropkin, 1983; Leudders, 1985). Although it may be argued that these data support the idea that different alleles at the same locus are responsible for parasitic ability (Leudders and Dropkin, 1983; Leudders, 1985), the lack of fixation of parasitism genes during primary selection casts doubts on that conclusion (Triantaphyllou, 1987). More recent experiments support the shift in reproductive abilities as being due to secondary selection, but still do not provide genetic evidence as to the mechanism (Young, 1984; Anand and Shumway, 1985).

Controlled crosses between various races of SCN have suggested that parasitic ability is inherited in a dominant fashion, but no specific ratios were observed (Triantaphyllou, 1975; Price et al., 1978). This was most likely because the populations used for these experiments represented a mixture of genotypes rather than pure strains. Inbred lines of SCN have been developed that have been selected over a seven year period for parasitic ability on a given soybean host (Dropkin and Halbrendt, 1986). Experiments with these lines revealed that nematode genes for parasitism on one soybean genotype may have a dominant effect over genes for parasitism on a different host genotype. Unfortunately, these studies ended at the F_1, and segregation patterns were not observed, but this type of data points out the need to utilize pure genetic lines for the study of SCN parasitic ability.

GENETIC MAPPING OF THE SCN GENOME

All of the previously described experiments have relied upon phenotypic analysis of genotypes. There are many factors that can mitigate expression of a particular genetic trait, including the environment, quantitative inheritance, or partial and complete dominance. These types of factors can have a significant impact on interpretation of genetic analyses based on phenotype. In the past 10 years, direct assay of genotype through use of DNA-based markers has greatly enhanced our abilities to perform genetic analysis. For most of that time, the technology of choice has been the use of restriction fragment length polymorphisms (RFLP) (Botstein et al., 1980; Burr et al., 1983). RFLPs reveal DNA polymorphisms by restriction endonuclease digestions combined with hybridizations. These assays are generally time consuming and labor intensive, not to mention costly. The use of the polymerase chain reaction (PCR) has greatly facilitated genetic analysis, and several assays based on selective amplifications have been developed (Innes et al., 1990). Although PCR assays for genetic analysis have greatly increased the numbers of individuals that can be analyzed compared to RFLP technology, the requirement for sequence information for primer synthesis limits applications. Even so, there are now a number of nematode labs around the world applying PCR-based DNA diagnostics to the study of population variation.

A significant advance in PCR technology for genetic analysis was the development of RAPD assays (random amplified polymorphic DNA) (Welsh and McClelland, 1990; Williams et al., 1990). The RAPD procedure obviates the need for prior sequence information. Instead, polymorphic DNA sequences are detected by amplification with a single random 10 base primer. If the primer binds to complementary DNA strands within an amplifiable distance, a discrete product is formed. Generally, RAPD primers may direct the amplification of multiple discrete products from a genomic DNA sample, making this an efficient way to screen for polymorphisms between individuals (Welsh and McClelland, 1990; Williams et al., 1990).

Our laboratory maintains a substantial collection of soybean cyst nematode populations from multiple locations in the United States. In addition to our field populations, we have over 70 populations that have been selected repeatedly on certain resistant soybean hosts. Although many of these lines still maintain a degree of heterogeneity due to the limited number of selection cycles that have been imposed, we have several highly inbred lines.

These lines have been repeatedly selected for several traits of interest, including enzyme phenotype and parasitism. In particular, we are working with three lines that have been carried forward by single cyst selection and are highly homozygous (Table 2). The three lines, OP20, OP25, and OP50, have been inbred for a minimum of 22 generations. Isozyme analysis has demonstrated that these three lines are homozygous for the esterase and glucose phosphate isomerase (GPI) loci. Genetic analysis of esterase pattern indicates that the three observable phenotypes correspond to three codominant alleles at a single locus (Esbenshade and Triantaphyllou, 1988). No maternal effects were detected in these studies. Host range testing using the standard soybean differential genotypes reveals that these lines are highly specific in their parasitic abilities. Unlike the standard race concept, parasitism of a differential by one of the inbred lines is either positive or negative. No cysts develop on resistant soybean lines. This is an extremely important factor for genetic analysis using phenotypic selection.

Table 2. Soybean cyst nematode inbred lines.

		Host Response				
line	esterase	Lee	Pickett	Peking	88788	90763
OP20	cc	+	-	-	+	-
OP25	aa	+	+	-	-	-
OP50	bb	+	+	+	+	+

Esterase phenotype determined on PhastSystem (Esbenshade and Traintaphyllou, 1988).
+ = reproduction on the host differential
- = no reproduction (ie., no cyst recovered)

Controlled crosses have been performed between these strains. Preliminary analysis of a cross between OP20 and OP25 has revealed several important factors . These strains each parasitize a single different resistant soybean genotype. Segregation patterns in the F_2 progeny of the cross suggest that inheritance of parasitic ability is Mendelian in nature. The genes controlling parasitic ability appear to be unlinked dominant loci that are inherited close to a typical 9:3:3:1 pattern. Additionally, neither of these two loci appear to be linked to esterase or GPI phenotype. The progeny lines carrying genes conferring parasitic abilities on both resistant soybean genotypes did not appear to have altered interactions compared to the parents, suggesting that these two loci do not interact, at least for parasitism of the two soybean genotypes tested. No additional host range was detected. Finally, reciprocal crosses revealed no pattern of maternal or sex-linked inheritance for parasitic abilities.

Given these factors, it should be possible to obtain RAPD markers linked to the parasitism loci through bulked segregant analysis (Michelmore et al., 1991). In this protocol, segregating F_2 populations can be screened for host preference and the individuals pooled into groups of compatible and incompatible lines. RAPD screening of these pools should identify markers linked to parasitism. In theory, this procedure is a quick and straightforward method to generate markers linked to parasitism genes. In practice, the necessity of screening individual progeny (i.e., single nematodes) introduces a level of variability that is unacceptable for genetic mapping. Therefore, we have chosen the alternative strategy of making recombinant inbred lines (RIL) from a cross between OP20 x OP50. In this approach, approximately 400 lines have been generated originating from the initial cross. The lines are being developed by full sib mating and single cyst descent through the F_8 generation. Large amounts of clean DNA can be obtained from each of these lines, markers linked to the parasitism loci can be quickly identified, and the necessary host range tests may be performed on a large population. This approach provides a great deal more certainty regarding data collected from the individual lines.

The most efficient way to identify markers linked to a particular area of the genome is to pool samples of DNA into two separate bulks composed of DNA from individual lines that are segregating for one of the parental parasitism phenotypes. All RAPD markers in these bulks should appear in linkage equilibrium except for those linked to the parasitism loci. Markers linked to these loci will appear polymorphic between the two pools of DNA. The use of a large number of segregating recombinant inbred lines should minimize the chances of identifying polymorphic markers that are unlinked.

Because of the large number of primers that can be screened, RAPDs are very useful for obtaining markers tightly linked to genes for which no mapping has previously been performed (Williams et al., 1990), such as is the case for *H. glycines*. The OP20 x OP50 cross was performed to generate recombinant inbred lines that would provide information regarding segregation of two independent parasitism loci, namely, those for parasitism of Peking and of PI 90763. More than 300 random 10-mer primers have been screened against the parental lines from the cross. Numerous polymorphic bands have been identified between the parental lines, and the stability of these bands is being verified by repeated assay from independent DNA extractions. Markers identified in this procedure will be analyzed for cosegregation with host preference phenotype and linkage groups assigned by a χ^2 analysis. Computer map construction software will be utilized to assign order of markers within the identified linkage group (Michelmore et al., 1991).

Figure 1 shows the typical amplification patterns from the three parental lines for several RAPD primers. Attempts to get consistent amplification patterns from individual nematodes have not been successful. Although a fair degree of reliability has been obtained, it is not sufficient for genetic analysis. Therefore, our approach has been to develop recombinant inbred lines from controlled crosses between SCN strains. These RILs can be carried for numerous generations to fix parasitism alleles. The OP20 x OP50 F_1 females were checked for heterozygous esterase phenotypes, eggs from these individuals were pooled and inoculated on Lee soybeans. Approximately 30 days later, 400 single cysts were selected at random from this inoculation and placed individually on the roots of Lee. Single cysts from each of these lines have been carried forward at each generation, and we are now entering the F_8 generation, when the 400 individual populations will be bulked (i.e., no more single cyst descent) and increased for RAPD analysis and host screening. The lack of selection pressure insures that representative lines will be available for mapping purposes. In this particular cross, segregation for two separate parasitism loci should have occurred, so it may be possible to map each of these in preparation for cloning.

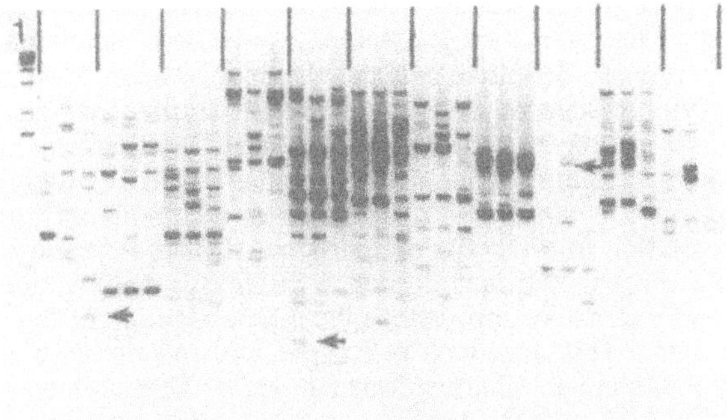

Figure 1. RAPD primer amplifications of three SCN inbred lines, OP20, OP25, and OP50. Size markers are in lane 1. This figures shows eleven primers run against each of the 3 lines in the order (left to right): OP20, OP25, OP50. Every three lanes there is a new primer. Arrows point to several representative polymorphisms.

A cDNA library has been constructed from second-stage juveniles of the OP50 strain. Characterization of the library indicates that it is representative and we have had no trouble isolating clones to low abundance transcripts. One hundred randomly selected cDNA clones are being sequenced. End sequencing runs have been completed on the 5' region of a number of clones. The sequences have been used in a database search and the majority have no known homology to anything previously characterized. However, eight clones have shown either significant homology to known peptide sequences or have apparent relationships to *C. elegans* clones (Table 3). These clones will be used as expressed sequence tags (EST) in the mapping project.

Table 3. cDNA sequences obtained as ESTs from SCN library

clone	BLAST score	peptide
OP#01	142	muscle specific protein
OP#06	76	homeiotic protein lab1
OP#08	103	extensin (tomato)
OP#10	79	keratin
OP#11	49	*C. elegans* cosmid CO7A9.11
OP#25	116	mouse DNA binding protein
OP#33	53	ceh-16-1 *C. elegans* homeobox
OP#36	51	*C. elegans* cosmid CEZK638

BLAST searches were performed using the BLAST algorythm (Altschul et al., 1990) via the BLAST Network Service of the National Center for Biotechnology Information.

The EST and RAPD markers obtained above can be used to probe an *H. glycines* genomic cosmid library in order to isolate large (~45 kb) DNA fragments overlapping the parasitism gene locus. Once overlapping cosmids are obtained, screening will be performed with the ESTs previously identified, as well as further screening with RAPD primers and other short oligonucleotide primers. In this way a saturated map of the region can be generated. Physical mapping of the clones should provide information on where the parasitism gene resides on the cosmid fragments. At this point, standard chromosome walking techniques can be used to isolate fragments containing the parasitism gene (Sambrook et al., 1989).

The investigations discussed above should lead to the identification of putative virulence genes. However, the only unambiguous way to determine whether or not a given sequence is related to parasitism is through nematode transformation. In this way, SCN strains unable to infect a particular host genotype could be transformed with genomic clones containing the putative gene. Because the parasitism genes appear to be dominant in nature, a screen of the R_1 generation would provide definitive evidence for involvement of a particular sequence in parasitism. Phenotypic rescue of *C. elegans* mutant strains by cosmid rescue is now a standard procedure. It may be possible to develop an SCN transformation system based on microinjection of the male gonad (Fire, 1986). One alternative strategy is microprojectile bombardment of developing females. A hydroponic or root explant culture system would be ideally suited to adaptation for this type of approach. This is necessary because the female gonad is not visible through the opaque nematode body, making injection of the ovary impossible. A second alternative approach is the direct injection of the genital primordia in second-stage juveniles. This approach is fraught with problems of its own and would certainly yield very low numbers of transformed animals. This strategy seems viable only as a last resort if the other two strategies fail.

Once a repeatable system is achieved to reliably transform SCN with the reporter constructs, it will be necessary to screen for identification of a selectable marker. This will be a key factor for transformations with putative parasitism genes. This particular facet of the project may take some amount of time. Selectable markers such as the roller phenotype commonly employed with *C. elegans* will not be appropriate for these purposes. Morphological or behavioral aberrations in plant-parasitic nematodes invariably interfere with mating and parasitic behavior. It will be critical to develop a reliable system to identify transformed second-stage juveniles prior to host range tests. Initially, cloned genes may be utilized from *C. elegans* that confer resistance to anthelminthics such as avermectin or to other toxins. Should this approach not yield a useful selectable marker, database searches can be conducted to identify potential genes from other organisms conferring resistance to toxins, antibiotics, or other compounds.

FUTURE CONSIDERATIONS

The mapping phase of the project is well under way. Within the next year, a linkage map should be generated for parasitism loci. The biggest hurdle is how tightly linked the markers obtained will be. We believe that the approach of using RAPD combined with expressed sequence tags will provide markers that map close to the target gene. The small size of the nematode genome and previous success with map based cloning in *C. elegans* (Emmons, 1988) provides confidence in this phase of the project. Due to the low level of repetitive DNA sequences in nematodes, the isolation of cosmid clones and chromosome walking should be very straightforward and no particular difficulties are anticipated with this portion of the project. Sequencing of an entire 50 kb cosmid may also be readily accomplished through automated sequencers, therefore cloning and sequencing of an SCN parasitism gene should be attainable in the near future.

Once a transformation protocol is in place, analysis can begin of cosmid clones isolated in the mapping part of the project. Clones containing markers linked to parasitism can be microinjected into males, and resulting transformed second-stage juveniles will be inoculated to resistant soybean cultivars to assay for parasitic abilities. Although this seems to be a relatively straightforward approach, numerous things could confound the analysis. For example, position effects may alter expression patterns of the parasitism gene or poor integration may cause the generation of mosaic animals that vary in their ability to parasitize the resistant host. For this reason, a large number of independent transformants must be evaluated. Although isolation of a cosmid clone containing putative parasitism genes is possible based on the mapping strategies, unambiguous identification of a gene requires the transformation of a non-parasitic strain and its subsequent alteration of parasitism. In addition, the development of a plant parasitic nematode transformation system will be of tremendous significance to future studies on a large number of traits.

Future areas to pursue with regard to this project are several. First, it will be possible to develop a linkage map of the entire genome of *H. glycines*. This will be an important resource for future studies of nematode biology and parasitic abilities. Of particular interest are sex determination genes in SCN, but there are many traits that can be studied including survival, diapause, hatching, and others. A genetic map will be useful to the entire community of researchers working on cyst nematode genetics and biology. Once a number of genes have been isolated from SCN, it will be possible to begin analysis of genome organization and function. In addition, the tremendous data base that exists from the *C. elegans* genome sequencing project will be very useful in analysis of SCN genetics.

In the short term, once a parasitism gene has been isolated its expression patterns and control elements may be characterized. Of particular interest is the function of the gene product in the interaction between SCN and soybean. Localization of the peptide product and its structure may provide important clues as to how it functions in the nematode. Finally, we are optimistic that new ideas for developing host resistance and nematode control strategies will be devised once an understanding of the interaction at the molecular level is achieved.

REFERENCES

Altschul, S.F., Gish, W., Miller, W., Myers, E.W., and Lipman, D.J., 1990, Basic local alignment search tool, *J. Mol. Biol.* 215:403.

Alston, D.G., and Schmitt, D.P., 1988, Development of *Heterodera glycines* life stages as influenced by temperature, *J. Nematol.* 20:366.

Anand, S.C., 1992, Registration of Hartwig soybean, *Crop Sci.* 32:1069.

Anand, S.C., and Brar, G.S., 1983, Response of soybean lines to differentially selected cultures of soybean cyst nematode *Heterodera glycines* Ichinoe, *J. Nematol.* 15:120.

Anand, S.C., and Gallo, K.M., 1984, Identification of additional soybean germplasm with resistance to race 3 of soybean cyst nematode, *Plant Dis.* 68:593.

Anand, S.C., Myers, G.O., and Rao-Arelli, A.P., 1988, Resistance to race 3 of soybean cyst nematode in PI 437654, *Crop Sci.* 28:563.

Anand, S.C., and Rao-Arelli, A.P., 1989, Genetic analysis of soybean genotypes resistant to soybean cyst nematode race 5, *Crop Sci.* 29:1181.

Anand, S.C., and Shumway, C.R., 1985, Response of secondary selection on soybean cyst nematode reproduction on resistant soybean lines, *Crop Prot.* 4:231.

Andersen, S., 1965, Heredity of race 1 or race 2 in *Heterodera avenae*, *Nematologica* 11:121.

Andersen, S., and Andersen, K., 1982, Suggestions for determination and terminology of pathotypes and genes for resistance in cyst-forming nematodes, especially *Heterodera avenae*, *OEPP/EPPO Bull.* 12:379

Botstein, D., White, R.L., Skolnick, M.H., and Davis, R.W., 1980, Construction of a genetic map in man using restriction fragment length polymorphisms, *Am. J. Hum. Genet.* 32:314.

Brim, C.A., and Ross, J.P., 1966, Registration of Pickett soybeans, *Crop Sci.* 6:305.

Burr, B., Evola, S.V., and Burr, F.A., 1983, The application of restriction fragment length polymorphism to plant breeding, *in*: "Genetic Engineering-Principles and Methods," J.K. Setlow and Hollander, A., eds., Plenum, New York.
Caldwell, B.E., Brim, C.A., and Ross, J.P., 1960, Inheritance of resistance of soybeans to cyst nematode, *Heterodera glycines, Agron. J.* 52635.
Caviness, C.E., 1992, Breeding for resistance to soybean cyst nematode, *in:* "Biology and Management of the Soybean Cyst Nematode," R.D Riggs and Wrather, J. A., eds., APS Press, St. Paul, MN.
Cook, R., and York, P.A., 1982, Resistance of cereals to *Heterodera avenae*: Methods of investigation, sources, and inheritance of resistance, *OEPP/EPPO Bull.* 12:423.
Dropkin, V.H., and J.M. Halbrendt, 1986, Inbreeding and hybridizing soybean cyst nematodes on pruned soybeans in petri plates, *J. Nematol.* 18:200.
Emmons, S., 1988, The genome, *in*: "The Nematode Caenorhabditis elegans," W.B. Wood, ed., Cold Spring Harbor Laboratory, Cold Spring Harbor.
Endo, B. Y., 1992, Cellular responses to infection, *in:* "Biology and Management of the Soybean Cyst Nematode," R.D Riggs and Wrather, J. A., eds., APS Press, St. Paul, MN.
Epps, J.M., and Hartwig, E.E., 1972, Reaction of soybean varieties and strains to race 4 of the soybean cyst nematode, *J. Nematol.* 4:222.
Esbenshade, P.R., and Triantaphyllou, A.C., 1988, Genetic analysis of esterase polymorphism in the soybena cyst nematode, *Heterodera glycines, J. Nematol.* 20:486.
Fire, A., 1986, Integrative transformation of *Caenorhabditis elegans, EMBO J.* 5:2673.
Goldstein, P., and Triantaphyllou, A.C., 1979, Karyotype analysis of the plant-parasitic nematode *Heterodera glycines* by electron microscopy. II. The tetraploid and an aneuploid hybrid, *J. Cell Sci.* 43:225.
Golden, A.M., Epps, J.M., Riggs, R.D., Duclos, L.A., Fox, J.A., and Bernard, R.L., 1970, Terminology and identity of infraspecific forms of the soybean cyst nematode (*Heterodera glycines*), *Plant Dis. Rep.* 54:544.
Hancock, J.A., Hancock, F.G., Caviness, C.E., and Riggs, R.D., 1987, Genetics of resistance to "Race X" of soybean cyst nematode, *Crop Sci.* 27:704.
Hartwig, E.E., 1985, Breeding productive soybeans with resistance to soybean cyst nematode, *in:* "World Soybean Research Conference III," R.A. Shibles, ed., Westview Press, Boulder.
Hussey, R.S., 1989, Disease-inducing secretions of plant parasitic nematodes, *Ann. Rev. Phytopath.* 27:123.
Innes, M.A., Gelfand, D.H., Snitsky, J.J., and White, T.J., 1990, "PCR Protocols," Academic Press, San Diego.
Janssen, R., Bakker, J., and Gommers, F.J., 1990, Selection of virulent and avirulent lines of *Globodera rostochiensis* for the H1 resistance gene in *Solanum tuberosum* ssp. *andigena* CPC 1673, *Rev. Nematol.* 13:265.
Janssen, R., Bakker, J., and Gommers, F.J., 1991, Mendelian proof for a gene-for-gene relationship between *Globodera rostochiensis* and the H1 resistance gene from *Solanumtuberosum* ssp. *andigena* CPC 1673, *Rev. Nematol.* 14:213.
Jones, M.G.K., 1981, The development and function of plant cells modified by endoparasitic nematodes, *in:* "Plant Parasitic Nematodes, Vol. 3," B. M. Zuckerman and Rohde, R.A., eds., Academic Press, New York.
Jones, M.G.K., and Dropkin, V.H., 1975, Scanning electron microscopy of syncytial transfer cells induced in roots by cyst-nematodes, *Phys. Plant Path.* 7:259.
Luedders, V.D., 1983, Genetics of the cyst nematode-soybean symbiosis, *Phytopath.* 73:944.
Luedders, V.D., 1985, Selection and inbreeding of *Heterodera glycines* on *Glycine max, J. Nematol.* 17:400.
Luedders, V.D., and Dropkin, V.D., 1983, Effects of secondary selection on cyst nematode reproduction on soybean, *Crop Sci.* 23:263.
Matson, A.L., and Williams, L.F., 1965, Evidence for a fourth gene for resistance to soybean cyst nematode, *Crop Sci.* 5:477.
McCann, J., Luedders, V.D., and Dropkin, V.D., 1982, Selection and reproduction of soybean cyst nematodes on resistant soybeans, *Crop Sci.* 22:78.
Miller, L.I., 1970, Differentiation of eleven isolates as races of the soybean cyst nematode. *Phytopath.* 60:1016
Michelmore, R.W., Paran, I., and Kesseli, R.V., 1991, Identification of markers linked to disease resistance genes by bulked segregant analysis: a rapid method to detect markers in specific genomic regions by using segregating populations, *Proc. Nat. Acad. Sci. USA* 88:9828.
Netscher, C., 1969, L'ovogénese et la reproduction chez *Heterodera oryzae* et *H. sacchari* (Nematodea: Heteroderidae), *Nematologica* 15:10.
Niblack, T. L., 1992, The race concept. *in:* "Biology and Management of the Soybean Cyst Nematode," R.D Riggs and Wrather, J. A., eds., APS Press, St. Paul, MN.
Noel, G. R., 1992, History, distribution, and economics, *in:*: "Biology and Management of the Soybean Cyst Nematode," R.D Riggs and Wrather, J. A., eds., APS Press, St. Paul, MN.
Person, F., and Rivoal, R., 1979, Hybridation entre les races Fr1 et Fr4 d'*Heterodera avenae* Wollenweber en France et étude du comportement d'agressivite des descendants F1, *Rev. Nematol.* 2:177.

Price, M., Caviness, C.E., and Riggs, R.D., 1978, Hybridization of races of *Heterodera glycines, J. Nematol.* 10:114.

Rao-Arelli, A.P., and Anand, S.C., 1988, Genetic relationships among soybean plant introductions for resistance to race 3 of soybean cyst nematode, *Crop Sci.* 28:650.

Rao-Arelli, A.P., Anand, S.C., and Wrather, J.A., 1992, Soybean resistance to soybean cyst nematode race 3 is conditioned by an additional dominant gene, *Crop Sci.* 32:862.

Riggs, R.D., Hamblen, M.L., and Rakes, L., 1981, Infraspecific variation in reaction to hosts in *Heterodera glycines* populations, *J. Nematol.* 13:171.

Riggs, R.D., and Schmitt, D.P., 1988, Complete characterization of the race scheme for *Heterodera glycines, J. Nematol.* 20:392.

Rodríguez-Kábana, R., 1992, "Chemical control," *in:* "Biology and Management of the Soybean Cyst Nematode," R.D Riggs and Wrather, J. A., eds., APS Press, St. Paul, MN.

Ross, J.P., 1962, Physiological strains of *Heterodera glycines,Plant Dis. Rep.* 46:766.

Ross, J.P. and Brim, C.A., 1957, Resistance of soybeans to the soybean cyst nematode as determined by the double row method, *Plant Dis. Rep.* 41:923.

Sambrook, J. E., Fritsch, E. F. and Maniatis, T., 1989, "Molecular Cloning: A Laboratory Manual, 2nd edition," Cold Spring Harbor Laboratory Press, Cold Spring Harbor.

Sasser, J.N., and Freckman, D.W., 1987, "A world perspective on nematology: The role of the society," *in:* "Vistas on Nematology," J.A. Veech and Dickson, D.W., eds., Society of Nematologists, Inc., Hyattsville.

Schmitt, D.P., and Noel, G.R., 1984, Nematode parasites of soybean. *in:* "Plant and Insect Nematodes," W. R. Nickle, ed., Marcel Dekker, Inc., New York.

Thomas, J.D., Caviness, C.E., Riggs, R.D., and Hartwig, E.E., 1975, Inheritance of reaction to race 4 of soybean cyst nematode, *Crop Sci.* 15:208.

Triantaphyllou, A.C., 1970, Oogenesis and reproduction of the birch cyst nematode, *Heterodera betulae, J. Nematol.* 2:399.

Triantaphyllou, A.C., 1975, Genetic structure of races of *Heterodera glycines* and inheritance of ability to reproduce on resistant soybean, *J. Nematol.* 7:356.

Triantaphyllou, A.C., 1982, Cyotgenetics and sexuality of root-knot and cyst nematodes. *South. Coop. Ser. Bull.* 276:71.

Triantaphyllou, A.C., 1987, Genetics of nematode parasitism on plants, *in:* "Vistas on Nematology," J.A. Veech and Dickson, D.W., eds., Society of Nematologists, Inc., Hyattsville.

Triantaphyllou, A.C., and Hirschmann, H., 1962, Oogenesis and mode of reproduction in the soybean cyst nematode, *Heterodera glycines, Nematologica* 7:235.

Triantaphyllou, A.C., and Hirschmann, H., 1978, Cytology of the *Heterodera trifolii* pathenogenetic species complex, *Nematologica* 24:418.

Triantaphyllou, A.C., and Riggs, R.D., 1979, Polploidy in an amphimictic population of *Heterodera glycines, J. Nematol.* 11:371.

Welsh, J., and McClelland, M., 1990, Fingerprinting genomes with PCR using arbitrary primers, *Nuc. Acids Res.* 18:7213.

Williams, J.G.K., Kubelik, A.R., Livak, K.J., Rifalski, J.A., and Tingey, S.V., 1990, DNA polymorphisms amplified by arbitrary primers are useful as genetic markers. *Nuc. Acids Res.* 18:6531.

Wyss, U., and Zunke, U., 1986, Observations on the behavior of second stage juveniles of *Heterodera schachtii* inside host roots, *Rev. Nematol.* 9:153.

Wyss, U., and Zunke, U., 1992, Observations on the feeding behavior of *Heterodera schachtii* throughout development, including events during molting, *Fund. appl. Nematol.* 15:75-89.

Young, L.D., 1982, Reproduction of differentially selected soybean cyst nematode populations on soybeans, *Crop Sci.* 22:385.

Young, L.D., 1984, Changes in the reproduction of *Heterodera glycines* on different lines of *Glycine max, J. Nematol.* 16:304.

Young, L.D., 1992, Epiphytology and life cycle, *in:* "Biology and Management of the Soybean Cyst Nematode," R.D Riggs and Wrather, J.A., eds., APS Press, St. Paul, MN.

MOLECULAR CHARACTERIZATION OF THE BURROWING NEMATODE SIBLING SPECIES, *RADOPHOLUS CITROPHILUS* AND *R. SIMILIS*

David T. Kaplan[1]

[1]Subtropical Plant Pathology Research Unit
U.S. Horticultural Research Laboratory
U.S.D.A.-Agricultural Research Service
Orlando, FL 32803

Burrowing nematodes, *Radopholus* spp. are migratory endophytoparasitic nematodes which are prevalent in many tropical and subtropical regions throughout the world. They are highly destructive and damage a wide range of plants by destroying the cortex as they feed in roots (Ducharme, 1959; Ford et al., 1960). *Radopholus* spp. are considered to be among the 10 most damaging nematodes worldwide (Sasser and Freckman, 1987).

Although there are 24 species described within the genus *Radopholus*, the two sibling species *R. citrophilus* (Huettel et al., 1984b) and *R. similis* (Cobb) Thorne are the most infamous because they are associated with significant economic loss. *Radopholus similis* has been detected in numerous countries throughout the tropics and subtropics (Gowen and Queneherve, 1990), but *R. citrophilus* has only been detected in Florida and Hawaii (Huettel et al., 1986). The diseases, spreading decline of citrus (Suit and Ducharme, 1953), black head toppling disease or toppling disease of banana (Loos and Loos, 1960), and slow wilt and yellows of black pepper (Van der Vecht, 1950) are caused by these two burrowing nematode species. Despite the presence of citrus in many countries where burrowing nematodes are commonly associated with a variety of crops, populations of *R. citrophilus* that attack citrus have only been detected in Florida (Holdeman, 1986).

Federal and state regulatory agencies have established quarantines to prevent the export of *R. citrophilus* from Florida. However, a fundamental obstacle to quarantine of *R. citrophilus* attacking citrus and to clarifying the taxonomy of *Radopholus* is the lack of knowledge of the biochemical or molecular basis that enabled burrowing nematodes in Florida to expand their host range to include citrus. To this end, Huettel and Dickson (1981a, 1981b) and Huettel et al. (1982, 1983a, 1983b, 1984a) identified biochemical, physiological, and karyotypic differences that distinguished the citrus and banana races of *R. similis*. These findings were used as criteria to establish the sibling species *R. similis* and *R. citrophilus* from the banana and citrus races of *R. similis*, respectively (Huettel et al., 1984b). More recently minor morphological differences in the tail region in males were identified (Huettel and Yaegashi, 1988). However, assays based on this information were considered impractical and wanting of sensitivity (Holdeman, 1986). While Siddiqi

(1986) classified the sibling species as subspecies *R. similis similis* and *R. similis citrophilus,* regulatory agencies disregarded the proposed sibling species concept and maintained the taxonomic status of these nematodes as races of *R. similis sensu lato* (Holdeman, 1986).

Technical capability to rapidly and accurately identify the distribution of burrowing nematode field populations that attack citrus or those that damage traditionally resistant rootstocks is lacking and is critical to grove, nursery site, and rootstock selection. Citrus growers interested in planting a resistant rootstock have no practical means of determining whether such a decision will improve production. This may also be relevant to banana production where burrowing nematode populations also appear to differ in their pathogenicity (Pinochet, 1979; Tarte et al., 1981; Sarah, et al., 1993).

There is a need to develop diagnostics and to characterize genetic variation in burrowing nematodes in order to provide decision-making processes for regulatory agencies and to facilitate agricultural production. Attempts to generate immunoassays to discriminate between the two sibling species were unsuccessful (Kaplan, unpublished data). Glycoproteins have proven useful in the identification of plant varieties (Niedz et al., 1991), but only 3 putative differences were detected in *R. similis* and *R. citrophilus* nematode homogenates (Kaplan and Gottwald, 1992), emphasizing the relatedness of the two sibling species and negating glycoprotein profiles as a plausible means of burrowing nematode identification.

Molecular characterization of the burrowing nematode genome should facilitate the development of sensitive and accurate diagnostics as well as provide insight into genome organization. This in turn should be helpful in understanding genetic variation in burrowing nematodes relative to host range. To this end, a collection of burrowing nematode strains maintained in carrot disk culture which differ in host range and geographic origin was assembled from Florida, Hawaii, the Carribean and Central America (Kaplan and Davis, 1990). A complimentary collection of burrowing nematode populations collected from Africa, Asia, and Indonesia was assembled and maintained at Commonwealth Agricultural Bureau, St. Albans, UK. These collections are essential to the analysis of the burrowing nematode genome and subsequent development of validated diagnostic assays.

From the Florida culture collection, two well-characterized strains of the burrowing nematode which differ in their ability to parasitize citrus were selected as subjects for the identification of randomly amplified polymorphic DNA (RAPD) (Williams et al., 1990). In RAPD studies involving 380 randomly generated single decamer primers and purified genomic DNA from two burrowing nematode populations, only 6 polymorphisms were identified. To develop a PCR-based procedure designed to specifically amplify these polymorphisms, each polymorphism was cloned and the terminal ends sequenced. A pair of primers was designed to complement the sequence of each end of the polymorphic DNA near the termini but excluding the original oligomer sequence (sequence-tagged site or STS) or including the oligomer sequence (sequence characterized amplified region or SCAR) (Olson et al., 1989; Michelmore et al., 1992). Currently, three sets of sequence specific primers that discriminate between the two strains of the burrowing nematode sibling species are under evaluation using the burrowing nematode culture collection.

Ribosomal DNA sequence has been studied to clarify the phylogenetic relationship of numerous organisms (Hillis and Dixon, 1991). Nematode rDNA is organized in much the same manner as that of other eukaryotic organisms by being comprised of multiple repeats of three ribosomal genes interspersed between spacer regions of DNA (Files and Hirsh, 1981). The extent of sequence divergence in rDNA is used as a parameter to estimate the relatedness of organisms. For *Caenorhabditis elegans*, the rDNA sequence is highly conserved between strains; but considerable differences were detected in the rDNA of the closely related species, *C. elegans* and *C. briggsae* (Files and Hirsh, 1981). Differences in rDNA restriction fragment length polymorphisms (RFLP) were also analysed for their ability to discriminate among *Xiphinema* spp. (Vrain et al., 1992). Interspecific differences in sequence appear to more pronounced in spacer regions that lie between genes in the tandem ribosomal RNA gene clusters in rDNA than in the gene themselves (Fedoroff, 1979). However, the alignment of the rDNA sequence for the internal transcribed spacer regions (ITS) and the sequence of the 5.8S rDNA for *R. similis* and *R. citrophilus* (Figure 1) implies a high level of sequence conservation between the two burrowing nematode sibling species (Table 1). Highly levels of conservation among the 5.8S and ITS regions within cyst nematode rDNAs (Ferris et al., 1993), and the 5.8S regions within some *Meloidogyne* species (T. Powers, personal communication) suggest that the ITS regions flanking the 5.8S region as well as the 5.8S gene will not prove useful for development of species-specific diagnostics for all plant parasitic nematodes.

Rc CTTACAAGTGAACCTATCCAATACGATTCCGTCCTTTGGTGGGCAGTGCC
Rs ....

Rc CTCAGGCATTGGCGAAACCCATCAAAAGTGACGCGAACGGCTGGGTTGGC
Rs -.........................T.......................

Rc GTCTGTGAGTCGTGGAGCAGTTGTAGTCCATGTCCGTGGCTGCGATGATG
Rs ....

Rc CGACTCGGTAGGGCTGTCATCGCCTTTGGCAGCTTAAGACTTGATGAGCG
Rs ....C.......

Rc CAGACCAAGCGCCGCCAACAACCATTTTTT-CCAATAAAACTTTTTCAAA
Rs ..............................T........-..........

Rc ATGCCATT-CAAGGCAAACAAGAATT**CTAGCCTTATCGGTGGATCACTCG**
Rs -.......T...

Rc **GCTCGTAGGTCGATGAAGAACGCAGCCAGCTGCGATATCTAGTGTGAACT**
Rs ....

Rc **GCAGAACCTTTGAACACTAAACATTCGAATGCACATTGCGCCATTGGAGT**
Rs ....

Rc **CACTTCCTCTGGCACGCCTGGTTCAGGTCGTTA**CCAAAAACGCAAGAC
Rs ....-.................

Rc AAATGCGTACATGAATGCGTGATAATTAATATTTTCATGCAATGGAAATT
Rs ....

Rc CAGAGTTTGCTAGTTTTTTACTTGGACATGAATTTCGTGAGTATTTGCCG
Rs ....

Rc TGTGGCGAAAGAAATTCACGTATGTTCATTGTGCAAA
Rs ....

Figure 1. Comparison of ITS-1, 5.8S and ITS-2 region of ribosomal DNA sequence from *Radopholus similis* (Rs) and *R. citrophilus* (Rc). The 5.8S gene is identified by bold font; ITS-1 (274 bp) precedes the 5.8S gene and the ITS-2 (152 bp) follows the 5.8S rDNA gene.

A comparison of similarity of *R. similis* and *R. citrophilus* with a variety of nematodes and other higher organisms suggests that intergenic spacer regions and the 5.8S region are conserved between genera (Tables 1 and 2). That is, the relative level of divergence in sequence appears similar for all of the organisms. Surprisingly, *Radopholus* spp.. demonstrates greater similarity with the 5.8S of *Meloidogyne* spp., *Xenopus laevis, Rattus norvegicus* and *Bombyx mori* than with *C. elegans*. Predictably, the ITS spacer regions are more greatly diverged than the 5.8S coding region (Federoff, 1979). A high degree of sequence similarity is expected in the 5.8S gene regardless of its phylogenetic origin because it has an explicit celluar function common to all living organisms. The gene can only tolerate changes to an extent that do not disrupt its cellular function. In contrast, the function of the ITS regions is uncertain, but they appear to tolerate high levels of nucleotide substitution as sequence is highly divergent between distant genera. Sequence homology in spacer regions for the two burrowing nematode species was slightly greater for ITS 2 which was 122 base pairs shorter than ITS 1. We are currently directing research on rDNA to explore the intergenic spacer region (IGS previously identified as nontranscribed spacer region or NTS) found between the 18S and the 28S tandem repeats of rDNA for sequence heterogeneity.

Table 1. Percent similarity of *Radopholus similis* 5.8S rDNA genetic sequence with that of *Caenorhabditis elegans*, *R. citrophilus*, *Meloidogyne* spp., *Xenopus laevis*, *Bombyx mori*, and *Rattus norvegicus*.

Organism[1]	***Radopholus similis******
*R. citrophilus**	—
Meloidogyne spp.**	75.3-82.2
Xenopus laevis	75.3
M. arenaria	75.3
Rattus norvegicus	73.6
Bombyx mori	73.2
Caenorhabditis elegans	67.1

[1]Source of sequence 5.8S genetic sequence: * = Kaplan, D.T. and Opperman, C.H., unpublished data; **Powers, T.O., unpublished data; *X. laevis* (XELRGB12); *M. arenaria* (MARR58SI); *R. norvegicus* (RNRRNA01); *B. mori* (BMORGA2); and *C. elegans* (CERDNA).

Table 2. Sequence similarity (%) of ITS 1 and ITS 2 of rDNA from *Radopholus citrophilus* and *R. similis* with sequence of *Caenorhabditis elegans* and *Heterodera* spp..

	ITS 1	ITS 1	ITS2
Nematode	*R. citrophilus**	*R. similis**	*R. similis/ R. citrophilus*[1]*
*R. citrophilus**	—	96.0	—
Heterodera spp.**	41.5-46.9	40.7-44.6	50.0-55.8
*C. elegans****	42.0	44.0	58.5

[1] The sequence of ITS 2 for *R. citrophilus* and *R. simlis* demonstrated complete identity.

Source of sequence ITS 1 and ITS 2 genetic sequence: * = Kaplan, D.T. and Opperman, C.H., unpublished data; **Ferris et al., 1993.

Our findings on rDNA and RAPD analysis suggest that the burrowing nematode genome is highly conserved. This concept is supported by the RAPD analysis of *R. similis* populations from Sri Lanka where 85% of the 14 populations collected from different geographic locations had common profiles (Hahn et al., 1994). A high level of conservation among burrowing nematodes is in sharp contrast to studies involving other plant parasitic nematodes where considerable variation in amplification products were detected with geographical isolates of *Meloidogyne* and *Heterodera* (Caswell-Chen et al., 1992; Ferris et al., 1993; T. Powers, personal communication).

It had been suggested that the limited genetic variation observed between species of *Radopholus* may be attributed to laboratory culture techniques (Hahn et al., 1994); however we believe genome conservation may be more strongly influenced by the reproductive mode (facultative parthenogenesis), manner of long-range nematode dissemination (through vegetative plant propagation), and the nematode's migratory parasitic habit. Of these, parthenogenesis may play a relatively minor role in stabilizing genome organization since considerable variation in RAPD analysis has been reported for other parthenogenetic species, i.e., *Meloidogyne* (Cenis, 1993; Xue et al., 1993). Movement of burrowing nematodes through the planting of infested nursery plants (banana, citrus, and ornamentals) may have significantly contributed to the apparent homogeneity of the burrowing nematode genome throughout the world. Furthermore, the migratory parasitic habit of burrowing nematodes in conjunction with a wide host range may also contribute to the apparent homogeneity of the genome. Burrowing nematodes (and other migratory plant parasitic nematodes) may avoid selection pressure through their ability to move away from roots of poor hosts or resistant plants. In contrast, sedentary plant parasitic nematode species infect roots of resistant and susceptible plants and attempt to establish feeding sites and to overcome resistance genes where necessary (or possible).

In general, burrowing nematode populations with polymorphic DNA appear to be those isolates with expanded host range (Hahn et al., 1994; Kaplan et al., unpublished data). If shifts in phenotype such as host range (e.g., citrus as a host), are associated with readily identifiable genetic markers and if sexual recombination occurs following mating, then the burrowing nematode could provide a model system for the study of the genetics of parasitism in migratory endoparasitic nematodes.

Application of *R. citrophilus*-specific primers developed to STS and SCARs are being evaluated for use as molecular markers to study genetic variation and for use as diagnositics. It is hoped that the information generated by this project will provide a rapid and practical method of identifying burrowing nematodes, clarify the taxonomic status of the burrowing nematode sibling species, elucidate the mode of reproduction of burrowing nematodes (facultative parthenogenesis?), and provide insight into the basis of genetic variation relative to plant resistance. These findings would significantly contribute to the development of innovative strategies to reduce losses associated with burrowing nematode infestations.

Acknowledgements

I wish to thank C. M. Vanderspool, D. L. Zies, D. M. Johnson, M.E. Hilf, and C. J. Nairn, USDA-ARS, S. Chang, C. Garrett, and C. H. Opperman, North Carolina State University; and A. G. Abbott, Clemson University, for their assistance and encouragement. Appreciation is expressed to T. Powers, University of Nebraska, for sequence information and W. Farmerie, Interdisciplinary Center for Biotechnology Research, University of Florida, for assistance with primer design.

REFERENCES

Caswell- Chen, E.P., Williamson, V.M., and Wu, F.F., 1992, Random amplified polymorphic DNA analysis of *Heterodera cruciferae* and *H. schachtii* populations, *J. Nematol.* 24:343.

Cenis, J.L., 1993, Identification of four major *Meloidogyne* spp.. by random amplified polymorphic DNA (RAPD-PCR), *Phytopathology* 83:76.

DuCharme, E.P., 1959, Morphogenesis and histopathology of lesions induced on citrus roots by Radopholus similis, *Phytopathology* 49:388.

Fedoroff, N.V., 1979, On Spacers, *Cell* 16: 697.

Ferris, V.R., Ferris, J. M., and Faghihi, J., 1993, Variation in spacer ribosomal DNA in some cyst-forming species of plant parasitic nematodes, *Fund. Appl. Nematol.* 16:177.

Files, J.G., and Hirsh, D., 1981, Ribosomal DNA of *Caenorhabditis elegans*, *J. Mol. Biol.* 149: 223.

Ford, H.W., Feder, W.A., and Hutchins, P.C., 1960, Citrus varieties, hybrids, species, and relatives evaluated for resistance to the burrowing nematode *Radopholus similis*. Citrus Experiment Station Mimeo Series, 60.

Gowen, S., and Queneherve, P., 1990, Nematode parasites of bananas, plantains, and abaca. *in:* "Plant Parasitic Nematodes in Subtropical and Tropical Agriculture," M. Luc, R.A. Sikora, and J. Bridge, eds., CAB, St. Albans, UK.

Hahn, M.L., Burrows, P.R. , Gnanapragasam, N.C., Bridge, J., Vines, N., and Wright, D.J., 1994, Molecular diversity amongst *Radopholus similis* populations from Sri Lanka detected by RAPD analysis, *Fund. Appl. Nematol.* (In press).

Hillis, D.M., and Dixon, M.T., 1991, Ribosomal DNA: Molecular evolution and phylogenetic inference, *Q. Rev. Biol.* 66:411.

Holdeman, Q.L., 1986, The burrowing nematode *Radopholus similis, sensu lato*. California Department of Food and Agriculture, Sacramento, CA.

Huettel, R. N., and Dickson. D.W., 1981a, Parthenogenisis in the two races of *Radopholus simlis* from Florida, *J. Nematol.* 13:13.

Huettel, R.N., and Dickson, D.W., 1981b, Karyotype and oogenesis of *Radopholus similis* (Cobb) Thorne, *J. Nematol.* 13:16.

Huettel, R.N., Dickson, D.W., and Kaplan, D.T., 1982, Sex attraction and behavior in the two races of *Radopholus similis* by starch gel electrophoresis, *Nematologica* 28:360.

Huettel, R.N., Dickson, D.W., and Kaplan, D.T., 1983a, Biochemical identification of two races of *Radopholus similis* by starch gel electrophoresis, *J. Nematol.* 15:338.

Huettel, R.N., Dickson, D.W., and Kaplan, D.T., 1983b, Biochemical identification of two races of *Radopholus similis* by polyacrylamide gel electrophoresis, *J. Nematol.* 15:345.

Huettel, R.N., Dickson, D.W., and Kaplan, D.T., 1984a, Chromosome number of populations of *Radopholus similis* from North, Central and South America, Hawaii and Indonesia, *Revue Nematol.* 7:113.

Huettel, R.N., Dickson, D.W., and Kaplan, D.T., 1984b, *Radopholus citrophilus* sp. n. Nematoda, a sibling species of *Radopholus similis*, *Proc. helminthol. Soc. Wash.* 51:32.

Huettel, R.N., Dickson, D.W., and Kaplan, D.T., 1986, Characterization of a new burrowing nematode population , *Radopholus citrophilus*, from Hawaii, *J. Nematol.* 18:50.

Huettel, R.N., and Yaegashi, T., 1988, Morphological differences between *Radopholus citrophilus* and *R. similis*, *J. Nematol.* 20:150.

Kaplan, D.T., and Davis, E.L., 1990, Improved nematode extraction from carrot disk culture, *J. Nematol.* 22:399.

Kaplan, D.T., and Gottwald, T.R., 1992, Lectin binding to *Radopholus citrophilus* and *R. similis* proteins, *J. Nematol.* 24:281.

Loos, C.A. and Loos, S.B., 1960, The blackhead disease of bananas, *Proc. helminthol. Soc. Wash.* 27:189.

Michelmore, R.W., Kesseli, R.V., Francis, D.M., Paran, I., Fortin, M.G., and Yang, D.H., 1992, Strategies for cloning plant disease resistance genes. *in:* "Molecular Plant Pathology 2. A Practical Approach". Oxford University Press Oxford, U.K.

Niedz, R.P., Bausher, M.G., and Hearn, C.J., 1991, Detection of citrus leaf and seed glycoproteins using biotinylated lectin probes, *Hort Sci.* 26:910.

Olson, M., Hood, L., Cantor, C., and Botstein, D., 1989, A common language for the physical mapping of the human genome. *Science* 248: 1434.

Pinochet, J., 1979., Comparison of four isolates of *Radopholus similis* from central america on valery bananas, *Nematropica,* 9:40.

Sarah, J.L., Sabatini, C., and Boisseau, M., 1993, Differences in pathogenicity to banana (*Musa* sp., cv. poyo) among isolates of *Radopholus similis* from different production areas of the world, *Nematropica* 23: 75.

Sasser, J.N., and Freckman, D.W., 1987, A world perspective on nematology: The role of the society, in: "Vistas on Nematology," J.A. Veech, and D.W. Dickson, eds. Society of Nematologists, Inc., Hyattsville, MD.

Siddiqi, M.R., 1986, "Tylenchida: Parasites of Plants and Insects," CAB, St. Albans, UK.

Suit, R.F., and Ducharme, E.P., 1953, The burrowing nematode and other parasitic nematodes in relation to spreading decline, *Citrus Leaves* 33:8.
Tarte, R.., Pinochet, J., Gabrielli, C., and Ventura, O., 1981, Differences in population increase, host preferences and frequency of morphological variants among isolates of the banana race of *Radopholus similis*, *Nematropica* 11:43.
Van der Vecht, J., 1950, Plant parasitic nematodes. *in:* "Diseases of Cultivated Plants in Indonesian Colonies", L.G.E. Karshoven, and J. Van der Vecht, eds., I.S'gravenhage, W. van Woeve.
Vrain, T.C., Wakarchuk, D.A., Levesque, A.C., and Hamilton, R.C., 1992, Intraspecific rDNA restriction fragment length polymorphism in the *Xiphinema americanum* group, *Fundam. Appl. Nematol.* 15:563.
Williams, J.G.K., Kubelik, A.R., Livak, K.J., Rafalski, J.A., and Tingey, S.V., 1990, DNA polymorphisms amplified by arbitrary primers are useful as genetic markers, *Nucl. Acid Res.*, 18:6531.
Xue, G., Baillie, D.L., and Webster, J.M., 1993, Amplified fragment length polymorphisms of *Meloidogyne* spp.. using oligonucleotide primers, *Fund. Appl. Nematol.* 16:481.

CUTICLIN GENE ORGANIZATION IN *MELOIDOGYNE ARTIELLIA*

Francesca De Luca,[1] Carla De Giorgi,[1] and Franco Lamberti[2]

[1]Dipartimento di Biochimica e Biologia Molecolare
University of Bari, Bari, Italy
[2]Istituto di Nematologia Agraria
Consiglio Nazionale delle Ricerche, Bari, Italy

INTRODUCTION

The cuticlins are insoluble, non-collagenous components of the nematode cuticle. They were first described in *Ascaris lumbricoides* by Fujimoto and Kanaya in 1973. Their insolubility, even in the presence of strong detergents and reducing agents, indicates that they are highly cross-linked by non reducible bonds. Recently, Sebastiano et al. (1991) have identified and cloned two distinct cuticlin genes from *Caenorhabditis elegans,* namely *cut-1* and *cut-2.*

The *cut-1* gene is expressed specifically during dauer larva formation. It codes for a protein that is localized under the alae and is absent in the cuticles of other stages. The *cut-2* gene, in contrast, is expressed in all developmental stages of *C. elegans* and is localized in the external cortical layer. Neither cuticlin proteins are exposed to the surface of the cuticle.

It is conceivable that many of the structural and developmental properties of the *C. elegans* cuticle can have a direct parallel in the parasitic nematode cuticle, despite the diversity of life styles to which nematode species have adapted. So far, the best example is the similarity in gene structure and organization of the collagens (Politz and Philipp, 1992).

In this communication, we report the isolation and characterization of the *cut-1* gene in the plant parasitic nematode *Meloidogyne artiellia* (Di Vito et al., 1985), by using the cuticlin gene isolated from *C. elegans* as a probe.

RESULTS AND DISCUSSION

Fig.1 shows a Southern blot of *M. artiellia* genomic DNA probed with *C. elegans cut-1.* This result clearly indicates that the *cut-1* gene is present in the genome of *M. artiellia.* Evidence was obtained that the *cut-2* gene is also present in the *M. artiellia* genome (result not shown).

In order to isolate the cuticlin gene, a genomic DNA library was made

Figure 1. Southern blot experiment of *Meloidogyne artiellia* total DNA.
Total genomic DNA was isolated by the proteinase K-SDS method (Emmons et al., 1979), digested with *EcoRI*. The 1.8 kb fragment, containing part of the *C.elegans cut-1*, was nick-translated and used as a probe. The filter was hybridized in 50% formamide, 5x Denhardt's solution, 0.1% SDS at 42° C for 14 hrs.

in lambdaGEM-11(Promega) and screened with the *C. elegans cut-1* probe. Two positive phages, presumably identical, showing the same restriction pattern, have been isolated. One of them has been analysed in detail.

A Southern analysis of the cloned fragment of *M. artiellia*, cleaved with different enzymes was hybridized using *C. elegans cut-1* (Fig.2A). A restriction site map is also reported in Fig. 2B. The 3 kb fragment derived from *Bam*HI/*Pst*I digestion and containing the *cut-1*, was subcloned and sequenced using the dideoxy chain-termination method (Sanger et al., 1977).

The comparison of the sequence of *M. artiellia* with that of *C. elegans* revealed that the fragment contains the first and second exons and part of the third exon, while the sequence at the 3' end of the *cut-1* gene is not yet available. However the gene organization in the two nematodes is quite different. The 5' end of the*cut-1* gene of *M. artiellia* is longer than the corresponding region*C. elegans*, the first exon is immediately followed by the second exon and the third exon is immediately adjacent to the second exon. The third exon is interrupted by a sequence of 200 nt, which is absent in the *C. elegans cut-1* gene (Fig.3).

We suspect that this novel sequence is an intron because we did not find any reading frame in phase with the rest of the protein. Moreover the sequence is A+T rich as are those of the introns inferred in *C. elegans*, and in addition the homology of the amino acid sequences between the two genes ends abruptly at the potential donor splice site, GT, and starts again immediately downstream of the potential acceptor splice site, AG.

The 5' flanking region of the *M. artiellia cut-1* gene contains the putative eukaryotic promoter elements TATA, at position -38, and CAAT, at position

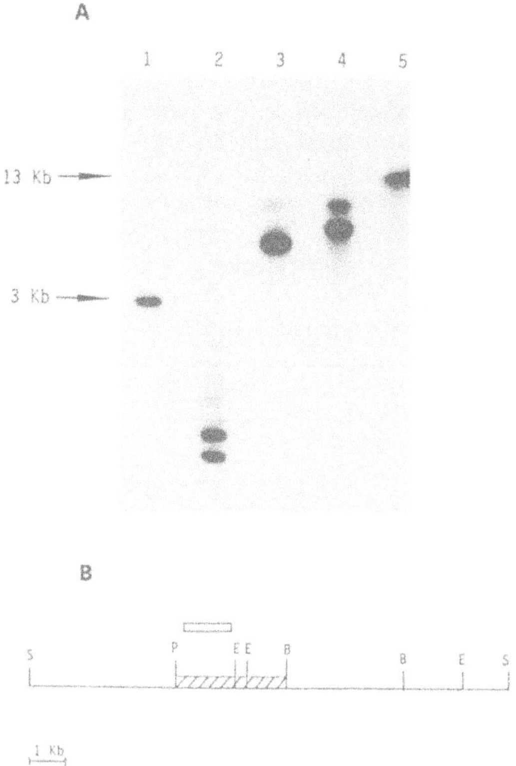

Figure 2. Identification of the 3 kb fragment containing *cut-1* in *M. artiellia*. A) The 13 kb insert was eluted from the positive recombinant phage, cleaved with different enzymes and hybridized with the *C. elegans cut-1* probe. Lane 1, *Bam*HI and *Pst*I; lane 2, *Bam*HI and *Hinf*I; lane 3, *Bam*HI and *Eco*RI; lane 4, *Bam*HI; lane 5, recombinant phage, uncut. B) Restriction map of the 13 kb fragment of *M. artiellia* Cleavage sites for each restriction endonuclease are indicated by vertical lines topped with the following symbols: S=*Sfi*I; B=*Bam*HI; E=*Eco*RI; P=*Pst*I. The hatched box represents the 3 kb fragment containing the *cut-1* gene. The upper box indicates the *C. elegans cut-1* probe.

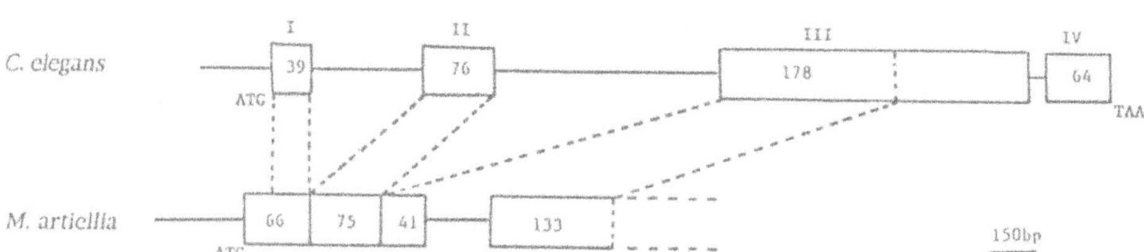

Figure 3. Schematic representation of the organization of the *M. artiellia cut-1* gene. The boxes indicate the exons and the continuous lines the introns. Numbers within the boxes indicate the amino acid length.

-64, that are absent in the corresponding flanking region of C. *elegans*. The sequence of the genomic *cut-1* gene of *M. artiellia* has not revealed the presence of the acceptor splice sequence immediately upstream from the initiation codon (TTCAG/ATG), as in *cut-1* of C. *elegans* (Krause and Hirsh, 1987). This finding suggests that the *cut-1* message in *M. artiellia* is probably not transpliced.

When the 5′ flanking regions in the two nematodes *C. elegans* and *M. artiellia* are aligned, a 13 bp region of homology is revealed. It is localized at -14 in *M. artiellia* and at -34 in *C. elegans*. This sequence, in which 11 out of 13 residues are conserved, is rich in pyrimidines. Moreover, a similar conserved sequence in the promoter regions of the dauer larva specific genes *col-2* and *col-6* (Cox et al., 1989), has also been detected.

Table 1. Sequence similarities in the 5′ flanking regions.

Nematode	Gene	Sequences	Positions
C. elegans	*col-2*	TATTCTTTCTCTTTTCTTTCT	-134
C. elegans	*cut-1*	TATTCCTTCC-TCTCAA	-34
M. artiellia	*cut-1*	TAATCCTTCC-TTCCCTTCAC	-14

The similarities were identified through computer searches for conserved sequences between the 5′ flanking regions of *cut-1* genes and for related sequences. The nucleotide sequence of *col-2* used for these analyses was from Kramer et al., (1985), while that of *C. elegans cut-1* from Sebastiano et al., (1991) (Table 1).

The correlation between the presence of these sequences and the expression profiles in C. *elegans*, strongly suggests that these sequences may be involved in the developmental regulation of these genes.

The codon usage is nearly the same in both nematodes. The most used base in the first position of the codon is G in both nematodes. In the second position the *M. artiellia* codon uses A, while the *C. elegans* codon indifferently uses A, C, or G. In the third position pyrimidines are present in both nematodes.

In conclusion, these preliminary data show that the *cut-1* gene is conserved in the plant parasitic nematode *M. artiellia* and probably has the same function as in *C. elegans*. However, the organization of the two genes, the differences in the splicing signals (encoded in the genomic sequences) and the different promoter regions, strongly indicate that the mechanisms of their expression are different in the two nematodes which diverged during evolution.

Acknowledgements

We wish to thank Dr. Paolo Bazzicalupo for generously providing the *C. elegans* cuticlin probes. We are also grateful to Dr. Mauro Di Vito for providing *M. artiellia* egg masses.

REFERENCES

Cox, G.N., Fields, C., Kramer, J.M., Rosenzweig, B., and Hirsh, D., 1989, Sequence comparisons of developmentally regulated collagen genes of *Caenorhabditis elegans*, *Gene* 76:331.

Di Vito, M., Greco, N., and Zaccheo, G., 1985, On the host range of *Meloidogyne artiellia.*, *Nematol. Medit.* 13:207.

Emmons, S.W., Klass, M.R., and Hirsh, D., 1979, Analysis of the constancy of DNA sequences during development and evolution of the nematode *Caenorhabditis elegans*, *Proc. Natl. Acad. Sci. USA* 76:1333.

Fujimoto, D., and Kanaya, S., 1973, Cuticlin: a noncollagen structural protein from *Ascaris* cuticle, *Arch. Biochem. Biophis.* 157:1.

Kramer, J.M., Cox, G.N., and Hirsh, D., 1985, Expression of the *Caenorhabditis elegans* collagen genes *col-1* and *col-2* is developmentally regulated, *J. Biol. Chem.* 260:1945.

Krause, M., and Hirsh, D., 1987, A trans-spliced leader sequence on Actin mRNA in *C.elegans* , *Cell* 49:753.

Politz, S.M., and Philipp, M., 1992, *Caenorhabditis elegans* as a model for parasitic nematodes: a focus on the cuticle, *Parasitol. Today* 8:6.

Sanger, F., Nicklen, S., and Coulson, A.R., 1977, DNA sequencing with chain-terminating inhibitors, *Proc. Natl. Acad. Sci. USA* 74:5463.

Sebastiano, M., Lassandro, F., and Bazzicalupo, P., 1991, *cut-1* a *Caenorhabditis elegans* gene coding for a dauer specific non-collagenous component of the cuticle, *Dev. Biol.* 146:519.

DISCUSSION

Georgi asked the percentage of amino acids of this putative cuticlin identical with the *C. elegans* sequence.

De Luca replied that the amino acid homology is higher than 75%.

GENETIC VARIATION IN TROPICAL *MELOIDOGYNE* SPECIES

Mireille Fargette[1], Vivian C. Blok[2], Mark S. Phillips[2] and David L. Trudgill[2]

[1] ORSTOM-Montpellier
2051 Av. du Val de Montferrand
BP 5045
34032 Montpellier Cedex 1
France

[2] Scottish Crop Research Institute
Invergowrie
Dundee DD2 5DA
Scotland

INTRODUCTION

In tropical regions the parthenogenetic *Meloidogyne* spp. are major crop pests. The presence of populations in West Africa which overcome resistance in cultivars as diverse as tomato cv. 'Rossol', sweet potato cv. 'CDH' and soya bean cv. 'Forrest' has been reported (Prot, 1984; Fargette and Braaksma, 1990). The relationship of these populations to other tropical *Meloidogyne* spp. was investigated as the nature of such resistance breaking populations has implications both for integrated control and for appropriate quarantine measures. At a broader level, variation within the various tropical *Meloidogyne* spp. is also of interest to gain an understanding of the extent of genetic variation in this important group of pests.

ESTERASE PHENOTYPES

Initial studies showed that the resistance breaking populations from West Africa have a distinct esterase phenotype (Fargette and Braaksma, 1990) which is also produced by *M. mayaguensis*, a species described from Mexico (Rammah and Hirschmann, 1988). This implies that these populations are biologically distinct from *M. incognita*, *M. javanica* and *M. arenaria*, the other major tropical root-knot nematode species. Whilst there is relatively little within species variation in the esterase patterns, there are indications particularly between *M. arenaria* populations that intraspecific genetic variation does occur.

Advances in Molecular Plant Nematology
Edited by F. Lamberti *et al.*, Plenum Press, New York, 1994

RFLP STUDIES

To assess this variation further and to confirm the identification of the resistance breaking populations from West Africa RFLP studies were undertaken. Clones were established from a single egg mass for the populations shown in Table 1. These lines were maintained on the susceptible tomato cv. Moneymaker and juveniles mass extracted from the roots 6–7 weeks after inoculation by placing washed roots in a mist unit. Nematodes were collected, stored at -70°C until used for DNA extraction.

Table 1. Specific identification and geographic origin of the 29 *Meloidogyne* lines.

Nematode species	Population code	Geographic origin
M. javanica	22	Burkina Faso, Sahelian climate
	23	Burkina Faso, South more humid climate
	24	Spain
	25	Portugal
M. arenaria	10	Ivory Coast
	26	Portugal
	28	French West Indies
	29	French West Indies (Ste Anne)
	31	French West Indies
	32	French West Indies
	34	French West Indies
M. mayaguensis	30	Burkina Faso
	1	Ivory Coast
	2	Ivory Coast
	3	Ivory Coast
	5	Ivory Coast
	7	Ivory Coast
	13	Puerto Rico
M. incognita	11	Louisiana, USA, race 3
	12	North Carolina, USA, race 4
	15	Thailand, race 1
	9	Ivory Coast
	17	Burkina Faso, Soudanian climate
	20	Guyana
	16	Senegal
	18	Chad
	19	French West Indies (Martinique)
	27	North Carolina, USA
M. hapla	33	The Netherlands

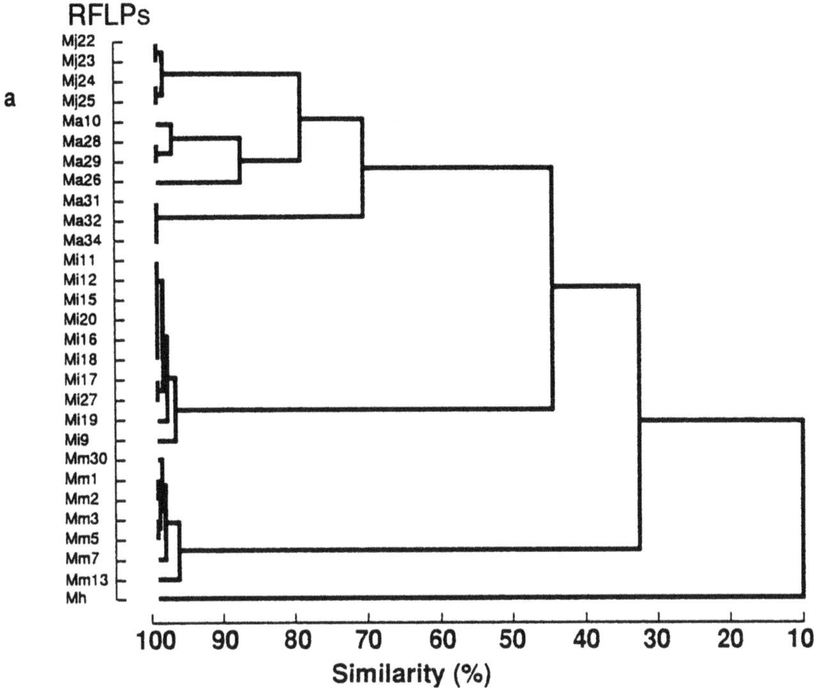

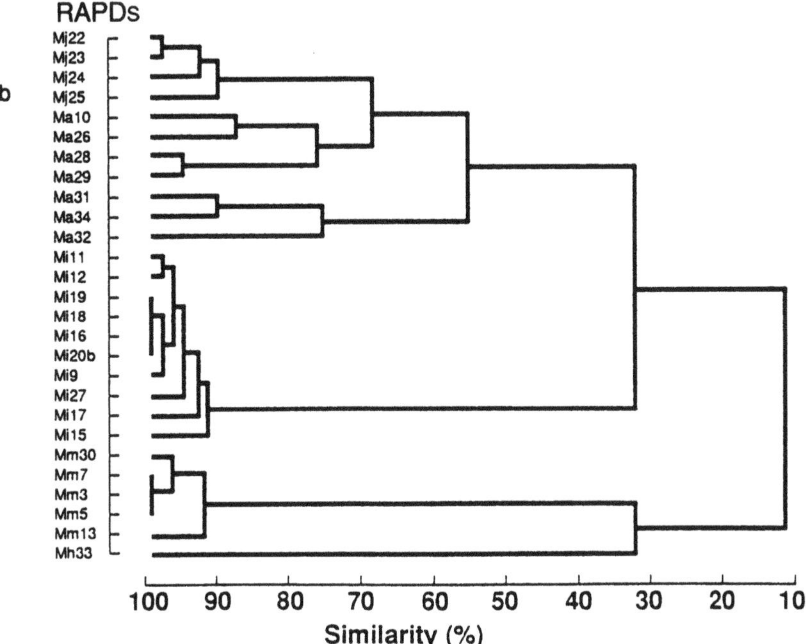

Figure 1. Dendrograms of similarities between populations of *Meloidogyne* species based on RFLPs (Fig. 1a) and RAPDs (Fig. 1b).

DNA was isolated from juveniles which were ground in liquid nitrogen. Extraction buffer (7 M urea, 0.35 M NaCl, 2% SDS, 10 mM EDTA, 0.1 M Tris–HCl pH 7.5) was then added and the mixture swirled for 15 min, then extracted two times with phenol:chloroform (1:1) and then chloroform followed by ethanol precipitation. Five restriction enzymes were used (BamHI, EcoRI, EcoRV, HindIII and BglII) to digest the DNA (200 ng per digest) in the presence of RNase. Samples were electrophoresed on 1% TBE agarose gels (Sambrook et al., 1989) and alkali blotted on to Hybond N^+ membrane. Inserts from four clones of low copy DNA prepared from a genomic library from *M. mayaguensis* line 5 were random labelled with ^{32}P–dCTP and used to probe the blots. Following hybridization, blots were washed at high stringency (0.1 x SSC, 0.1% SDS) 65°C. Twenty different autoradiograms were produced with the four probes and five endonucleases.

The RFLP patterns readily discriminated between the species. Two of the probes produced simple patterns which were easy to score whereas the others gave more complicated patterns. Polymorphisms within species, particularly *M. arenaria* were found. A total of 383 bands were scored and these data were used to produce a similarity matrix using the formula of Nei and Li (1979) and from which a dendrogram was produced (Figure 1a).

The populations of *M. incognita, M. javanica* and *M. mayaguensis* were all very homogeneous with average within group similarities of 98.5% or higher (Table 2). However, each group was clearly distinct with intergroup similarities of less than 45%. The *M. arenaria* group was most variable and was more similar to the *M. javanica* group than to any of the others. Within the *M. arenaria* group two subgroups consisting of lines 10, 26, 27 and 28, and lines 31, 32 and 34 occur with the first group showing more similarity to *M. javanica* than to the second group. *M. hapla*, only represented by one line, appears to be quite distinct from the tropical root–knot nematode species. Sequence divergence of *M. hapla* from *M. mayaguensis*, from which the probes were derived probably accounts for the lower number of bands scored for *M. hapla* due to poor hybridization of the probes in the high stringency washes of the blots.

Table 2. Between and within species similarity matrix derived from RFLP data; Mj = *M. javanica*, Ma = *M. arenaria*, Mm = *M. mayaguensis*, Mi = *M. incognita*, Mh = *M. hapla*.

	Mj	Ma	Mm	Mi	Mh
Mj	99.7				
Ma	73.5	83.9			
Mm	35.7	35.2	98.7		
Mi	43.3	44.4	30.9	99.0	
Mh	6.3	9.6	11.0	14.5	100

RAPD STUDIES

It was desirable to find a PCR based method as an alternative to the RFLP technique as the latter requires relatively large amounts of high quality DNA. Initially the internal transcribed ribosomal cistron was examined using primers designed by Vrain et al. (1992). However very little polymorphism between species was found in this region so was not expected to be informative for the assessment of intraspecific variation. Next single random primers 10 nucleotides in length were used as described by Williams et al. (1990) and Welsh and McClelland (1990) in the random amplified polymorphic DNA-PCR technique. Reactions included 10 ng of DNA in a volume of 50 µl with standard amounts of buffer, dNTPs and *Taq* polymerase (Boehringer Hannheim) as indicated by the manufacturer. Amplification for 45 cycles with an annealing temperature of 38°C was performed in a Perkin-Elmer 480 thermal cycler. The products from these PCR reactions were separated by electrophoresis in TBE buffered 1.5% agarose gels and the products visualised with ethidium bromide and UV illumination.

Primers differed in their ability to produce clear, reproducible patterns but several patterns were produced which showed both inter- and intraspecific variation. Reproducibility was assessed by repeating the amplifications on three separate occasions, taking considerable care to maintain consistent reaction conditions as this technique is very sensitive to minor changes. The three replicates were generally very similar although this was not always so. With some primers the replicates did not show such good reproducibility and consequently it was difficult to determine whether bands are equivalent between different lines. Only reaction products which were consistently produced were scored. Data from nine primers were scored (112 bands) and used to generate a similarity matrix and dendrogram using the same methods as for the RFLP data (Figure 1b).

This dendrogram shows many similarities to that produced from the RFLP data. The resistance breaking lines from West Africa were grouped with the *M. mayaguensis* isolate from Puerto Rico; the *M. javanica, M. incognita* and *M. mayaguensis* groups were less variable than *M. arenaria* and the subgroup of *M. arenaria* (lines 10, 26, 28 and 29) were more similar to the *M. javanica* group than to the other *M. arenaria* lines (Table 3). The RAPD data revealed more intraspecific variation than RFLPs which is partly explained by the greater proportion of monomorphic bands observed within groups using the latter technique. Overall the dendrograms produced from the RFLP and RAPD analyses were very similar. Some differences in the within species groupings revealed by the two techniques are apparent but it is not surprising given the small amount of discrimination found between lines within groups such as *M. mayaguensis* and *M. incognita*.

Table 3. Between and within species similarity matrix derived from RAPD data; Mj = *M. javanica*, Ma = *M. arenaria*, Mm = *M. mayaguensis*, Mi = *M. incognita*, Mh = *M. hapla*.

	Mj	Ma	Mm	Mi	Mh
Mj	94.7				
Ma	72.3	76.5			
Mm	6.9	8.2	95.9		
Mi	51.4	44.9	18.1	96.1	
Mh	13.4	19.8	20.5	17.6	100

CONCLUSIONS

Carpenter et al. 1992 reported the relatively high levels of variation in the *M. arenaria* group even with populations which were geographically closely related.

Garate et al. (1991) and Xue et al. (1992) observed that *M. arenaria* and *M. javanica* were closely related and could possible by considered as one group. These conclusions are supported by these RFLP and RAPD studies. Studies of mitochondrial DNA, however, show that *M. arenaria* is distinct (Powers and Sandall, 1988; Hyman and Powers, 1991) from other *Meloidogyne* spp. and thus at some point the maternal inheritance, as indicated by mtDNA, of these groups diverged.

The RFLP and RAPD analyses gave the same species groupings of the lines and in general showed similar relationships of the species to each other. More within species variation was detected with the RAPD analysis and less DNA was required to generate this information. It is not yet clear whether discrepancies in the intraspecies groupings obtained with the two types of analysis would be diminished with more data. The data obtained so far are limited in the amount of genomic information revealed and hence more data would give a more accurate indication of the relationships. Clearly, despite reproduction by mitotic parthenogenesis intraspecific variation in the tropical root knot nematodes does occur.

M. mayaguensis appears to be a distinct and highly virulent biological group and the extent of its global distribution is of importance to assess the risk of this potentially dangerous pest and to implement appropriate quarantine measures.

Acknowledgements

The authors acknowledge the technical skills of Jane Roberts and Kirsty Murray and funding by the Scottish Office Agriculture and Fisheries Department and EEC Project No. TS3-CT92-0098.

REFERENCES

Carpenter, A.S., Hiatt, E.E., Lewis, S.A., and Abbot, A.G., 1992, Genomic RFLP analysis of *Meloidogyne arenaria* race 2 populations, *J. Nematol.* 24:23.

Fargette, M., and Braaksma, R., 1990, Use of the esterase phenotype in the taxonomy to the genus *Meloidogyne*. 3. A study of some "B" race lines and their taxonomic position, *Revue Nématol.* 13:375.

Garate, T., Robinson, M.P., Chanon, M.R., and Parkhouse, R.M.E., 1991, Characterization of species and races of *Meloidogyne* by DNA restriction enzyme analysis, *J. Nematol.* 23:414.

Hyman, B.C., and Powers, T.O., 1991, Integration of molecular data with systematics of plant parasitic nematodes, *Ann. Rev. Phytopath.* 29:89.

Nei, M., and Li, W., 1979, Mathematical model for studying genetic variation in terms of restriction endonucleases, *Proc. Natl. Acad. Sci.* 76:5269.

Powers, T.O., and Sandall, L.J., 1988, Estimation of genetic divergence in *Meloidogyne* mitochondrial DNA, *J. Nematol.* 20:505.

Prot, J.C., 1984, A naturally occurring resistance breaking biotype of *Meloidogyne arenaria* on tomato. Reproduction and pathogenicity on tomato cultivars Roma and Rossol, *Revue Nématol.* 7:23.

Rammah, A., and Hirschmann, H., 1988, *Meloidogyne mayaguensis* n. sp. (Meloidogynidae), a root-knot nematode from Puerto Rico, *J. Nematol.* 20:58.

Sambrook, J., Fritsch, E.F., and Maniatis, T., 1989, Molecular cloning. A laboratory manual, Cold Harbor Laboratory Press, Vol. 1, 2 and 3.

Vrain, T.C., Wararchuk, D.A., Lévesque, A.C., and Hamilton, R.L., 1992, Intraspecific rDNA restriction fragment length polymorphism in the *Xiphinema americanum* group, *Fundam. appl. Nematol.* 15:563.

Welsh, J., Petersen, C., and McClelland, M., 1990, Polymorphisms generated by arbitrarily primed PCR in the mouse: application to strain identification and genetic mapping, *Nucl. Acid. Res.* 19:303.

Williams, G.K., Kubelik, A.R., Livak, K.J., Rafalski, J.A., and Tingey, S.V., 1990, DNA polymorphisms amplified by arbitrary primers are useful genetic markers, *Nucl. Acid. Res.* 18:6531.

Xue, B., Baille, D.L., Beckenbach, K., and Webster, J.M., 1992, DNA hybridization probes for studying affinities of three *Meloidogyne* populations, *Fundam. appl. Nematol.* 15:35.

PART II

IDENTIFICATION OF NEMATODES

NUCLEOTIDE SEQUENCES IN NEMATODE SYSTEMATICS

Thomas O. Powers and Byron J. Adams

Department of Plant Pathology
University of Nebraska
Lincoln, NE 68583 USA

INTRODUCTION

The addition of nucleotide sequence data to systematics has dramatically altered the study of relationships among organisms. In some respects, it has merely intensified the debate involving congruence between morphological and nonmorphological data sets (Patterson et al., 1993; Swofford, 1991; Hillis, 1987). In other regards, it has shifted the focus of the debate to methods of handling large data sets comprised of nonmorphological characters (Felsenstein, 1988; Hillis and Huelsenbeck,1992; Simon, 1991; Swofford and Olsen, 1990). Nucleotide sequence alignment, assessments of homology, tree building protocols, and tree optimization and evaluation procedures are all recognized as critical components in contemporary systematic analysis. It is indisputable that molecular methods will have an impact in nematode systematics. Already there is a rapidly accumulating literature in molecular systematics, albeit some of it rather obtuse and difficult to interpret. Some fields, such as mammalian systematics, have vigorously embraced the new technologies. Nematode molecular systematics is in its infancy, with less than a dozen technical papers using nucleotide sequence data to assess relationships among nematodes. Yet it could be argued that molecular systematics will have its greatest impact among the lesser understood taxa, those that have received scant attention due to small size of the organisms and conservation of morphological characters. Molecular systematics can put nematodes on the same footing as better understood organisms. The same genes that are being used to resolve relationships among mammals can be used to resolve relationships among nematodes. For that reason, it is a good time to examine some of the questions facing taxonomists wishing to incorporate molecular data into their systematic studies, and to give a general overview of the initial efforts in nematode molecular systematics. For those interested in a more detailed analysis of the field of molecular systematics, several thorough reviews and texts are listed below.

A REVOLUTION IN SYSTEMATICS

Systematics is currently undergoing a revolution driven by the accessibility of molecular data. There are new journals such as *Molecular Biology and Evolution*, *Molecular Phylogenetics and Evolution*, and *Molecular Ecology* which publish studies of relationships based on molecular data. An increasing percentage of molecular systematic studies are included in established journals like *Evolution*, *Systematic Biology* (formerly *Systematic Zoology*), and *Genetics*. Recent textbooks reflect the rate of growth in the field of molecular systematics (Dutta, 1986; Nei, 1987; Patterson, 1987; Fernholm et al., 1989; Doolittle, 1990; Hillis and Moritz, 1990; Brooks and McLennan, 1991; Gillespie, 1991; Harvey and Pagel, 1991; Miyamoto and Cracraft, 1991; Selander et al. 1991; Soltis et al. 1992; Quicke, 1993; Avise, 1994). Special mention should be made of Molecular Systematics by Hillis and Moritz (1990), since this is probably the most comprehensive of the methods books and presents techniques in a detailed "cookbook" style as well as providing an in-depth discussion of theoretical issues.

Foremost among the factors which have contributed to the rapid increase of molecular data sets in systematics is the development of the Polymerase Chain Reaction (PCR) (Mullis et al., 1986). Taxonomists are no longer constrained by the size of the organism in obtaining a sufficient amount of biological material for analysis. PCR amplifications may be conducted upon a single juvenile nematode or an individual egg in a relatively crude assay (Harris et al., 1990). The concept of "universal primers" has reduced the need to initiate each taxonomic project with a time-consuming hunt for primer sequences that will amplify the desired product. Universal amplification primers are oligonucleotide sequences which anneal to genomic regions of such high evolutionary conservation, that they will serve as amplification primers for a wide taxonomic range of organisms. Kocher et al., (1989) demonstrated how primers designed to amplify certain portions of the mitochondrial genome could successfully amplify DNA from more than 100 organisms tested ranging from spiders to humans. Ironically, these same primers have not been useful for the amplification of nematode DNA (Powers, unpublished; C. De Giorgi, pers. comm.). Nonetheless, several primer sets have been published that work for numerous nematode taxa (Table 1).

Table 1. Primer Sets Used in PCR Amplification of Nematode DNA

PRIMER-5'->3'		REGION AMPLIFIED	NEMATODE	REFERENCE
C2F3 1108	GGTCAATGTTCAGAAATTTGTGG TACCTTTGACCAATCACGCT	3' of COII to 16 S mitochondrial genes	most Tylenchida	Powers & Harris (1993)
18s 28S	TTGATTACGTCCCTGCCCTTT TTTCACTCGCCGTTACTAAGG	ITS region of ribosomal repeat	most nematodes	Vrain et al. (1992)
VRF1 VRF2	CGTAACAAGGTAGCTGTAG TCCTCCGCTAAATGATATG	ITS region of ribosomal repeat	most nematodes	Ferris et al. (1993)
HSPA1 HSPA2	GACACCGAGCGTCTAATCGGAG GTACCACCTCCAAGATCGAAG	heat shock 70A	*Bursaphelenchus* spp.	Beckenbach et al. (1992)
CEGLY CEHIS	TTCAGTATGTTTGACTTCCA GCTCTATATTCTCTTACACCACA	Entire COII subunit of mitochondrial gene	*Caenorhabditis* *Steinernema*	Thomas & Wilson (1991)
CECAL1 CECAL2	GGCAATCCGAGCAGAAAGAATG GGGTTCTGCCAATGCGAGAGAAGT	Calmodulin-like gene	*Caenorhabditis*	Thomas & Wilson (1991)

Another feature that has stimulated research in systematics has been the availability of software packages for computational analysis. Programs for data base searching, sequence editing, data conversions, alignment, and phylogenetic analysis can be obtained for many computer configurations. A listing of common computer packages for phylogenetic analysis, and the addresses where they may be obtained can be found in Hillis and Moritz (1990). A more complete listing (over twenty phylogenetic packages) can be found in the documentation for PHYLIP version 3.5, which together with the programs is available for free via anonymous ftp from: evolution.genetics. washington.edu. in directory pub/phylip. Basically, many systematic studies have either used the programs found in PHYLIP (Phylogeny Inference Package) distributed by Joseph Felsenstein, Department of Genetics SK-50, University of Washington, Seattle, WA, 98195 USA [E-mail: joe@genetics.washington.edu] or PAUP (Phylogenetic Analysis Using Parsimony) distributed by David Swofford, Laboratory of Molecular Systematics, MRC 534, MSC, Smithsonian Institution, Washington D.C. 20560 USA [E-mail: swofford@onyx.si.edu]. MacClade, a program currently only available for Macintosh computers, is finding considerable use in the analysis of character evolution and the evaluation and testing of phylogenetic trees. It is commercially available from Sinauer Associates, Sunderland, MA, 01375 USA. Another software package, recently made available for IBM compatible computers is MEGA (Molecular Evolutionary Genetics Analysis). This package contains methods for estimating genetic distance as well as several methods for phylogenetic inference including UPGMA, neighbor-joining, and maximum parsimony. Information concerning this package can be obtained through Joyce White, Institute of Molecular Evolutionary Genetics, The Pennsylvania State University, 328 Mueller Laboratory, University Park, PA 18602 USA [E-mail: imeg@psuvm.psu].

Currently there appears to be a small cottage industry writing and testing computer programs for phylogenetic inference. Needless to say, the selection of a program and the conditions under which it is run can influence the outcome of a systematic investigation. This topic is too large for a general review, and at best it can be suggested to try the common phylogenetic inference packages mentioned above under a range of conditions. Fortunately the documentation for the major programs is quite extensive in scope and the programs are very user friendly.

ASSUMPTIONS AND CONCERNS

Given the convenience of software packages for computational analysis, it is tempting to insert data into the programs and uncritically accept the first tree that is produced. Yet many of the programs have parameter settings that could dramatically alter the results. Consider the comparison of several sequences of ribosomal RNA, where alignments cannot be aided by protein coding information (Figure 1). Often in comparisons of ribosomal sequence, it is necessary to insert gaps or deletions to improve alignments among multiple sequences. In the computer alignment program Pileup in the Sequence Analysis Software Package by Genetics Computer Group, University of Wisconsin, default values are provided which have a specified penalty for creating a gap or widening a gap. Depending on the nature of the data, slight alterations in these penalties could result in a different alignment that supports a significantly different tree topology. Some alignment programs are sensitive to the input order of the sequence data. There is probably no single "best alignment", only optimal alignments for the given set of parameters (D. Davidson, pers. comm.). It is suggested to try various parameters and always carefully examine alignments by eye after computer analysis. A detailed discussion of computer alignment can be found in Waterman et al., (1991) and concerns of weighing and homology assessment are expressed by Mindell (1991).

```
CTAGCA        CTAGCA
CTT-CA        CT-TCA
  ↑             ↑
```

Figure 1. The insertion of a gap to improve alignment may occur at two possible positions. In computer alignments, highly divergent sequences may require the insertion of numerous gaps for optimal alignment. The phylogenetic treatment of gaps is problematic and is discussed by Swofford and Olson (1990).

Once a sequence alignment is selected, researchers often construct tables comparing genetic distance among the representative sequences. Although not recommended for phylogenetic inference by character based methods, genetic distances do provide a convenient means of species comparisons and could aid in the study of evolutionary processes. These distances are estimates of the number of nucleotide substitutions per site between two sequences, and may incorporate different assumptions in their computation. The simplest calculation is the observed proportion of nucleotides which differ between two sequences. However, as distance between two sequences increases, or other factors indicate that observed distance will give a biased estimate, then it may become necessary to use more complex calculations. For example, in animal mitochondrial it has been observed that transitions (A<->G, C<->T substitutions) occur more frequently than transversions (C<->A, C<->G, T<->A, T<->G) (Brown et al., 1982). This also appears to be true for nematodes (Thomas and Wilson, 1991). Methods such as the Kimura 2-Parameter Distance Model (Kimura, 1980) and the Tajima-Nei Distance Model (Tajima and Nei, 1984) allow for corrections for transition/transversion bias. Similarly, corrections may be made for nucleotide frequencies that strongly deviate from 0.25 or show unequal rates of mutation. A strong bias toward A-T nucleotide has already been observed for nematode mitochondrial DNA (Okimoto et al., 1992; Powers et al., 1993). The Tamura-Nei Model (1993) works well when there is a nucleotide content bias and a transition/transversion bias. A good description of distance methods can be found in the MEGA Package documentation and in Nei (1987).

There are instances when genetic distance can be misleading. If the loci selected for comparison are saturated with mutations, where mutations have occurred at most individual nucleotide sites repeatedly during the course of evolution, then the data will essentially be comprised of random noise. Mutational saturation will compromise all methods of phylogenetic inference (Simon, 1991). Figure 2 illustrates how repeated rounds of mutation at a single nucleotide site can obscure genetic relationships and produce a misleading result. In this example, the current state root-knot 1 shares a T with cyst 1, whereas cyst 2 and lesion 1 share an A. Convergence in the state of the nucleotide site after the second and third round of mutation produces two different but equally erroneous views of relationships. This situation may be encountered fairly frequently in nematology since nematodes are an ancient group and very long branch lengths may extend from the common ancestor of two extant plant parasitic nematodes. Rapidly evolving mitochondrial genes and noncoding DNA such as the internally transcribed spacer region of the nuclear ribosomal genes may be saturated with mutations in comparisons between nematodes of different genera or distantly related congeneric species. Ultimately convergence is a problem for all forms of analysis in systematics, but special care must be taken in selecting genetic loci for comparison, particularly in light of possible bias in the nature of nucleotide substitutions.

Nematode	Ancestral State	1st Mutation	2nd Mutation	3rd mutation/ Current State
root-knot 1	A →	T →	C →	T
cyst 1	A →	T →	C →	T
cyst 2	A →	T →	G →	A
lesion 1	A →	G →	C →	A

Figure 2. Molecular convergence at a single nucleotide site produces a misleading interpretation of relationships.

EXPECTATIONS

In nematology, many systematics inquiries are initially diagnostic studies, generally attempts to uncover some new diagnostic characteristic that will facilitate identification of a target pest species. They develop into systematic studies because, before it can be claimed that a character, or a "probe" in the case of many molecular studies is species-specific, some evaluation of closely related species must be undertaken. Species-specificity implies we know something about species boundaries. Molecular characteristics have been particularly appealing in this regard because of the virtually limitless number of characters that are available for evaluation. However, when we make the conceptual step from genetic marker to genetic relatedness, we must be careful not to bias our view of nematode relationships with expectations of well-delimited species boundaries. It would be convenient if nematode species existed at the terminal branches of a well-pruned phylogenetic tree. The reality may be that some nematode groups exist at the terminal branches of a tangled hedge, with rather indistinct species boundaries. The parthenogenetic plant-parasitic species in agriculturally important genera such as *Meloidogyne*, *Xiphinema*, and *Pratylenchus* could pose special taxonomic problems for phylogenetic inference.

In a recent review of molecular changes and speciation, Harrison (1991) evaluated patterns of nucleotide polymorphism subsequent to speciation, and posed the following two questions. 1) Does speciation leave a distinctive signature of patterns of molecular genetic variation? 2) Can we use genetic distances to make judgements about species status? Unfortunately, the answers appear to be (1) sometimes a distinctive signature is left following speciation (but generally, not), and (2) it depends on several historical factors.

Harrison (1991) cites a number of examples where distinct genetic breaks occur between species that appear to have evolved in allopatry. In these cases, the "genetic signature" is particularly strong and can generally be correlated with a former or existing barrier to gene flow. However, there are many examples cited where "good" species have been examined and been found to display remarkably little divergence, as well as cases where a single species exhibits a great range of genetic variation. These include organisms as diverse as American eels (Avise et al., 1986), red-winged blackbirds (Ball et al., 1988), monarch butterflies (Brower and Boyce, 1991), and sea urchins (Palumbi and Wilson, 1990).

The pattern of genetic variation in a species depends on a number of factors. These include the genetic population structure of the ancestral population, and how the genetic variation was partitioned geographically at the time the ancestral population began to split into differentiated lineages. Also important is the amount of gene flow that continues to exist between the ancestral population and its lineages, and the history of the populations since splitting relative to evolutionary forces such as selection and genetic drift. While the relative importance of factors during speciation are the focus of many studies in evolutionary biology, one consequence has significant relevance to molecular systematics. Species trees are not necessarily the same as gene trees (Pamilo and Nei, 1988; Harrison, 1991). This means that any gene tree constructed from a single locus may show a lack of congruence or topological dissimilarity with a species tree, especially if the time since splitting has not been long. This incongruence not only applies to trees derived from organelles (which are considered a single locus), but any single locus in the genome. A dramatic example of incongruent trees comes from studies of the major histocompatibility complex (MHC) in primates (Klein et al., 1993). Some MHC loci have alleles that display much greater similarity between humans and chimpanzee, than they do between other conspecific allelic comparisons. It has been estimated that some of the loci in this complex started diverging as long as 65 million years ago, predating splits in the human-chimp or human-gorilla lineages.

To avoid the difficulty posed by gene trees versus species trees, the recommendation seems to be to view single gene trees with caution, and when possible use several unlinked loci for the construction of phylogenetic trees.

NEMATODE EXAMPLES

As might be expected at this the early stage in nematode molecular systematics, more questions have been generated from the initial studies than answers. Nonetheless, a few observations can be made concerning nematode relationships. First, it is apparent that there is extreme genetic diversity in Nematoda. In comparisons among *Caenorhabitis* species, the large genetic differences indicated by allozyme studies (Butler et al., 1981) have been measured by sequence comparisons of mitochondrial and nuclear genes (Thomas and Wilson, 1991). In an early study, *C. elegans* and *C. briggsae* exhibited differences in 22 of 24 allozymes, but no variation in the 5 S rRNA gene (Butler et al., 1981). Recently, Thomas and Wilson (1991) compared sequence of the complete mitochondrial cytochrome oxidase subunit II gene (COII), and the nuclear single copy gene, calmodulin (*cal-1*). They calculated an approximate 50% divergence in the COII gene between the same two species, when nucleotide substitutions were corrected for multiple hits. They point out that the divergence observed in this congeneric comparison of nematodes greatly exceeds that seen between rats and mice, a divergence estimated by molecular data to have occurred 20-35 million years ago (Wilson et al., 1987).

The most extensive molecular comparison to date is that of the entire mitochondrial genome of *C. elegans* to the *Ascaris suum* genome (Okimoto et al., 1992). In addition to providing an excellent structural framework for future mitochondrial comparisons, the evolutionary data supported a divergence approximately 80 million years ago, roughly concurrent with the splitting of the human and cow ancestral lineages. Still, these divergences may be considered recent in comparison with other nematodes. Sequence comparison of the mitochondrial genes for NADH dehydrogenase subunit 3, the large (16 S) rRNA, and cytochrome b, among *C. elegans, Ascaris suum, Meloidogyne incognita* and *Romanomermis culicivorax*, indicate that except for the *Caenorhabditis-Ascaris* comparison, these genes are virtually saturated with multiple mutations obscuring any phylogenetic inference from these data (Powers et al., 1993). Considering the estimated rate of ribosomal RNA (Mindall and Honeycutt, 1990), it is possible that the ancestral lineages of *Caenorhabditis* and plant-parasitic nematodes diverged during or before the Cambrian period, over 500 million years ago. Similar conclusions about the large genetic divergences between plant-parasitic nematodes and *C. elegans* have been drawn from a comparison of major sperm protein genes (Novitski et al., 1993).

In light of the extreme divergence in the phylum Nematoda, it is clear that careful consideration must accompany the selection of genetic loci for systematic studies.

Nuclear ribosomal genes have been used for diagnostic (Vrain et al., 1992) as well as systematic analyses (Ferris et al., 1993). This repeating array of genes and spacer regions has the desirable characteristic of highly conserved genes (18S, 5.8S, 28S) flanking the rapidly evolving spacer DNA. Well-placed PCR primers can amplify DNA from a wide taxonomic range of nematodes and yet provide systematic information at or below the species level. Ferris et al. (1993) compared sequences from cyst-forming species over a portion of the two internally transcribed spacer regions and found surprisingly little difference between *Heterodera schachtii, H. glycines*, and *H. trifolii*. Less than 1% divergence was observed for approximately 600 bp between the three species, compared to 28% when these species were compared to *H. avenae*. The low divergence of the internal transcribed spacer region contrasts with the near 15% divergence estimated for *H. glycinces* from *H. schachtii* based on mitochondrial RFLP data (Radice et al., 1988). In *Meloidogyne*, the internal transcribed spacer region of the major parthenogenetic polyploid species, *M. arenaria, M. incognita*, and *M. javanica*, is virtually identical, but comparisons with *M. hapla* and *M. chitwoodi* show extensive divergence (Powers, unpublished). Examinations of the ribosomal repeats in *Meloidogyne* are interesting phylogenetically because in *M. arenaria* it has been demonstrated that a 5 S rRNA gene is located in the nontranscribed spacer region between the 18 S and 28 S genes (Vahidi et al., 1988; Vahidi et al., 1991; Vahidi and Honda, 1991). This genomic structure is unusual in higher eucaryotes, and differs from that of *Ascaris* and *Caenorhabditis*.

To date, few attempts at phylogenetic inference have been made using nematode nucleotide sequences. Beckenbach et al. (1992) used sequence from the 70A heat shock gene to construct a neighbor-joining tree of *Bursaphelechus* isolates. Their data supported recognition of three broad groupings of *Bursaphelenchus* species, *B. xylophilus, B. mucronatus* from Japan, and a putative B. *mucronatus* from Europe that was genetically distant from the Japanese isolate. Ferris et al. (1993) presented a maximum parsimony analysis of ribosomal DNA for isolates of *Heterodera glycines, H. schachtii*, and *H. trifolii*. Hugall et al. (1994) used sequence data from mitochondrial DNA and restriction site polymorphism to demonstrate that there was as much sequence divergence between lineages of *Meloidogyne arenaria* as there was between *M. arenaria* and *M. javanica*. This study helps explain the apparent discordances between nuclear and organelle phylogenies in *Meloidogyne* (Castagnone-Sereno et al. 1993).

Clearly there is much to be done in nematode systematics. As we move diagnostically-oriented studies toward those which target relationships between species or "races" of uncertain taxonomic status, we will discover the phylogenetic value of molecular approaches to systematics. We may finally be able to lay to rest the tired cliché that nematode systematics is in a state of flux.

REFERENCES

Avice, J.C., 1994, Molecular Markers, Natural History, and Evolution, Chapman and Hall, New York.

Avise, J.C., Helfman, G.S., Saunders, N.C., and Hales, L.S., 1986, Mitochondrial DNA differentiation in North Atlantic eels: Population genetic consequences of an unusual life history pattern, *Proc. Natl. Acad. Sci. USA* 83:4350.

Ball, R.M. Jr., Freeman, S., James, F.C., Bermingham, E., and Avise, J.C., 1988, Phylogeographic population structure of Red-winged Blackbirds assessed by mitochondrial DNA, *Proc. Natl. Acad. Sci. USA* 85:1558.

Beckenbach, K., Smith, M.J., and Webster, J.M., 1992, Taxonomic affinities and intra- and interspecific variation in *Bursaphelenchus* spp. as determined by polymerase chain reaction, *J. Nematol.* 24:140.

Brooks, D.R., and McLennan, D.A. 1991, Phylogeny, Ecology and Behavior: A Research Program in Comparative Biology, University of Chicago Press, Chicago
Brower, A.V.Z., and Boyce, T.M., 1991, Mitochondrial DNA variation in monarch butterflies, *Evolution* 45:1281.
Brown, W.M., Prager, E.M., Wang, A., and Wilson, A.C., 1982, Mitochondrial DNA sequences of primates: Tempo and mode of evolution, *J. Mol. Evol.* 18:225.
Butler, M.H., Wall, S.M., Luehrsen, K.R., Fox, G.E., and Hecht, R.M., 1981, Molecular relationships between closely related strains and species of nematodes, *J. Mol. Evol.* 18:18.
Castagnone-Sereno, P., Potte, C., Uijthof, J., Abad, P., Wajnberg, E., Vanlerberghe-Masutti, F., Bongiovanni, M., and Dalmasso, A., 1993, Phylogenetic relationships between amphimictic and parthenogenetic nematodes of the genus *Meloidogyne* as inferred from repetitive DNA analysis, *Heredity* 70:195.
Doolittle, R.F., ed., 1990, Molecular Evolution: Computer Analysis of Protein and Nucleic Acid Sequences. Methods in Enzymology, Vol. 183, Academic Press, New York.
Dutta, S.K., ed., 1986, DNA Systematics. CRC Press, Boca Raton.
Felsenstein, J., 1988, Phylogenies from molecular sequences: inferences and reliability, *Annu. Rev. Genet.* 22:521.
Ferris, V.R., Ferris, J.M., and Faghihi, J., 1993, Variation in spacer ribosomal DNA in some cyst-forming species of plant parasitic nematodes, *Fundam. Appl. Nematol.* 16:177.
Fernholm, B., Bremer, K., and Jornvall, H., eds., 1989, The Hierarchy of Life. Molecules and Morphology in Phylogenetic Analysis, Elsevier, Amsterdam.
Gillespie, J., 1991, The Causes of Molecular Evolution, Oxford University Press, Oxford.
Harris, T.S., Sandall, L.J., and Powers, T.O., 1990, Identification of single *Meloidogyne* juveniles by polymerase chain reaction amplification of mitochondrial DNA, *J. Nematol.* 22:518.
Harrison, R.G., 1991, Molecular changes at speciation, *Annu. Rev. Ecol. Syst.* 22:281.
Harvey, P.H., and Pagel, M.D., 1991, The Comparative Method in Evolutionary Biology, Oxford University Press, Oxford.
Hillis, D.M., 1987, Molecules versus morphological approaches to systematics, *Annu. Rev. Ecol. Syst.* 18:23.
Hillis, D.M., and Huelsenbock, J.P., 1992, Signal, noise, and reliability in molecular phylogenetic analyses, *J. Hered.* 83:189.
Hillis, D.M., and Moritz, C., eds., 1990, Molecular Systematics. Sinauer Associates, Sunderland.
Hugall, A., Moritz, C., Stanton, J., and Wolstenholme, D.R., 1994, Low, but strongly structured mitochondrial DNA diversity in root knot nematodes (*Meloidogyne*), *Genetics* (in press).
Kimura, M., 1980, A simple method for estimating evolutionary rate of base substitutions through comparative studies of nucleotide sequences, *J. Mol. Evol.* 16:111.
Klein, J., Takahata, N., and Ayala, F.J., 1993, MHC polymorphism and human origins, *Scientific American* Dec. 78.
Kocher, T.D., Thomas, W.K., Meyer, A., Edwards, S.V., Paabo, S., Villablanca, F.X., and Wilson, A.C., 1989, Dynamics of mitochondrial DNA evolution in animals: Amplification and sequencing with conserved primers, *Proc. Natl. Acad. Sci. USA* 86:6196.
Mindell, D.P., 1991, Aligning DNA sequences: Homology and phylogenetic weighing, in: "Phylogenetic Analysis of DNA Sequences", Miyamoto, M.M. and Cracraft, J., eds., Oxford University Press, New York.
Mindell, D.P. and Honeycutt, R.L. 1990, Ribosomal RNA in vertebrates: Evolution and phylogenetic applications, *Annu. Rev. Ecol. Syst.* 21:541.
Miyamoto, M.M., and Cracraft, J., eds., 1991, Phylogenetic Analysis of DNA Sequence. Oxford University Press, New York.
Nei, M., 1987, Molecular Evolutionary Genetics, Columbia University Press, New York.
Novitski, C.E., Brown, S., Chen, R., Corner, A.S., Atkinson, H.S., and McPherson, M.J., 1993, Major Sperm Protein Genes from *Globodera rostochiensis*, *J. Nematol.* 25:548.
Mullis, K., Faloona, F., Scharf, S., Saiki, R., Horn, G., and Erlich, H., 1986, Specific enzymatic amplification of DNA in vitro: the polymerase chain reaction. Cold Spring Harbor, *Symp. Quant. Biol.* 51:263.
Okimoto, R., Macfarlane, J.L., Clary, D.O., and Wolstenholme, D.R., 1992, The mitochondrial genomes of two nematodes, *Caenorhabditis elegans* and *Ascaris suum*, *Genetics* 130:471.
Palumbi, S.R., and Wilson, A.C., 1990, Mitochondrial DNA diversity in the sea urchins *Strongylocentrotus purpuratus* and *S. droebachiensis*, *Evolution* 44:403.
Pamilo, P., and Nei, M., 1988, Relationships between gene trees and species trees, *Mol. Biol. Evol.* 5:568.

Patterson, C., ed., 1987, Molecules and Morphology in Evolution: Conflict or Compromise?, Cambridge University Press, Cambridge.
Patterson, C., Williams, D.M., and Humphries, C.J., 1993, Congruence between molecular and morphological phylogenies, *Annu. Rev. Ecol. Syst.* 24:153.
Powers, T.O., and Harris, T.S., 1993, A polymerase chain reaction method for identification of five major *Meloidogyne* species, *J. Nematol.* 25:1
Powers, T.O., Harris, T.S., and Hyman, B.C., 1993, Mitochondrial DNA divergence among *Meloidogyne incognita, Romanomermis culicivorax , Ascaris suum,* and *Caenorhabditis elegans, J. Nematol.* 25:563.
Quicke, D.L.J., 1993, Principles and Techniques of Contemporary Taxonomy, Blackie Academic and Professional, London.
Radice, A.D., Powers, T.O., Sandall, L.J., and Riggs, R.D., 1988, Comparisons of mitochondrial DNA from the sibling species *Heterodera glycines* and *H. schachtii, J. Nematol.* 20:443.
Selander, R.K., Clark, A.G., and Whittam, T.S., 1991, Evolution at the Molecular Level, Sinauer Associates, Sunderland.
Simon, C., 1991, Molecular systematics at the species boundary: Exploiting conserved and variable regions of the mitochondrial genome of animals via direct sequencing from amplified DNA, in: "Molecular Techniques in Taxonomy", Hewitt, G.M., Johnson, A.W.B., and Young, J.P.W., eds., Springer-Verlag, Berlin.
Soltis, P., Soltis, D., and Doyle, J.J., eds., 1992, Molecular Systematics of Plants, Chapman and Hall, New York.
Swofford, D.L., 1991, When are phylogeny estimates from molecular and morhological data incongruent?, in: "Phylogenetic Analysis of DNA Sequences", Miyamoto, M.M. and Cracraft, J., eds., Oxford Univ. Press, New York.
Swofford, D.L., and Olsen, G.L., 1990, Phylogeny reconstruction, in: Molecular Systematics, D.M. Hillis and C. Moritz, eds., Sinauer, Sunderland, Mass.
Tajima, F., and Nei, M., 1984, Estimation of evolutionary distance between nucleotide sequences, *Mol. Biol. Evol.* 1:269.
Tamura, K., and Nei, M., 1993, Estimation of the number of nucleotide substitutions in the control region of mitochondrial DNA in humans and chimpanzees, *Mol. Biol. Evol.* 10:512.
Thomas, W.K. and Wilson, A.C., 1991, Mode and tempo of molecular evolution in the nematode *Caenorhabditis*: cytochrome oxidase II and calmodulin sequences, *Genetics* 128:269.
Vahidi, H., Curran, J., Nelson, D.W., Webster, J.M., McClure, M.A., and Honda, B.M., 1988, Unusual sequences, homologous to 5 S RNA, in ribosomal DNA repeats of the nematode *Meloidogyne arenaria, J. Mol. Evol.* 27:222.
Vahidi, H., and Honda, B.M., 1991, Repeats and subrepeats in the intergenic spacer of rDNA from the nematode *Meloidogyne arenaria, Mol. Gen. Genet.* 227:334.
Vahidi, H., Purac, A., LeBlanc, J.M., and Honda, B.M., 1991, Characterization of potentially functional 5 S rRNA-encoding genes within ribosomal DNA repeats of the nematode *Meloidogyne arenaria, Gene* 108: 281.
Vrain, T.C., Wakarchuk, D.A., Levesque, A.C., and Hamilton, R.I., 1992, Intraspecific rDNA restriction fragment length polymorphism in the *Xiphinema americanum* group, *Fundam. appl. Nematol.* 15:563.
Waterman, M.S., Joyce,J., and Eggert, M., 1991, Computer alignment of sequences, in: "Phylogenetic Analysis of DNA Sequences", Miyamoto, M.M. and Cracraft, J., eds., Oxford University Press, New York.
Wilson, A.C., Ochman, H., and Prager, E.M., 1987, Molecular time scale for evolution, *Trends Genet.* 3:241.

DISCUSSION

BAILLIE stated that the mitochondrial lineages can't be pushed beyond 100myr, but POWERS noted that there were analytic methods that can be exploited to examine more ancient divergences. BAILLIE claimed to be terrified by ribosomal spacer sequences because there are no suitable mathematical methods for analyzing them. MARSHALL noted that, in general, the phylogenetic analysis programs were easy to use; EDDY noted that there are occurrences of ancient nematodes embedded in amber, and wondered if someone was looking at their genome. POWERS answered that George Poinar, Jr. has talked about doing just that. WILLIAMSON asked if *Meloidogyne arenaria* should be more than one species. POWERS responded by saying that he didn't like the concept of species for these parthenogenetic variants, and argued that more phylogenetic data were required. WILLIAMSON noted that distinct nucleotide haplotypes appear to be falling out of the molecular analyses. WILLIAMSON asked what would be suitable genes to examine to help understand higher order systematics. POWERS replied that ribosomal sequences have been applied for divergences of up to 500myr. ATKINSON commented that according to the Maynard Smith view of evolution, lack of sex should be an evolutionary blind alley. In light of that, how can worms have lived without sex for 100myr? BAILLIE argued that the notion of "no sex" is false, and suggested that the genes for sex should be present and therefore, sex should have occurred in evolutionary time. BIRD concurred, and noted that acquisition and loss of sex is a common evolutionary occurrence, i.e., sex is evolutionary plastic.

THE USE OF mtDNA FOR THE IDENTIFICATION OF PLANT PARASITIC NEMATODES

Carla De Giorgi,[1] Mariella Finetti Sialer,[1] Franco Lamberti[2]

[1]Dipartimento di Biochimica e Biologia Molecolare
University of Bari, Bari, Italy
[2] Istituto di Nematologia Agraria
Consiglio Nazionale delle Ricerche, Bari, Italy

INTRODUCTION

Mitochondria from all organisms studied to date have their own mitochondrial DNA (mtDNA), distinct from that of the nucleus. Animal mtDNA carries the genetic information for two rRNAs, a complete set of tRNAs and only a few essential components of the respiratory complexes: three subunits of the cytochrome oxidase (COI, COII and COIII), the cytochrome *b* (Cyt b), the subunits 6 and 8 of the F_O ATPase complex and several subunits of the NADH dehydrogenase (ND1-ND6 and ND4L). In addition, each molecule contains a control region in which signals of DNA duplication and transcription are included (for a review see Wolstenholme, 1993).

Several peculiarities of mtDNA have made this molecule a powerful molecular tool for revealing the genealogical history of many different animal species.

Animal mtDNA is a circular double-stranded closed structure, relatively small. The typical mt genes are accomodated in molecules ranging in size from about 14 to about 17 kb. Smaller molecules have also been found, but they are not functional (Holt et al., 1988; Hayashi et al., 1991). Larger molecules, resulting from tandem duplication of specific regions, most often of non coding sequences within the control region, have been observed (Moritz and Brown, 1986; Snyder et al., 1987). Both characteristics, the small size and the circular form, allowed simple procedures for the extraction of mtDNA.

One striking peculiarity of mt genomes is that mtDNA from a broad range of invertebrates and vertebrates including humans, have the same gene content. However there is a great variation in the gene organization, but the mitochondrial-peculiar compact gene arrangement is invariably conserved. The coding genes are not interrupted by introns, as frequently occurs in the nuclear genome. There are few or no nucleotides between genes. Finally there is a single non coding region regarded as the control region, which is characterized by a large variability in the size and in the sequence among different organisms.

Advances in Molecular Plant Nematology
Edited by F. Lamberti *et al.*, Plenum Press, New York, 1994

Table 1. Mitochondrial DNAs completely sequenced

Organism	Size (bp)	Gene content conservation	Two-directional Transcription	ATG as translation initiation codon	TTG as translation initiation codon	References
Mammalia						
Homo sapiens	16,569	Yes	Yes	Yes	No	Anderson et al., 1981
Mus musculus	16,295	Yes	Yes	Yes	No	Bibb et al., 1981
Bos taurus	16,338	Yes	Yes	Yes	No	Anderson et al., 1982
Rattus norvegicus	16,298	Yes	Yes	Yes	No	Gadaleta et al., 1989
Balaenoptera physalus	16,398	Yes	Yes	Yes	No	Arnason et al., 1991
Aves						
Gallus domesticus	16,775	Yes	Yes	Yes	No	Desjardins and Morais, 1990
Amphibia						
Xenopus laevis	17,553	Yes	Yes	Yes	No	Roe et al., 1985
Echinodermata						
Paracentrotus lividus	15,697	Yes	Yes	Yes	No	Cantatore et al., 1989
Strongylocentrotus purpuratus	15,650	Yes	Yes	Yes	No	Jacobs et al., 1988
Arbacia lixula	15,722	Yes	Yes	Yes	No	De Giorgi et al., unpublished
Asterina pectinifera	16,260	Yes	Yes	Yes	No	Asakawa et al., 1991
Arthropoda						
Drosophila yacuba	16,019	Yes	Yes	Yes	No	Clary and Wolstenholme, 1985
Apis mellifera	16,343	Yes	Yes	Yes	No	Crozier and Crozier, 1993
Nematoda						
Caenorhabditis elegans	13,794	No	No	No	Yes	Okimoto et al., 1992
Ascaris suum	14,284	No	No	No	Yes	Okimoto et al., 1992
Meloidogyne javanica	20,500	No	No	No	Yes	Okimoto et al., 1991
Cnidaria						
Metridium senile	17,443	Yes	No	Yes	No	Beagley et al., unpublished

During the evolution of animal mitochondrial DNA, three different kinds of change have been described as occurring: base substitutions, insertions or deletions and gene rearrangements. Consequently estimation of phylogenetic relationships can be based on these different criteria. However differences existing in mtDNA are primarily a result of nucleotide substitutions.

Nucleotide changes in mtDNA occurr at a very high rate, therefore mtDNA evolves rapidly and exhibits extensive polymorphism within most species. Sequence divergence usually averages from 1 to 3% within species, but can be as high as 10% (Hoeh et al., 1991).

Finally, mtDNA molecules are maternally inherited. Only a few cases of biparental inheritance have been reported so far (Hoeh et al., 1991), and evidence for recombination between molecules is lacking. Therefore mtDNA molecules from the same mother are inherited as separate clones. It is now clear that also within interbreeding populations, mtDNA lineages are genetically isolated from one another, therefore any observed homologies in structure presumably result from historical connection in a matriarchal genealogy.

MITOCHONDRIAL DNA IN NEMATODES

The availability of fast sequencing techniques has revealed the detailed structure of mtDNA from very different organisms and in several cases the nucleotide sequence of the entire mt molecule has been obtained. Table I shows the list of organisms in which the mtDNA has been completely sequenced. Some molecular features have also been reported.

It can be seen that the nematodes show characteristics not shared with other taxa. Nematode mt genomes display broader variation in size, gene content and arrangement. The size of mtDNA shows a trend to homogeneity within the same phylum. In the case of nematodes, *Meloidogyne javanica* is characterized by a bigger size, for the presence of multiple repeat regions and all the genes are located in the same strand. Moreover, while the general rule of animal mtDNA is the conservation of the gene content, nematode mtDNA represents an exception, as illustrated below. Finally, the translation initiation codon of a protein gene TTG is never used except by nematodes, while the universal initiation codon ATG is not used by the nematode mitochondrial system (Wolstenholme, 1993).

Therefore, although further data are necessary to substantiate this hypothesis, it seems that specific changes occurring in the nematode mtDNA have been accumulated during evolution.

In Fig. 1, the maps summarizing gene content and organization of the nematode mtDNAs completely sequenced, show some peculiarities found in the mtDNA within this phylum. The size is relatively different, and the ATPase 8 gene, while present in all the sequenced mtDNA, has not been located in any of these mtDNA molecules. Only the *M. javanica* mtDNA contains an open reading frame of 116 codons downstream from the COII gene that would be transcribed in the same direction as all other genes.

Furthermore these maps clearly show that while the mtDNA molecule of *Ascaris suum* differs from that of *Caenorhabditis elegans* only in the location of the AT region, extensive gene rearrangement has occurred in the mtDNA of the *M. javanica* molecule. The most evident variation between the three nematode mtDNAs, is the presence in *M. javanica* of repeated sequence sets that

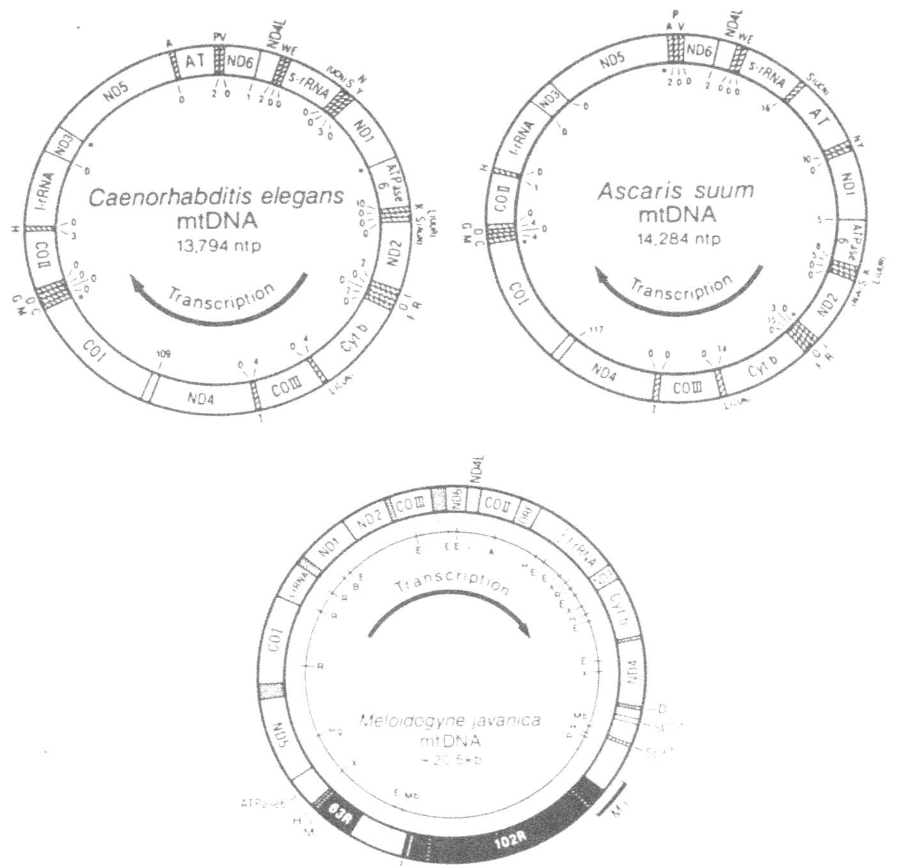

Figure 1. Gene maps of nematode DNAs completely sequenced. The *Caenorhabditis elegans* and the *Ascaris suum* maps are taken from Okimoto et al., (1992), while that of the *Meloidogyne javanica* from Okimoto et al., (1991) with permissions.

lack standard mt genes; approximately 36 copies of a 102 nt sequence, 11 copies of a 63 nt sequence and 5 copies of an 8 nt sequence are present (Okimoto et al., 1991). It is important to stress that the presence of long reiterated DNA sequences whose functions are still unknown, have as yet only been reported in two diverse nematodes, *Romanomermis culicivorax* (Hyman et al., 1988) and *Meloidogyne* spp. but their localization, size and coding potential are different in the two nematode genera.

The localization of genes on the map of *C. elegans* and of *A. suum*, demonstrate that the COI and COII genes are adjacent, a common feature also for other mtDNA . In the case of *R. culicivorax* , however the two genes are separated by approximately 8 kb (Hyman et al., 1988). The phylum Nematoda has been traditionally divided into two main classes Secernentea and Adenophorea. Since both *A. suum* and *C. elegans* are Secernentea, while *R. culicivorax* belongs to the class of Adenophorea, it was proposed that genome rearrangment of mtDNA could be used as important tool in nematode systematics (Hyman, 1988). However, the observation that a similar localization of the COI and COII genes has also been found in the *Meloidogyne*, not only renders this approach less atractive, but strongly indicates that much more information is required on the mt genome organization and sequences before a unique key for the identification and classification of nematodes on the basis of mtDNA can be established.

The three sets of tandemly arranged, direct repeat sequences described in the case of *M. javanica* mtDNA, have also been detected in *M. hapla*- Race A, and Race B, *M. incognita*-Race 1, 3 and 4, and *M. arenaria*- Race 1 and 2, but differences were seen either in the repeat number or in the location of the restriction enzyme used for the detection. Therefore it has been pointed out that differences in migration distances and in the number of repeat-containing bands of restricted mtDNA are sufficient to distinguish the different host-races of each species (Okimoto et al., 1991).

Differences in nucleotide sequences between mtDNA molecules of different populations of the same species, as well as mtDNA molecules of different species of a genus have been greatly exploited in recent years to gain information regarding population structure and evolutionary history of a wide variety of organisms.

The relationship between *C. elegans* and its close relatives was established by Thomas and Wilson (1991), by comparing a mtDNA fragment, sequenced after amplification. The primers used for the amplification of the cytochrome oxidase subunit II (COII) were designed on the basis of the published tRNA sequences bordering the COII gene in *C. elegans*. The nematodes used were different strains of *C. elegans* and two other *Caenorhabditis* species, *C. remanei* and *C. briggsae*, compared to a distantly related nematode *Steinernema intermedii* The results demonstrated that the intra-specific differences in the COII gene are low (<2%). However, the percent divergences among the COII sequences of the three *Caenorhabditis* species are very high at almost 50%. The increase in the divergence between genera is consistent with the change in the pattern of nucleotide substitutions. The ratio of transitions to transversions falls by a factor of more than ten as one shifts from intra- to inter-specific comparison. Interestingly, the higher proportion of transversions is due to an increase in the T-A transversion. The percentage nucleotide composition of the sense strand in *C. elegans* , and partially in *A. suum*, demonstrates a strong bias toward T and A and a bias against C. In contrast mtDNA of higher vertebrates is strongly biased against G. As a consequence, significant differences have been detected in the usage of the two leucin codons between nematode and higher vertebrates.

Table 2. Different leucin codon usage between nematode and mammalian in the entire mtDNA

Codon	*C. elegans*	*A. suum*	Human	Bovine
UUR	437	446	89	126
CUN	93	67	553	471

Table 2 shows that the leucin codons in nematodes are biased against C with the UUR codon being the most used instead of the CUN codon. Although further data are required, it is conceivable that differences in codon usage could be used in order to discriminate between mtDNA of nematodes and other eukaryotic organisms.

MOLECULAR DIAGNOSTIC ANALYSES

The indication that mtDNA is a suitable probe for nematode molecular diagnostics, is based on two main peculiarities, namely its high copy number and the occurrence of polymorphic forms.

One of the first examples that demonstrated the use of mtDNA as a diagnostic tool for the rapid detection of *Meloidogyne* species used the technique of DNA hybridization (Powers et al., 1986). Furthermore, it has been possible to detect mtDNA of different *Meloidogyne* species not only in the presence of exogenous DNA, but even in a crude field sample (Hyman et al., 1990)

On the other hand, the fast evolving sequences of mtDNA represent a useful taxonomic fingerprint among closely related species, because changes in nucleotide sequences are likely to occur also in endonuclease restriction sites, therefore differentiating molecules with different endonuclease cleavage sites.

This strategy has been followed to demonstrate the existence of readily scorable differences between *M. incognita, M. hapla , M. arenaria* and *M. javanica* mtDNAs. Furthermore, when mtDNAs from individual *M. incognita* host races were used, the extent of polymorphism decreased, indicating that the host-race development is a relatively recent evolutionary acquisition (Powers et al., 1986; Powers and Sandall, 1988).

Restriction fragment polymorphisms of mtDNA are also easily observable between sibling species of cyst nematodes *Heterodera glycines* and *H. schachtii* (Radice et al., 1988).

The most powerful approach for the isolation and characterization of a mtDNA fragment is PCR amplification. The succesful application of this technique is demonstrated by the nematode identification on a single egg or juvenile when the amplified segment contains diagnostic restriction site polymorphisms (Harris et al., 1990). The most detailed analysis of mtDNA differences within *Meloidogyne* species has been recently obtained (Powers and Harris, 1993). In this work, the amplified region encompasses the region between the 3′ portion of the CO II gene and the large RNA gene. In some cases the amplification products showed size differences, in some other cases discrimination was made possible by the restriction endonuclease pattern analyses.

mtDNA Amplification in *Xiphinema index*

The possibility of amplifying a segment of mtDNA for use in the identification and taxonomic studies of *Xiphinema* species has been investigated in our laboratory. One subunit of the cytochrome oxidase was selected for study, because the comparison, even between distantly related organisms, demonstrated that in this gene, the amino acid sequence is very conserved because of constraints on the protein function, whereas the nucleotide sequence shows a lower degree of homology. Therefore intra-specific differences could be easily detected at the nucleotide level.

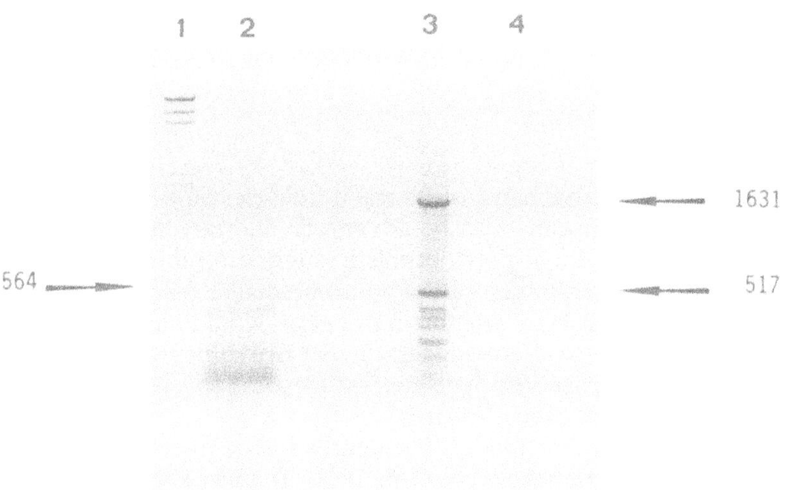

Figure 2. Amplification of COIII fragment from *Xiphinema index* single individual. Individual nematode cut in pieces was lysed in 100 mM Tris buffer containing 200 mM NaCl, 5 mM EDTA, 1% SDS and 2 mg/ml proteinase K at 55° C for 4 hours. The DNA was isolated by using glassmilk (Geneclean II, Bio 101, Inc.,) and electrophoresed. Lane 1: *Hind* III cleaved Lambda DNA: lane 2: amplification product; Lane 3 : *Hinf* I cleaved pBR322 DNA; Lane 4: corresponding product amplified from C. *elegans* . The arrows indicate specific DNA markers fragments.

The pair of primers were designed on the COIII gene of *C. elegans* . However in order to avoid the annealing of the primers to other contaminating eukaryotic organisms, the oligonucleotides selected on the *C. elegans* were those with the lowest homology with the corresponding sequence of other organisms present in a database.

The experiment was carried out by using a single *X. index* nematode and the experiments had to be repeated several times before the amplification was achieved. The results of the amplification are reported in Fig. 2. The figure clearly shows that a faint band of the expected size, 650 nt, is visible on the gel. When the gel was transferred and hybridized with the *C. elegans* labelled probe, the results (Fig. 3) indicate that an authentic nematode COIII gene had been amplified Nevertheless, two bands were present which were found to be 650 and 560 nt in size respectively.

Figure 3. Hybridization of the amplified product from *X. index* single individual with the 650 nt long fragment amplified from *C. elegans* single individual. The arrows indicate the two different bands hybridized.

Obviously, the smaller band could result from cleavage from the upper band, but other explanations can also be advanced. The results could indicate the presence of intra-individual polymorphism suggesting the presence of at least two functional active molecules, one of which is smaller than the other. However this seems unlikely because a deletion in this gene would most probably alter the function of mtDNA. Instead, the suggestion that two different molecules of which only one is functional, is more appealing. The localization of the deleted molecule can be mitochondrial or even nuclear.

The presence of mitochondrial sequences in the nuclear genome of other organisms has been reported (Jacobs et al., 1983). Therefore it is not surprising that both the authentic gene and a modified homologous gene have been amplified in our experiments. This point will be clarified by sequencing both bands.

In conclusion, mtDNA seems to be a good candidate as a tool for identification and taxonomic studies of plant parasitic nematodes. However, much more information is required on the gene organization and particularly on the nucleotide sequences in the different plant parasitic nematodes. A recent report (Peloquin et al., 1993) demonstrated the existence of at least two distinct mitochondrial genomes in *M. hapla* Therefore the degree of intra-specific variation must be considered. One final concern regards the presence of heteroplasmy i. e. the existence of different mtDNA molecules in the same individual. This phenomenon seems to be a peculiar characteristic of the mtDNA population, but the extent of heteroplasmy may be species-specific.

Therefore the unequivocal identification of nematode based on mtDNA, requires the analysis to be carried out on single individuals, but it must be stressed that the knowledge on the intra-individual variability has yet to be determined.

Acknowledgments

The authors thank Dr Mauro Di Vito for supplying *Xiphinema index* adult single individuals.

REFERENCES

Anderson, S., Bankier, A.T., Barrel, B.G., de Bruijn, M.H.L., Coulson, A.R., Drouin, J., Eperon, I.C., Nierlich, D.P., Roe, B.A., Sanger, F., Schreier, P.H., Smith, A.J.H., Staden, R. and Young, I.G., 1981, Sequence and organization of the human mitochondrial genome,*Nature* 290: 457.

Anderson, S., de Bruijn, M.H.L., Coulson, A.R., Eperon, I.C.,Sanger, F. and Young, I.G., 1982, Complete sequence of bovine mitochondrial DNA. Conserved features of the mammalian mitochondrial genome, J.*Mol. Biol.* 156: 683.

Arnason, U., Gullberg, A. and Widegren, B., 1991, The complete nucleotide sequence of the mitochondrial DNA of the fin whale *Balaenoptera physalus*, *J. Mol. Evol.* 33: 556.

Asakawa, S., Kumazawa, Y., Araki, T., Himeno, H., Miura, K., and Watanabe, K., 1991,Strand-specific nucleotide composition bias in echinoderm and vertebrate mitochondrial genomes, *J. Mol. Evol.* 32:511.

Bibb, M.J., Van Etten, R.A., Wright, C.T., Walberg, M.W. and Clayton, D.A., 1981, Sequence and gene organization of mouse mitochondrial DNA, *Cell* 26: 167.

Cantatore, P., Roberti, M., Rainaldi, G., Gadaleta, M.N. and Saccone, C., 1989, The complete nucleotide sequence, gene organization, and genetic code of the mitochondrial genome of *Paracentrotus lividus*, *J. Biol. Chem.* 264: 10965.

Clary, D.O. and Wolstenholme, D.R., 1985, The mitochondrial DNA molecule of *Drosophila yacuba:* nucleotide sequence, gene organization and genetic code, *J. Mol. Evol.* 22: 252.

Crozier, R.H. and Crozier, Y.C., 1993, The mitochondrial genome of the honeybee *Apis mellifera* : complete sequence and genome organization, *Genetics* 133:97.

Desjardins, P. and Morais, R., 1990, Sequence and gene organization of the chicken mitochondrial genome. A novel gene order in higher vertebrates, *J. Mol. Biol.* 212:599.

Gadaleta, G., Pepe, G., De Candia, G., Quagliariello, C., Sbisà, E, and Saccone, C., 1989, The complete nucleotide sequences of the *Rattus norvegicus* mitochondrial genome: cryptic signals revealed by comparative analysis between vertebrates, *J. Mol. Evol.* 28: 497.

Harris, T.S., Sandall, L.J. and Powers, T.O., 1990, Identification of single *Meloidogyne* juveniles by polymerase chain reaction amplification of mitochondrial DNA, *J. Nematol.* 22:518.

Hayashi, J.L., Ohta, S., Kikuchi, A., Takemiten, M., Goto, Y. I., and Nonaka, I., 1991, Introduction of disease-related mitochondrial deletions into He La cells lacking mitochondrial DNA results in mitochondrial dysfunction, *Proc Natl. Acad. Sci. USA* 88:10614.

Hyman, B. C., 1988, Nematode mitochondrial DNA: anomalies and applications, *J. Nematol.* 20:523.

Hyman, B. C., Beck, J.L. and Weiss, K.C.,1988, Sequence amplification and gene rearrangement in parasitic nematode mitochondrial DNA, *Genetics* 120:707.

Hyman, B.C., Peloquin, J.J. and Platzer E.G., 1990, Optimization of mitochondrial DNA-based hybridization assays to diagnostics in soil, *J. Nematol.* 22: 273.
Hoeh, W. R., Blakley, K.H. and Brown, W.M., 1991, Heteroplasmy suggests limited biparental inheritance of *Mytilus* mitochondrial DNA, *Science* 251: 1488.
Holt,I.J., Harding, A.E. and Morgan-Hughes, J.A., 1988, Deletion of muscle mitochondrial DNA in patients with mitochondrial myopathies, *Nature* 331:717.
Jacobs, H.J., Elliot, D.J., Math , V.B. and Farquharson, A., 1988, Nucleotide sequence and gene organization of sea urchin mitochondrial DNA, *J. Mol. Biol.* 202: 185.
Jacobs, H.J., Posakony, J.W., Grula, J.W., Roberts, J.W., Xin, J.H., Britten, R.J.and Davidson, E.H., 1983, Mitochondrial DNA sequences in the nuclear genome of *Strongylocentrotus purpuratus*, *J.Mol. Biol.* 165: 609.
Moritz, C. and Brown, W.M., 1986, Tandem duplication of D-loop and ribosomal RNA sequences in lizard mitochondrial DNA , *Science* 233:1425.
Okimoto, R., Chamberlin, H. M., Macfarlane , J.L. and Wolstenholme, D.R., 1991,Repeated sequence sets in mitochondrial DNA molecules of root knot nematodes (*Meloidogyne):* nucleotide sequences, genome location and potential for host-race identification, *Nucleic Acids Res.* 19:1619.
Okimoto, R., Macfarlane,J.L., Clary, D. O. and Wolstenholme, D.R., 1992, The mitochondrial genomes of two nematodes *Caenorhabditis elegans* and *Ascaris suum, Genetics* 130:474.
Peloquin, J.J., Bird, D.M., Kaloshian, I. and Matthews, W.C., 1993, Isolates of *Meloidogyne hapla* with distinct mitochondrial genomes, *J. Nematol.* 25: 239.
Powers, T.O.and Harris, T.S., 1993, A polymerase chain reaction method for identification of five major *Meloidogyne* species, *J. Nematol.* 25:1.
Powers, T.O., Platzer, E.G. and Hyman, B.C., 1986, Species-specific restriction site polymorphism in root-knot nematode mitochondrial DNA, *J.Nematol.* 18:288.
Powers, T.O. and Sandall, L.J., 1988, Estimation of genetic divergence in*Meloidogyne* mitochondrial DNA, *J. Nematol.* 20:505.
Radice, A.D. Powers, T.O., Sandall, L.J. and Riggs, R.D.,1988, Comparison of mitochondrial DNA from the sibling species *Heterodera glycines* and *H. schactii, J. Nematol.* 20:443.
Roe, B.A., Ma, D.P., Wilson, R.K. and Wong, J.F.H., 1985, The complete nucleotide sequence of the *Xenopus laevis* mitochondrial genome, *J. Biol. Chem.* 260: 9759.
Snyder, M., Fraser, A.R., LaRoche, J., Gartner-Kepkay, K.E. and Zouros, E., 1987,Atypical mitochondrial DNA from the deep sea scallop *Placopecten magellanicus* , *Proc. Natl. Acad. Sci. USA* 84: 7595.
Thomas, K.W. and Wilson, A.C., 1991, Mode and tempo of molecular evolution in the nematode Caenorhabditis: cytochrome oxidase II and calmodulin sequences, *Genetics* 128:269.
Wolstenholme, D.R., 1993, Animal mitochondrial DNA: structure and evolution, *Int. Rev. of Cytology* 141:173.

DISCUSSION

Eddy observed that the *C. elegans* genome project has examined the possibility of using the codon usage to distinguish *C. elegans* clone from contaminating yeast DNA. In practice, one finds that really only highly expressed genes show strong codon bias. For random DNA fragments we have found that about the best we can do is assign a 70:30 chance of being yeast vs. worm, not good enough to be useful.
Georgi asked whether the predominant product might actually not be related to the probe (apart from the primers) and the genuine COIII product is invisible on staining.
De Giorgi replied that the two bands show differences in the intensity of the hybridization signal. The predominant band coincides with the stained one.
Bird asked if there is any evidence for editing of mtDNA transcripts.
De Giorgi pointed out that RNA editing does not occur in animal mtDNA, while it does occur in mtDNA of plants and trypanosomes.

PCR FOR NEMATODE IDENTIFICATION

Valerie M. Williamson, Edward P. Caswell-Chen, Frances F. Wu, and Doris Hanson

Department of Nematology
University of California
Davis, CA 95616
USA

INTRODUCTION

Accurate identification of plant pathogenic nematode species is a vital component of many areas of nematology, from systematic research to regulatory nematology and management. It is especially important for management that does not rely on chemical nematicides. In many cases it is desirable to identify to species single females or juveniles. In other cases, it is desirable to identify the presence of small numbers of juveniles in a complex mixture. A major constraint to developing identification techniques for single females and juveniles has been the small size of the nematodes. The availability of the polymerase chain reaction (PCR) amplification technique has revolutionized many aspects of biology by allowing amplification of specific sequences from minute quantities of DNA. This procedure opens new tools for nematology. For example it is now possible to differentiate juveniles of the major root-knot nematode species based on differences in their mitochondrial spacer regions (Powers and Harris, 1993).

A novel type of genetic marker which is based on DNA amplification has been developed (Williams et al., 1990; Welsh and McClelland, 1990). These markers, commonly called RAPD (for Random Amplified Polymorphic DNA) markers are generated by the amplification of random DNA segments in the target genome with single primers of arbitrary nucleotide sequence. RAPD markers provide a large spectrum of markers distributed throughout the target genome and have been shown to have wide applicability for genetic diagnostics. Because of the tremendous number of oligomer primers that can be generated, the number of DNA markers that can potentially be identified is enormous. RAPD markers are generally inherited in a Mendelian fashion and thus are valuable for generating genetic maps in variety of organisms. In many cases, these markers have been used to obtain a measure of the relatedness of species or isolates of an organism (Williams et al., 1990; Goodwin and Annis, 1991; Smith et al., 1992; Hedrick, 1992) and there have been several reports of using this technique, or related techniques, for comparison of nematode species and populations (Caswell-Chen et al., 1992; Cenis, 1993; Baum et al., 1994; Fargette et al., 1994). The RAPD procedure does not require use of isotopes, a major advantage in small laboratories, and requires minimal knowledge about the genome of the target organism, an important consideration for the study of plant pathogenic nematodes.

To perform a RAPD assay, a single oligonucleotide of arbitrary sequence, generally 10 nucleotides in length, is mixed with the genomic DNA and amplification is carried out using PCR conditions. Because primers are short, annealing to target DNA must be carried out at low temperatures. Amplification occurs when the primer anneals to opposite strands of the DNA within an amplifiable distance (usually 2 kb or less). In nematodes, several bands are

Advances in Molecular Plant Nematology
Edited by F. Lamberti *et al.*, Plenum Press, New York, 1994

generally amplified for each primer. These bands are not likely to be primed by exact matches of 10-mers as there would theoretically be too few occurrences of exact matches the right distance apart to even amplify a single band from each primer from a simple genome such as that of root-knot nematodes (Williams et al., 1993; Pableo and Triantaphyllou, 1989). The banding pattern can be very consistent from run to run if assay conditions are kept constant. However, fluctuations in the pattern produced can result from differences in DNA concentration, sample impurities, *Taq* polymerase source, and differences in operator technique or thermocycler (Williams et al., 1993). Thus, although this technique is likely to be important in research laboratories for work on systematics or nematode genetics it may not be applicable for routine identification where crude DNA extracts and small samples (single nematodes) are used.

We describe here our experience using RAPDs to compare cyst and root-knot nematode species and populations and present our progress toward using this technique for identification. We propose ideas for applying and extending PCR technology to developing more robust primers and novel techniques applicable for routine identification of single juveniles and juveniles in soil samples. While some of our results are preliminary and our strategies are speculative, we hope to evoke ideas on development of new methods in nematology that prior to PCR would not have been feasible.

RAPD ANALYSIS ON *HETERODERA SCHACHTII* AND *H. CRUCIFERAE*

We had multiple motives for investigating the RAPD patterns of *Heterodera schachtii* and *H. cruciferae*. The sensitivity of the technique suggested that RAPD markers could be useful tools for studying the dynamics of mixed populations of *H. schachtii* and *H. cruciferae*. These two species are sympatric in California and frequently occur in the same field on the same host. Our goal was to be able to distinguish individual females in mixed populations. Currently, species are identified by vulval cone morphology and host range tests (Mulvey, 1972; Baldwin and Mundo-Ocampo, 1991), imprecise and tedious methods. In addition, RAPDs were a potential tool for assessing the variability between populations and individuals of *H. schachtii*. Intraspecific polymorphic markers could also be used to follow genetic crosses between *H. schachtii* isolates that differed in host preference and in ability to reproduce on resistant sugar beets (Baldwin and Mundo-Ocampo, 1991; Lange et al., 1993).

Initial experiments were carried out using DNA preparations from about 150 brown cysts of *H. schachtii* or *H. cruciferae* (Caswell-Chen et al., 1992). With each of six primers tested, a different array of bands was produced for each of the two species (see for example, Figure 1). Typically 2 to 12 fragments ranging from 0.2 to 1.5 kb were obtained for each

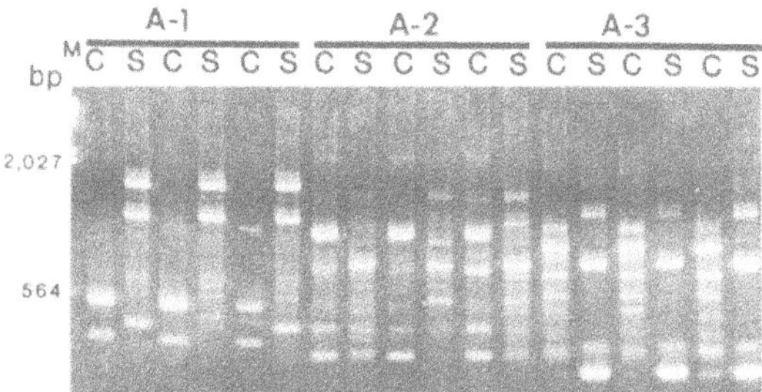

Figure 1. RAPD markers from DNA isolated from brown cysts of *Heterodera cruciferae* (C) and *H. schachtii* (S) produced using decamer primers OPA01 (A-1), OPA02 (A-2) and OPA03 (A-3). Nematodes of both species originated from the same field in Half Moon Bay, California. Each species was maintained on cabbage in the greenhouse and species identities were confirmed by vulval cone morphology. Three replicate amplifications were carried out with each DNA for each primer. Lane M contains molecular size standards. Electrophoresis was at 70 V for approximately 1 hr in a 1.7% agarose gel. Figure reproduced from Caswell-Chen et al., 1992.

primer. Replicate PCR reactions with the same DNA preparations yielded reproducible fragment patterns. RAPD analysis was also carried out on six California populations of *H. schachtii* from different locations in the state. In each case, several common bands were produced, but populations could easily be distinguished (Figure 2).

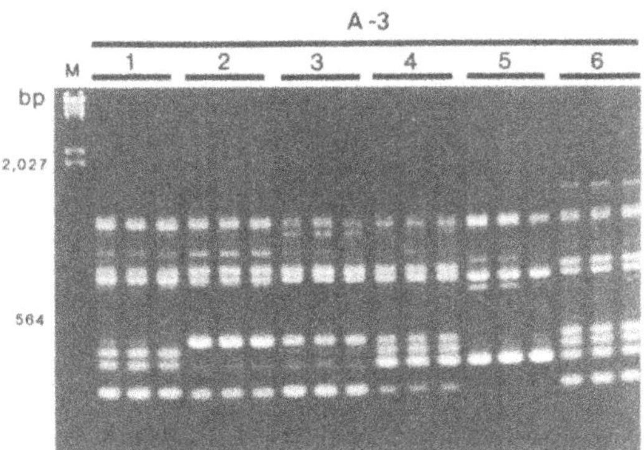

Figure 2. RAPD markers generated with decamer primer OPA03 and DNA from *H. schachtii* originating from different geographical locations in California (1 from Salinas; 2 from Lodi; 3 and 5 from different locations in Imperial Valley; 4 and 6 from different fields in Clarksburg). Each population is represented by three replicate lanes. Lane M contains molecular size standards. Figure reproduced from Caswell-Chen et al., 1992.

To test the applicability of this technique for measuring the genetic variability within a population, DNA was isolated from single gravid, brown cysts by a simple technique that used Chelex 100 resin to remove impurities from the crude extract that inhibit the PCR reaction (Walsh et al., 1991; Caswell-Chen et al., 1992). RAPD patterns were obtained from single females, but some variability in pattern was seen, even when inbred lines were used (Figure 3). This pattern variability could reflect genetic differences between the females in the inbred line or differences in DNA amount in the extracts of single females. RAPD patterns are subject to changes with template DNA concentration as spurious patterns have been reported to occur at very low concentrations (Williams et al., 1993). Thus, we conclude that although strong, reproducible RAPD bands may have some utility, RAPDs may not be generally useful as diagnostic tools for routine identification.

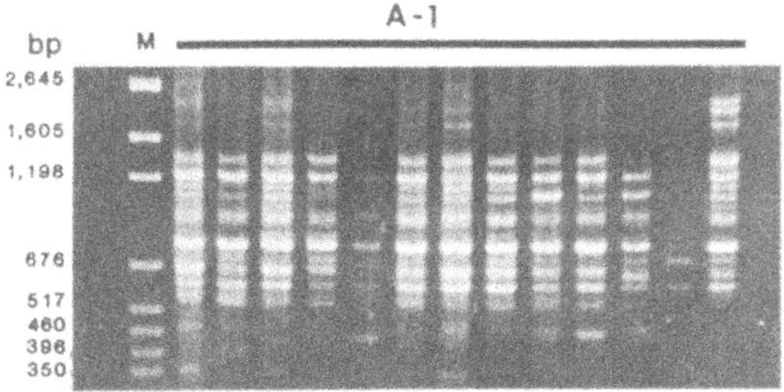

Figure 3. RAPD markers generated with primer OPA01 and DNA extracted from single cysts of an inbred line of *Heterodera schachtii* (line 5027-1-2-1 from P. Esbenshade, North Carolina State University) maintained in the greenhouse (lanes 2-13) through serial transfers of single cysts. In lane 14 is amplified DNA using the same primer and DNA extracted from a pool of 150 cysts as a template. Figure reproduced from Caswell-Chen et al., 1992.

Comparisons of RAPD patterns have been used to estimate the relationships between nematode species and populations. We used the presence or absence of 78 markers generated by 10 different primers to estimate the relationships among California populations of *H. schachtii* (Caswell-Chen et al., 1992). To assess the validity of this approach, it was necessary to determine whether bands of the same size are homologous to each other. A band of 480 bp that was amplified from *H. schachtii* with primer OPA03 (band OPA03-480) was cloned into a plasmid vector to generate clone p38. A Southern blot of *H. schachtii* DNA probed with clone p38 produced a pattern indicating that the RAPD band OPA03-480 corresponds to a single copy or low copy number gene in the *H. schachtii* genome (Caswell-Chen et al., unpublished). A Southern blot of a gel containing primer OPA03-amplified RAPD bands generated using DNA from different lines of *H. schachtii* and from related nematode species was probed with p38. Hybridization occurred to all bands that were the same size as the probe insert as well as additional, larger bands for DNA from all *H. schachtii* lines. No hybridization occurred to DNA from *H. cruciferae*, *Globodera tabacum*, or *H. glycines*. In other words, hybridization of probe p38 to DNA amplified by primer OPA03 was species specific, but the marker shared some homology to several RAPD bands from *H. schachtii* indicating that individual bands are not necessarily unique.

The apparent specificity of clone OPA03 to *H. schachtii* suggests a possible strategy for determining ratios of species in a mixed population. In this strategy, DNA from single females is isolated, using Chelex 100 resin, then the extracted DNA amplified with RAPD primer OPA03. The amplified DNA is spotted onto nitrocellulose in a grid, then probed with the species specific probe p38. The blot could then be reprobed with a probe specific for the other species being tested.

In sum, many RAPDs are useful tools for distinguishing *Heterodera* species and populations. Reproducibility of RAPDs on single females is not likely to be robust enough for routine analysis and thus the technique may be of limited diagnostic value. However RAPDs have the potential to be good starting points for developing diagnostic probes. An example is our proposed use of clone p38 as a species-specific probe on single female DNA amplified with OPA03. This probe together with other species specific probes could be used as tools to study the population dynamics of plant pathogenic nematodes.

PCR AND ROOT-KNOT NEMATODE IDENTIFICATION

Root-knot nematodes (*Meloidogyne* spp.) are broadly distributed, have wide host ranges and probably cause the most important nematode problems in agriculture. Because each species, as well as races or biotypes within each species, differ in host preference, it is essential that they be correctly identified for many integrated pest management strategies. Techniques for species identification have in the past included perinneal pattern observation and other morphological criteria, host range tests and isozyme electrophoresis (Eisenback and Triantaphyllou, 1991). Morphological criteria have been problematic due to variability and the need for specialized personnel to perform the determinations. Isozymes have proved useful for routine species identification (Esbenshade and Triantaphyllou, 1990; Cenis et al., 1992). The strengths of this technique include reliability and the ability to determine the species of single females. Thus isozymes have utility for studying mixed populations. Constraints are that isozyme electrophoresis does not work well for single second-stage juveniles (J2s) and does not differentiate subspecies populations that differ in host preference. In addition, samples must contain relatively young, healthy females for successful identification. In practice, many field samples contain only juveniles or eggs, and thus must be grown in greenhouse culture to produce females for the isozyme analysis.

A PCR method for identification of the species of single *Meloidogyne* juveniles has been developed (Powers and Harris, 1993). This technique involves amplification of the intergenic spacer region between the cytochrome oxidase subunit II gene and the 16S rRNA gene in the mitochondrial genome of the nematode. A 1.7 kb band is amplified from *M. javanica* and *M. incognita*, a 1.1 kb band is amplified for *M. arenaria* and a 0.5 kb band is amplified for *M. hapla* and *M. chitwoodi*. We have reproduced this result in our laboratory and are currently using this technique for routine identification of root-knot nematode juveniles. In general, we successfully obtain an amplified band for about 80% of the individual J2s.

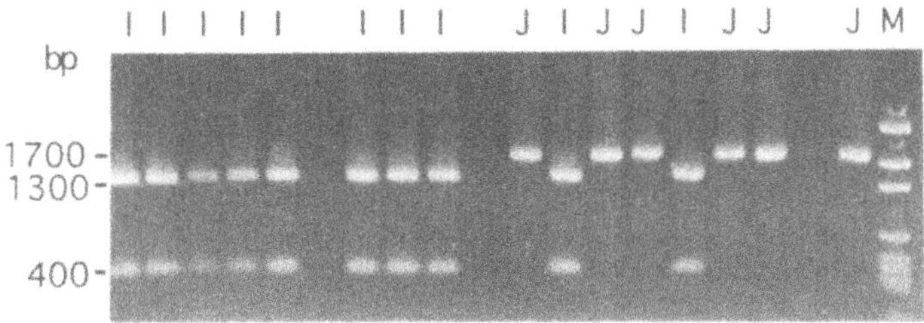

Figure 4. PCR amplification of mitochondrial DNA from single juveniles in a mixed population. Single juveniles from a mixed population of *Meloidogyne incognita* and *M. javanica* were amplified using PCR primers and conditions described in Powers and Harris (1993) except that individuals were picked with an insect pick (minuetin) and disrupted with the pick in the PCR tube before buffer was added. DNA was digested with *Hin*fI and electrophoresed on a 1.5% agarose gel. Lanes marked "J" were identified as *M. javanica* and those marked "I" were identified as *M. incognita*

To distinguish *M. incognita* from *M. javanica*, it is necessary to cut the amplified DNA with the restriction enzyme *Hin*fI, and size the cleaved product after electrophoresis on an agarose gel. According to Powers and Harris (1993), this should result in two bands (1.0 and 0.7 kb) for *M. javanica* and three bands (1.0, 0.4 and 0.3 kb) for *M. incognita*. In our hands, only a single band (1.7 kb) was present in *M. javanica* lanes, indicating no digestion, and two bands (1.3 and 0.4 kb) were present in *M. incognita* digests (Figure 4). This result has been reproduced on numerous occasions and on several different isolates of each nematode. The reason for the discrepancy with the results of Powers and Harris (1993) is not known but may be due to differences in PCR enzyme sources in the two laboratories. Nevertheless, in both laboratories, *M. javanica* and *M. incognita* are easily differentiated. The technique is useful for studies of mixed populations, as individual J2s can be analyzed (Figure 4). Disadvantages of the technique are the need for restriction enzyme digestion of the PCR products followed by electrophoresis. In addition, this technique does not allow identification of races or biotypes that differ in their host range.

The finding that RAPDs distinguish populations of the species *Heterodera schachtii* prompted us to explore this technique for distinguishing root-knot nematode species and populations. Using DNA extracted from eggs (Sulston and Hodgekin, 1988) we found that RAPD primers distinguish the major species when used to amplify purified DNA (see Figure 5 for example). We have tested 38 different primers so far. With a number of primers, *M. javanica and M. arenaria*, and sometimes *M. incognita* give similar or indistinguishable patterns, whereas *M. hapla* and *M. chitwoodi* are easily distinguished from this group with nearly all primers. Similar results were obtained by Cenis (1993). Cenis (1993) found frequent differences between races A and B of *M. hapla* as well as between isolates of *M. arenaria* but no clear polymorphisms that distinguish the host races of *M. incognita*. We have so far found no clear polymorphisms that distinguish populations of *M. incognita* that are avirulent in the presence of the tomato resistance gene *Mi* from those that are virulent on *Mi*-containing tomato lines. The few minor band differences that were seen between isolates differing in virulence on *Mi* did not correlate with virulence when different populations were compared. In general the RAPD patterns between *M. incognita* and *M. arenaria* populations are much more similar than among *H. schachtii* populations. This is not surprising, as we expected much less intraspecific genetic variability in the parthogenetic *Meloidogyne* species.

By a related technique called DNA Amplification Fingerprinting (DAF) that uses 8 mer primers and displays amplified DNA on acrylamide gels, Baum et al. (1994) found consistent polymorphisms that distinguished *M. incognita* and *M. arenaria* populations, but host races did not cluster in parsimonious trees. From this they speculate that host races do not originate from a common ancestor. If this is true, it may not be possible to identify race specific primers. However, population specific markers may be useful in specific geographic areas where a particular host range biotype has originated from a single ancestor.

We have amplified DNA from single *Meloidogyne* J2s using RAPD primers. However, the pattern is somewhat variable even when juveniles from a single egg mass are examined (results not shown). Cenis (1993) also amplified single juveniles and saw some variability in minor bands though the major bands were reproducible. The inconsistency in banding pattern is likely to be due to the low and variable amount of DNA that is extracted from the single J2. Williams et al. (1993) noted that too little DNA gives non-reproducible patterns under RAPD conditions. Our results suggest that RAPDs of single juveniles are not likely to be consistent enough to use for routine identification of single J2s. However, RAPDs can be converted into more robust markers that anneal to specific genomic sequences at higher temperatures (Paran and Michelmore, 1993). To do this, specific RAPD bands that are amplified in only one species are cloned. The DNA sequence of the ends of these clones is then determined and used to design oligonucleotide primers that prime amplification under stringent conditions to yield a single band representing one locus in the target genome. With the higher annealing temperature used for the amplification, DNA concentration is less critical, and reproducible amplification of tiny and variable amounts of DNA can be carried out. Such conditions have been shown to be sensitive enough to reproducibly amplify single copy genes from individual human sperm (Cui et al., 1989). These markers, called SCARs (sequence characterized amplified regions) or STSs (sequence tagged sites) have also been used successfully in lettuce and tomato (Paran and Michelmore, 1993; Williamson et al., 1994).

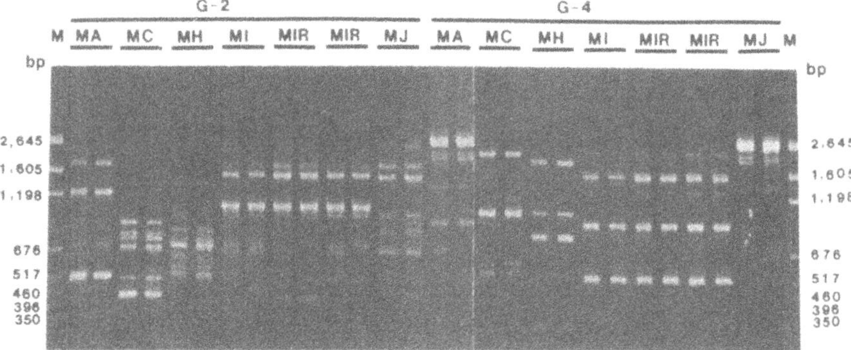

Figure 5. RAPD patterns of root knot nematode species. Primers OPG02 and OPG04 were used to amplify DNA prepared from eggs of *Meloidogyne arenaria* (MA), *M. chitwoodi* (MC), *M. hapla* (MH), *M. javanica* (MJ), and three populations of *M. incognita* (a culture avirulent on tomato with the *Mi* gene, and two cultures that are virulent on *Mi*, both labeled MIR). Each sample was run in duplicate.

We have carried out a preliminary study to determine whether RAPD-derived STS markers can be applied to developing species-specific PCR probes for nematode identification. In our laboratory, a strong PCR band of 1.1 kb is amplified from *M. hapla* (California isolate VC1R) using primer OPA01. The same size band was amplified by Cenis (1993) with primer OPA01 in *M. hapla* Race A and Race B isolates from Virginia and Chile, respectively. A large scale amplification was carried out and the 1.1 kb band was isolated from the gel, then cloned using a TA cloning kit (Invitrogen Corp., San Diego, CA) to generate plasmid pMH1. DNA sequence of the ends of the insert were determined and primers were designed by criteria that we have used to generate STS markers for genome mapping in tomato. Generally, our primers are 20 nucleotides in length and are designed such that primer pairs have approximately equal melting temperature and are approximately 50% GC. In addition, primer sequences are checked visually to be sure that they lack obvious secondary structure or complementary 3' ends. The primers do not necessarily include the original RAPD primer sequence. For *M. hapla,* we designed two primers, MH1F (5'-CTCTGTTTGGAGAGTAGGCA-3') that hybridized to one end of the insert in pMH1 and MH1R (5'-CTTCGTTGGGGAACTGAAGA-3') that hybridized to the opposite strand at the other end of the insert. We tested these primers for their ability to amplify DNA from *M. hapla*, *M. incognita*, *M. javanica*, and *M. arenaria*. A single band of the expected size was amplified only from the *M. hapla* DNA (Figure 6). A band of the appropriate size was also amplified from single J2 of *M. hapla* but not *M. chitwoodi* (Figure 6). This result needs to be extended to other isolates of *M. hapla* as well as with other nematode species.

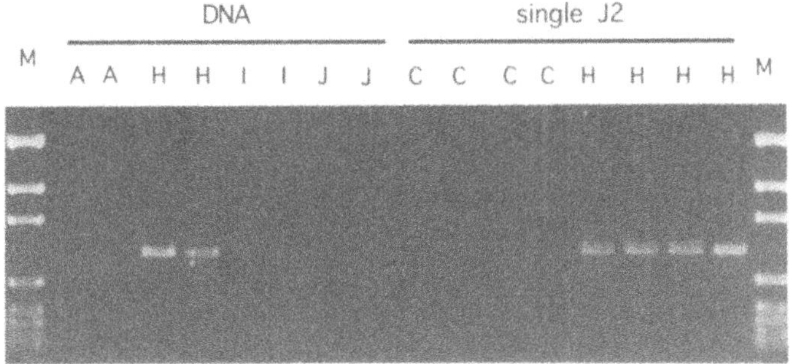

Figure 6. *Meloidogyne hapla*-specific probe. DNA preparations from *M. arenaria* (A), *M. hapla* (H), *M. incognita* (I), and *M. javanica* (J) were amplified using primers MH1F and MH1R using amplification conditions described in Williamson et al., 1994. Single J2 of *M. chitwoodi* (C) or *M. hapla* (H) were disrupted as for the mitochondrial spacer amplification (Figure 4), then subject to PCR with *M. hapla*-specific primer pairs.

If, indeed we are able to extend this procedure to convert RAPD bands into species-specific probes for each of the major root-knot nematode species, we still need to develop these into a format that will be useful for routine identification. One strategy is to generate primer pairs for each species of interest, such that each amplifies a species-specific band of a different size. For example, a mixture of 8 primers, two for each of the four major species could be amplified in a "multiplex" format, then the species would be identified after agarose gel electrophoresis to distinguish band sizes. Multiplex approaches have been used in human genetic studies (see for example, Cui et al., 1989). Alternatively, we may be able to use a hybrid procedure, mixing mitochondrial primers with the species-specific primers for *M. incognita*. This mixture of primers should allow us to distinguish the four major species with our restriction enzyme digestion of the amplified DNA.

Another possible use for species-specific primers is for identification of root-knot nematodes (eggs, J2s) in soil samples or in root tissue. Often it is difficult to obtain sufficient sensitivity. One strategy is to extract a nematode-containing fraction from the soil using classical techniques (elutriation, sieves, sucrose flotation). The sample would then be treated to release the DNA, extracted, perhaps with Chelex-100, then subjected to PCR amplification. Highly specific primers should amplify DNA from the nematode of interest and not other species or other contaminating microorganisms. Precise quantitation may be difficult but the method is likely to have a higher sensitivity than current procedures. Sensitivity and inhibition by contaminants in soil extracts can be monitored by addition of nematodes of interest (for example, *M. hapla*) to soil samples prior to extraction.

In summary, many RAPDs can be used to distinguish *Meloidogyne* species using DNA preparations of each nematode. The ability of RAPDs to distinguish host race differences is unclear. RAPDs appear also to be useful starting points for developing species-specific primers that can be used for identification of single juveniles and perhaps to determine the presence of nematodes in soil. Thus RAPDs are a good starting point for PCR marker development. The value of PCR markers for developing nematode identification techniques that are uncomplicated, reproducible, simple and safe is clear. At this stage, validation and testing of techniques is in progress and routine use is increasing. The potential for applying PCR to previously intractable areas of nematode identification and population biology is just beginning to be explored and applications are only limited by our creativity.

Acknowledgments

We thank Tom Baum for a preprint of an unpublished manuscript and Miriam Kaplan for comments on the manuscript. This research was funded by the California Statewide Integrated Pest Management Project.

REFERENCES

Baldwin, J.G., and Mundo-Ocampo, M., 1991, Heteroderinae, cyst- and non-cyst-forming nematodes, *in:* "Manual of Agricultural Nematology," W. R. Nickle, ed., Marcel Dekker, New York.

Baum, T.J., Gresshoff, P.M., Lewis, S.A., and Dean, R.A., 1994, Characterization and phylogenetic analysis of four root-knot nematode species using DNA amplification fingerprinting and automated polyacrylamide gel electrophoresis, *Molec. Plant Microbe Interact.*, in press.

Caswell-Chen, E.P., Williamson, V.M., and Wu, F.F., 1992, Random amplified polymorphic DNA analysis of *Heterodera cruciferae* and *H. schachtii* populations, *J. Nematol.* 24:343.

Cenis, J.L., 1993, Identification of four major *Meloidogyne* spp. by random amplified polymorphic DNA (RAPD-PCR), *Phytopathology* 83:76.

Cenis, J.L., Opperman, C.H., and Triantaphyllou, A.C., 1992, Cytogenetic, enzymatic, and restriction fragment length polymorphism variation of *Meloidogyne* spp. from Spain, *Phytopathology* 82:527.

Cui, X., Li, H., Goradia, T.M., Lange, K., Kazazian, Jr., H.H., Galas, D., and Arnheim, N., 1989, Single-sperm typing: determination of genetic distance between the γg-globin and parathyroid hormone loci by using the polymerase chain reaction and allele-specific oligomers, *Proc. Natl. Acad. Sci. USA* 86:9389.

Eisenback, J.D., and Triantaphyllou, H.H., 1991, Root-knot nematodes: *Meloidogyne* species and races, *in:* "Manual of Agricultural Nematology," W. R. Nickle, ed., Marcel Dekker, New York.

Esbenshade, P.R., and Triantaphyllou, A.C., 1990, Isozyme phenotypes for the identification of *Meloidogyne* species, *J. Nematol.* 22:10.

Fargette, M., Block, V.C., Phillips, M.S., and Trudgill, D.L., 1994, Genetic variation in tropical *Meloidogyne* species, *in*: "NATO ARW: Advances in Molecular Plant Nematology," F. Lamberti, De Georgi, C., and Bird, D. McK., eds., Plenum Press, New York.

Goodwin, P.H., and Annis, S.L., 1991, Rapid identification of genetic variation and pathotype of *Leptosphaeria maculans* by random amplified polymorphic DNA assay, *Appl. Environ. Microbiol.* 57:2482.

Hedrick, P., 1992, Shooting the RAPDs, *Nature* 355:679.

Lange, W., Muller, J., and DeBock, T.S.M., 1993, Virulence in the beet cyst nematode (*Heterodera schachtii*) versus some alien genes for resistance in beet, *Fundam. appl. Nematol.* 16:447.

Mulvey, R.H., 1972, Identification of *Heterodera* cysts by terminal and cone top structures, *Can. J. of Zool.* 50:1277.

Pableo, E.C., and Triantaphyllou, A.C., 1989, DNA complexity of the root-knot nematode (*Meloidogyne* spp.) genome, *J. Nematol.* 21:260.

Paran, I., and Michelmore, R.W., 1993, Development of reliable PCR-based markers linked to downy mildew resistance genes in lettuce, *Theor. appl. Genet.* 85:985.

Powers, T.O., and Harris, T.S., 1993, A polymerase chain reaction method for identification of five major *Meloidogyne* species, *J. Nematol.* 25:1.

Smith, M.L., Bruhn, J.N., and Anderson, J.B., 1992, The fungus *Armillaria bulbosa* is among the largest and oldest living organism, *Nature* 356:428.

Sulston, J., and Hodgekin, J., 1988, Methods, *in:* "The Nematode *Caenorhabditis elegans*," W.B. Wood, ed., Cold Spring Harbor Laboratory Press, Cold Spring Harbor, NY.

Walsh, P.S., Metzinger, D.A., and Higuchi, R., 1991, Chelex 100 as a medium for simple extraction of DNA for PCR-based typing from forensic material, *BioTechniques* 10:506.

Welsh, J., and McClelland, M., 1990, Fingerprinting genomes using PCR with arbitrary primers, *Nucl. Acids Res.* 18:7213.

Williams, G.G.K., Kubelik, A.R., Livak, K.J., Rafalski, J.A., and Tingey, S.V., 1990, DNA polymorphisms amplified by arbitrary primers are useful as genetic markers, *Nucl. Acids Res.* 18:6531.

Williams, J.G.K., Rafalski, J.A., and Tingey, S.V., 1993, Genetic analysis using RAPD markers. *Methods Enzymol.* 218:704.

Williamson, V.M., Ho, J.Y., Wu, F.F., Miller, N., and Kaloshian, I., 1994, A PCR-based marker tightly linked to the nematode resistance gene, *Mi*, in tomato, *Theor. appl. Genet.*, in press.

PCR FOR NEMATODE IDENTIFICATION, DISCUSSION

OPPERMAN noted a plus or minus score for a *Meloidogyne hapla* band, and wondered how a diagnostic based on absence of a character would work. WILLIAMSON replied that this type of diagnostic would require appropriate controls. For some experiments the mitochondrial primers could be used as a positive control for amplification as they amplify a band from most root-knot nematode species. A second control could be to add *M. hapla* extract or DNA to a duplicate reaction to that being assayed. OPPERMAN also asked whether the differences in mitochondrial PCR patterns after cleavage with restriction enzymes might be restriction enzyme brand specific. WILLIAMSON replied that this was unlikely as several brands of restriction enzyme were tested, including the one used by Powers. BLOK asked WILLIAMSON to elaborate on the skepticism of obtaining race-specific primers. WILLIAMSON replied that root-knot nematode host races are defined on the basis of differences in their ability to reproduce on a small number of hosts. She did not think that a nematode originating from North Carolina that reproduces on the test hosts was likely to be genetically identical to one from Europe or Africa that displays the same limited host range. Several research groups have found that adding new hosts to the test, will display differences between populations of the same race. She further suggested that it is likely that host range evolved independently several times, and thus it may be difficult or impossible to find markers common to all nematodes of a particular host race. KAPLAN wondered why Imperial Valley populations 1 and 2 differed given the historical cropping procedures in Imperial Valley where tare soils were recovered at beet processing centers and returned to different fields. WILLIAMSON replied that the practice of returning tare soil to fields was discontinued in approximately 1959, soon after the cyst nematode was discovered in the Imperial Valley. During the last 30 years there have undoubtedly been many opportunities for the introduction of new founder populations on, for example, machinery brought into the Valley from other areas in California such as Ventura or Salinas. WILLIAMSON also noted that in reality we don't really understand the dynamics of cyst nematode movement in the Imperial Valley, or anywhere else for that matter. There is very little information on the invasion and spread of plant parasitic nematodes. She suggested that genetic markers hold promise for helping to evaluate this question. BAKKER noted that his group was trying to use molecular markers (2-D gels) to examine "gene pools." After assaying approximately 150 pools, he found that the variance was saturated. PHILLIPS noted that discussions at a meeting in Portugal had concluded with the suggestion that if one uses two types of resistance, two types of race will emerge. BAKKER noted that the number of loci that contribute to a pathogenic trait are likely to be small and probably polymorphic. Therefore, the approach of using a small number of markers to try to follow these traits is a poor one. BAILLIE agreed, and used the human apo^R locus as a parallel: trying to map this locus in humans by mapping general polymorphisms between races is unlikely to work. BAKKER noted that there is limited gene flow in nematodes and that markers may be useful in a characterized local where introduction was relatively recent. WILLIAMSON agreed and suggested that markers that distinguish root-knot nematodes with different host ranges from a particular geographic area where they are likely to have a common origin, can be found and will be useful. POWERS asked if the diversity of worms in the polder was evidence for gene flow. BAKKER pointed out that differences likely were due to migration into the polder.

PERSPECTIVES FOR GENETICALLY ENGINEERED ANTIBODIES FOR THE IDENTIFICATION OF NEMATODES

Arjen Schots[1] and Jaap Bakker[2]

[1]Laboratory for Monoclonal Antibodies
P.O. Box 9060
6700 GW Wageningen
The Netherlands

[2]Wageningen Agricultural University
Department of Nematology
P.O. Box 8123
6700 ES Wageningen
The Netherlands

INTRODUCTION

Over the past decade monoclonal antibody (MA) based immunoassays have been developed for routine identification of plant-parasitic nematode species. The information obtained from such immunoassays can be used for advisory systems with regard to crop rotation, pesticide application and certification. Initially, the development of immunoassays was hindered by difficulties with standardization and raising of specific antibodies. Standardization was mainly hampered by the complexity of the samples which usually contain soil and by difficulties with the homogenization of nematodes. Obtaining specific antibodies was impeded by the great similarities in protein composition of closely related nematode species.

The hybridoma technology developed by Köhler and Milstein (1975) offers possibilities to overcome the problems associated with the development of immunoassays. With this technique, antibody-producing B-cells are isolated from an immunized animal and fused *in vitro* with a lymphoid tumor cell (myeloma). In the resulting hybrid cell (hybridoma), the characteristics of both parental cell types are combined, *i.e.* the production of specific (monoclonal) antibodies and the ability to multiply *in vitro*. In this way, the humoral immune response against complex molecules containing many antigenic sites, each of which gives rise to an individual antibody, can be dissected into separate component antibodies. Moreover, rare antibody specificities, whose potential reactivities never become manifest *in vivo* might be generated *in vitro*

Advances in Molecular Plant Nematology
Edited by F. Lamberti *et al.*, Plenum Press, New York, 1994

by the hybridoma technique (Metzger et al., 1984). This has resulted in monoclonal antibodies discriminating between iso-enzymes and other closely related proteins (Berzofsky et al., 1980; Hollander and Katchalski-Katzir, 1986).

Monoclonal antibodies have been developed against several nematode species including *Meloidogyne incognita* (Jones et al., 1988; Hussey et al., 1989), *Heterodera glycines* (Atkinson et al., 1988) and the potato cyst nematode species *Globodera rostochiensis* and *G. pallida* (Schots et al., 1989). MAs against the latter two species are used for identification by a number of inspection services. The outcome of these assays is used for advice in relation to crop rotation, soil disinfection and certification of seed potatoes.

The major drawback of the hybridoma technology is that it is only suited for the generation of murine antibodies, whereas the main interest has always been in human (like) antibodies for medical applications, e.g. tumor killing. Murine monoclonal antibodies offer no alternative, because they give rise to an undesirable immune response in humans (Winter and Milstein, 1991). Therefore other alternatives have been developed by using DNA-technology. Initially, variable (V) domains of murine hybridomas were grafted onto human constant domains (see Fig. 1; Orlandi et al., 1989). Later, mouse complementarity determining regions, the parts of the V domains responsible for binding the antigen, were put in a human framework (Fig. 1; Jones et al., 1986). More recently, human antibodies have been developed by using antibody phage display libraries (Clackson et al., 1991).

Although these techniques have been developed for medical use, plant pathologists can benefit from them. Antibodies developed with these new techniques will undoubtedly be of great value for diagnostic purposes. Another application is the expression of antibodies in plants, which offers possibilities to endow plants with new properties like resistance against pathogens (Schots *et al.*, 1992).

IMMUNE SELECTION

The ability to respond to an apparently limitless array of foreign antigens is one of the most remarkable features of the vertebrate immune system. Almost all antibody molecules contain a unique stretch of amino acids in their variable region. It is estimated that the immune system is capable of generating more than 10^8 different antibody molecules. This diversity is possible because during B-cell differentiation in the bone marrow, immunoglobulin gene segments are randomly shuffled by a dynamic genetic system. This process is carefully regulated. B-cell differentiation from an immature pre-B-cell to a mature cell involves an ordered progression of immunoglobulin gene rearrangements. The result is a mature immunocompetent B-cell which contains a <u>single</u> functional variable-region DNA sequence for its heavy and its light chain. After antigenic stimulation, further rearrangement of constant-region gene segments can generate changes in the isotype expressed and consequently change the associated biological effector functions without changing the specificity of the immunoglobulin molecule. Antigenic stimulation also results in the formation of plasma and memory cells. Upon repeated contact with antigen, memory cells produce antibodies with increased affinity which is a consequence of the affinity maturation process leading to somatic mutations in the variable regions of the heavy and light chain.

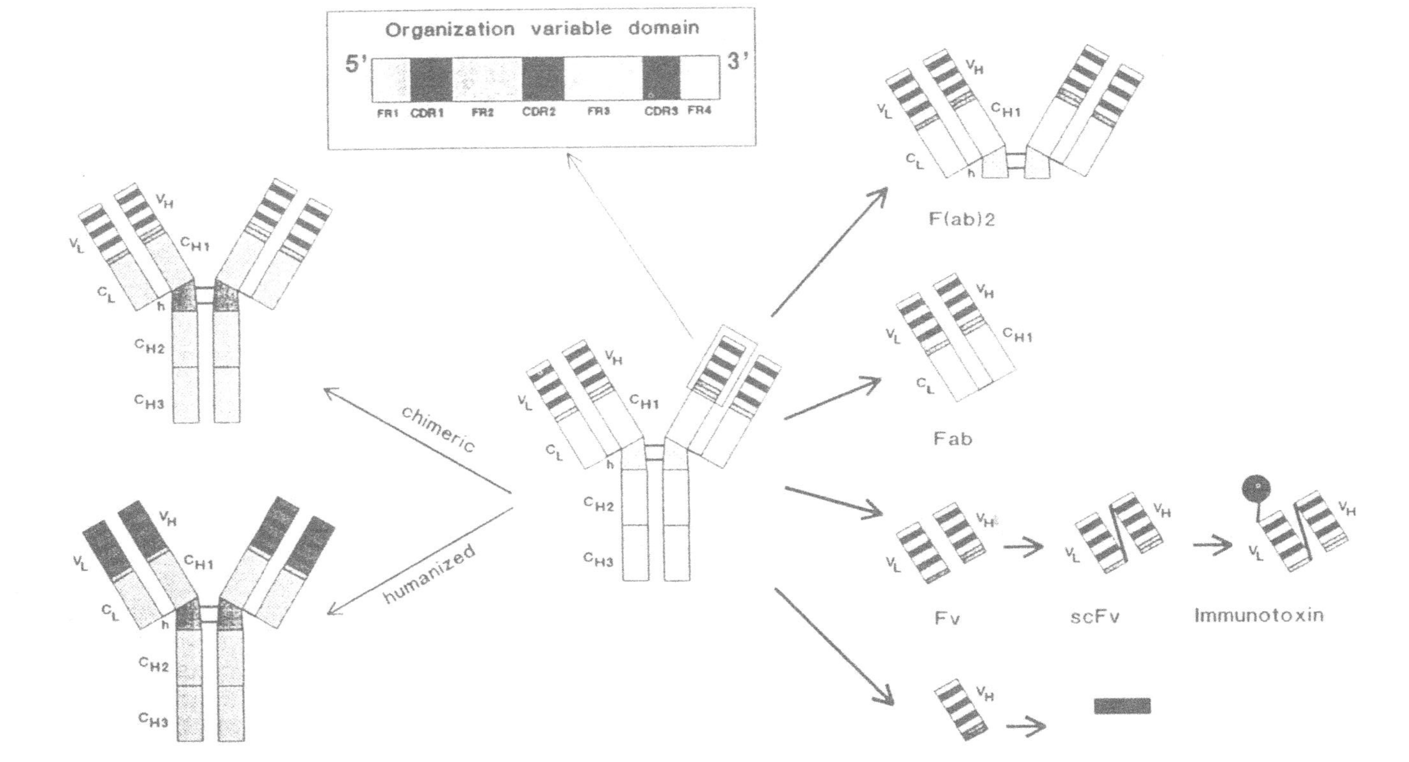

Figure 1. The domain structure of the antibody molecule facilitates antibody engineering. The different possibilities are indicated. $F(ab)_2$ and Fab fragments can be obtained through proteolytic cleavage of antibody molecules or through gene technology. Gene technology must also be used to obtain Fv-fragments, single chain Fv-fragments (both polypeptide chains linked by a 15-mer peptide), single domain antibodies (dAbs), chimeric (humanized) antibodies or reshaped antibodies. A minimal recognition unit (m.r.u.) is a synthetic peptide resembling the amino acid sequence of a hypervariable region on the heavy or light chain V-region. Abbreviations used: V, variable domain; C, constant domain; L, light chain; H, heavy chain; h, hinge region; FR, framework region; CDR, complementarity determining regions.

ANTIBODY ENGINEERING

Since 1985 gene technology has revolutionized hybridoma technology. Until then only few functional properties of monoclonal antibodies could be changed, for instance by switching heavy chain constant regions (Kipps, 1985) or by making bispecific antibodies (Milstein and Cuello, 1984). Antibody genes from hybridoma cell lines or lymphocytes can be cloned into plasmid vectors and expressed in bacteria, yeast, mammalian cells or plants. An antibody (IgG) is a Y-shaped molecule (Fig. 1), in which the domains (V_L, V_H) forming the tips of the arms bind to antigen. The domains forming the stem (Fc fragment) are responsible for triggering effector functions to eliminate the antigen from the animal. An IgG molecule consists of four polypeptide chains, two heavy (H) and two light (L) chains. Each domain consists of two ß sheets which pack together to form a sandwich, with exposed loops at the ends of the strands.

The domain structure in combination with the polymerase chain reaction facilitates protein engineering of antibodies (Orlandi et al., 1989; Winter and Milstein, 1991; Fig. 1):

- Fv and Fab fragments can be used separately from the Fc-fragments without loss of affinity. By using PCR for forced cloning, Fv and Fab fragments and single chain Fvs (scFvs) can be expressed in heterologous systems. In a scfv the variable parts of the light (V_L) and heavy chain (V_H) are linked by a peptide. Fv-fragments, Fab-fragments and scFvs have been successfully expressed in bacteria, yeast and murine myeloma cell lines.
- Toxins or enzymes can be fused to the antigen-binding domains to endow them with new properties. For example immunotoxins have been constructed as Fv fusion proteins (Batra et al., 1990, Chaudhary et al., 1990) and expressed in *Escherichia coli.*
- Murine antibodies can be humanized by grafting the variable domains on the constant domains of human IgG isotypes, chimeric antibodies can be made which are expressed in myeloma cells (Orlandi et al., 1989).

The humanization of rodent antibodies shows that engineering of antibodies is not restricted to the use of whole domains. The antigen-binding loops in a V-region framework of a human antibody can be replaced by those of a rodent antibody, thus transferring the antigen binding-site (Jones et al., 1986).

EXPRESSION OF ANTIBODIES IN BACTERIA

Bacterial production of antibody fragments (Better et al., 1988; Skerra and Plückthun, 1988) can be achieved in three ways. The first involves the expression of proteins intracellularly, the second as proteins secreted into the periplasmic space or culture medium and the third as fusion proteins expressed on the surface of phages.

Intracellular production of antibodies leads to nonfunctional proteins sequestered in inclusion bodies. As a result, they need to be refolded *in vitro*, procedures which often have to be individually tailored to a particular antibody, and the folded products always have to be separated from misfolded forms for accurate quantitative measurements. Secretion, on the other hand, offers several advantages compared to intracellular production: efficient purification, theoretically higher yields, no aggregation of the product, correctly folded proteins, the possibility for disulfide bond formation and the

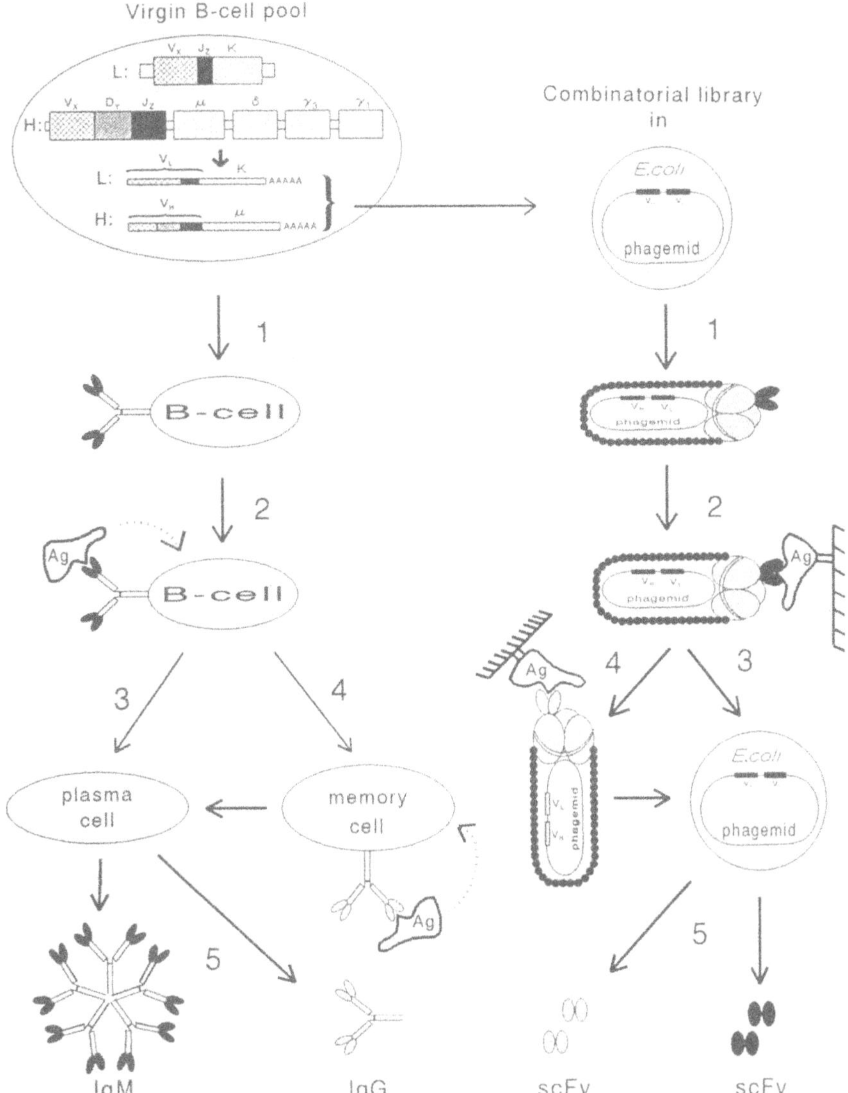

Figure 2. Comparison of the strategy of the immune system *in vivo* and using phage. V-genes are rearranged or assembled (step 1) and displayed on the surface (step 2). Antigen driven or affinity selection then takes place (step 3). The selected cells or phages then undergo affinity maturation (step 4) or produce soluble antibody (step 5).

possibility for continuous cultivation and production. The third possibility is the display on the surface of phages as fusion protein with a coat protein (McCafferty et al., 1990; Chang et al., 1991). Mainly the filamentous phage fd has been used. Fd is a non-lytic bacteriophage which has been used to display peptides, enzymes and antibodies. This is usually done as fusions with gene III, encoding the minor coat protein. Fusions with gene VIII, encoding the major coat protein, have also been described (Chang et al., 1991).

Although many antibody fragments have been successfully produced in *E. coli* by means of secretion, the production levels are not optimal. The causes of the low production levels can mainly be found in the use of suboptimal promoters, inefficient translation, folding and assembly and the inability to use a suitable host. Attempts to improve folding and assembly included, among others, the co-expression of disulfide isomerase, cis-trans isomerase and chaperones (Knappik et al., 1993; Söderlind et al., 1993). In these experiments, the helper proteins were produced with success. However, the effect on antibody production was minimal. Another host bacterium has also been tried. A scFv was successfully produced in *Bacillus subtilis* which possibly is a more suitable host than *E. coli* (Wu et al., 1993). The production levels obtained were similar to those from *E. coli*, but, the functional yield was much better, about 98% of the yield. Although in these experiments no real improvement of antibody production levels were observed, this type of research should continue as it will sooner or later reveal what factor hinders the production of heterologous proteins in bacteria.

ANTIBODY PHAGE DISPLAY LIBRARIES

Library development

The natural antibody selection system can be mimicked by displaying the immune repertoire on phage and selection with antigen (Clackson et al., 1991). To achieve this, a library of rearranged V-genes, representing the immune repertoire, can be constructed with genomic DNA or MRNA of B-lymphocytes or by making a synthetic library (Fig. 2). Such a library is constructed by amplifying a range of heavy and light chain V-genes using "universal" PCR primers, followed by direct cloning into expression vectors. Universal 3' PCR primers can be based on the sequences of the J-segments to copy the sense strand, while they can also be used for first strand cDNA synthesis. The nucleotide sequence at the 5' end of the V-exon proved to be sufficiently conserved to design a "degenerate" primer to copy the antisense strand. Thus, 'natural' mouse and human V-gene libraries have been constructed. In addition, synthetic V-gene repertoires have been made by randomizing CDR3 of a V_H segment of an bacterially expressed antibody fragment (Hoogenboom and Winter, 1992; Barbas et al., 1992). This was achieved by designing a highly degenerate PCR primer bridging CDR3 which randomly changed the sequence of the amino acid residues. The randomization could only result in one (amber) stop codon, TAG, suppressible in *supE E. coli* strains.

Starting from "natural" V-gene libraries it is necessary to link the V_H and V_L genes together in order to obtain functional antibody fragments. This results in a chimeric gene encoding either a scFv-fragment or Fab-fragment. The genes can be linked together by recombination by cloning, for instance, the light and heavy chain gene repertoire separately into the left and right arms of a modified lambda zapII vector and create a random combinatorial library by recombining the arms (Huse et al., 1989). An alternative and more general way of linking the heavy and light chains is "splicing by overlap extension" (Horton et al., 1989). After amplification of the V-genes by PCR, the V_H and V_L repertoires are combined with "linker" DNA which has regions of sequence homology with the 3' end of the amplified V_H gene and the 5' end of the amplified V_L gene or *vice versa*. The 'linker' can be either the linker-peptide in the case of scFv-fragments or a DNA fragment comprising the ribosomal binding site and leader sequence in the case of Fab-fragments. A further PCR amplification with outer flanking PCR primers splices together the two repertoires. Including restriction sites appended to the 5' ends of the flanking primers, enables direct cloning into an expression vector, usually a phagemid vector with the antibody fragments fused to the gene III protein.

Selection of binders

Phages displaying antibody fragments showing affinity can be selected by direct binding of the phage to immobilized antigen (McCafferty et al., 1990; Clackson et al., 1991; Marks et al., 1991). Antigens can be immobilized on columns; on dishes or tubes; or on the surface of cells. Alternatively, phages can be allowed to bind to biotinylated antigen in solution followed by capture on paramagnetic beads (Hawkins et al., 1992). After washing to remove non-binding phage, binders are eluted with antigen, acid or alkali buffers or with DTT when the antigen is biotinylated with a cleavable disulphide biotin. Eluted phage are used to reinfect *E. coli* cells, whereafter the whole procedure is repeated once or a few times more to enrich the fraction of binders.

V-gene source

Rearranged V-gene repertoires for phage display libraries can be obtained from either immunized or unimmunized donors. If, an immunized donor is used a relatively large number of binders is found. Consequently, the size of the library can be relatively small (10^6-10^7). In contrast, the number of binders obtained from a naive library is low. Therefore, the library has to be as large as possible, preferably larger than 10^8. It is difficult but nevertheless worthwhile to construct such a library. Large naive libraries contain the whole, expressed, immune repertoire and can, in principle, be used to select antibodies against almost any molecular structure. Recently such a library was successfully used to select antibody fragments against 12 different antigens, haptens, proteins and cellular antigens (Hoogenboom et al., 1992).

To obtain antibodies with higher affinities, it will be necessary to construct larger libraries. At present, the transformation efficiency of bacteria limits the construction of larger libraries ($>10^8$ combinations). At present there is an alternative method using *in vivo* recombination which is made possible by the *lox*-Cre site specific recombination system of bacteriophage P1 (Waterhouse et al., 1993). Within an infected bacterium this system locks together the heavy and light chain genes from two different replicons.

Increasing the affinity

The affinity of an antibody selected from an antibody phage display library is usually low and can be improved by introducing mutations. Three systems can be used to generate point mutations. The first possibility is to use reaction conditions which decrease the reliability of Taq polymerase. A reamplified, mutant, scFv was cloned and expressed on phage showing a 4-fold improved affinity (Hawkins et al., 1992). A second alternative is growth of the phage in an *E. coli mutD* strain in which the spontaneous mutation frequency is 10^3-10^5 times higher than in wild-type strains (Gram et al., 1992). The third system allows the introduction of multiple point mutations. This system is based on the use of spiked oligonucleotides (Hermes et al., 1989).

An alternative way to increase the affinity is to use the somatic mutations introduced by the immune system. Therefore, certain loops or entire heavy or light chains are shuffled with a repertoire of *in vivo* somatically-mutated V-genes (Marks et al., 1992). For instance, the light chain of a scFv is replaced by light chains from the library followed by selection. In a similar way parts of the heavy chain can be shuffled. However, care needs to be taken to avoid disrupting key features of antigen binding. In general, V_H-CDR3 should be left undisturbed.

PLANTIBODIES

Functional IgG molecules, as well as functional Fab and scFv fragments can also be produced in plants. These findings offer possibilities to endow plants with new properties like resistance against diseases, alteration of metabolic routes or introduction of new metabolic routes using catalytic antibodies.

Düring (1988) and Hiatt et al. (1989) were the first to describe the production of antibodies in plants. Constructs of coding-length cDNAs of the γ- and κ-chain, with and without leader sequences, were ligated into a plant expression vector. To transform tobacco plants Hiatt et al. (1989) constructed plasmids containing the gene for either the heavy or light chain, both with and without an immunoglobulin leader sequence, under control of the constitutive 35 S promoter. The transformants expressing individual immunoglobulin chains were then sexually crossed to produce progenies expressing both chains. Only plants expressing immunoglobulin chains with the original leader peptide contained assembled gamma-kappa complexes. Functional immunoglobulins, assessed using ELISA, were reported to accumulate up to 1.3% of the total leaf protein. However, nothing was reported on correct assembly and the *in situ* localization of assembled antibody molecules.

Alternatively, Düring et al. (1990) constructed a vector containing both heavy and light chain genes and the barley aleurone α-amylase leader peptide coding sequence under control of the pT_R and pNOS promoters. The barley aleurone α-amylase leader peptide is thought to direct proteins to the intercellular spaces. The *in situ* localization and correct assembly of functional antibodies was studied using anti-idiotype antibodies. Assembled antibodies were detected in the endoplasmic reticulum (ER) and, surprisingly, also inside chloroplasts. Nothing was reported on the presence of antibodies in the intercellular spaces.

Recently, scFv (Owen et al., 1992) and Fab fragments (De Neve et al., 1993) have been successfully expressed. The scFv described by Owen et al. (1992) was directed against phytochrome which was inhibited intracellularly. Seeds from transgenic progeny displayed aberrant phytochrome-dependent germination. The scFv is likely active in the cytoplasm as the chimeric gene contained no leader peptide. This might also explain the low expression levels ($\leq$ 0.1% of total protein) obtained. Fab-fragments were expressed up to 1.3% of the total protein in *Arabidopsis* leaves (De Neve et al., 1993). However, the same fragment accumulated only to around 0.05% in *Nicotiana* leaves. No satisfactory explanation can be given, although this feature may be related to the plant species.

To endow plants with new properties it is essential that several aspects concerning antibody expression in plants need optimization. Striking differences are observed in the amount of accumulated antibody. Hiatt et al. (1989), using *Nicotiana*, and De Neve et al (1993), using *Arabidopsis*, report expression levels up to 1.3% of total leaf protein. Much lower amounts (< 0.1%) were found by Düring et al. (1990) and again De Neve et al. (1993), both using tobacco. This discrepancy could be the result of different promoters (35S versus pT_R and pNOS) used possibly in combination with different leader peptides. Both aspects as well as the differences in the amount of protein expressed in different plant species have to be studied in more detail. Furthermore, De Neve et al. (1993) report aberrant assembly or degradation patterns of expressed whole IgG molecules. In the range between 100 kDa and 150 kDa a series of bands is observed on an SDS-PAGE gel under non-reducing conditions in both *Nicotiana* and *Arabidopsis*. In addition some bands with a molecular mass between 40 kDa and 65 kDa are observed in *Nicotiana*. Immunoblotting experiments revealed that all bands contain parts of the immunoglobulin molecule.

PROSPECTS

The hybridoma technology has solved many of the problems associated with polyclonal antibodies, such as standardization and specificity. However, certain questions remain, especially in those cases where the purification of the antigen is difficult and the impurities have an immunodominant character. These problems can be solved using antibody phage display libraries. A proper antibody phage display library consists of up to 10^8 different antibody molecules. The immune repertoire displayed with the hybridoma technology is limited to a maximum of a few thousand V-gene combinations i.e., the number of different hybridomas obtained. Rare specificities are only occasionally observed. This is in contrast to what is observed using phage display libraries. Moreover, the random association of V_H and V_L genes may lead to an even larger number of combinations than observed in nature, especially, when the *in vivo* recombination techniques (Waterhouse et al., 1993) come into practice for the construction of libraries.

One of the most promising spin-offs of antibody technology is the expression of antibodies in plants. Plantibodies are suitable to obtain resistance against pathogens. For instance resistance against viruses can be foreseen in analogy to the work described by Marasco et al. (1993). They found that intracellular expression, in eukaryotic cells, of an scFv derived from a monoclonal antibody recognizing the CD4 binding region of the human immunodeficiency virus (HIV) led to reduced viral infectivity. Furthermore, plants can be endowed with new properties in analogy with the recently described *in vivo* catalysis of a metabolic reaction by an antibody (Tang et al, 1991). A cytoplasmically expressed Fab fragment derived from a catalytic monoclonal antibody that displays chorismate mutase activity was shown to confer a growth advantage, under auxotrophic conditions, on a chorismate-mutase deficient mutant of yeast.

Antibody technology has an enormous potential and progress is being made rapidly. Practical applications in the medical field and biological sciences are likely to emerge soon. Developments which will lead to *in vivo* applications, e.g. in cancer therapy to the construction of new, improved, immunoassays and to better heterologous expression systems are likely to emerge. These findings will be beneficial to many aspects of agricultural research and practice.

ACKNOWLEDGEMENT

This paper is a revised version of a chapter appearing in *New Diagnostics in Crop Sciences*, edited by J.H. Skerrit and R. Appels (CAB International, 1994) reproduced with permission.

REFERENCES

Atkinson, H.J., Harris, P.D., Halk, E.J., Novitski, C., Leighton-Sands, J., and Fox, P.C., 1988, Monoclonal antibodies to the soya bean cyst nematode, *Heterodera glycines*, *Ann. Appl. Biol.* 112:459.

Barbas III, C.F., Bain, J.D., Hoekstra, D.M., and Lerner, R.A., 1992, Semisynthetic combinatorial antibody libraries: A chemical solution to the diversity problem, *Proc. Natl. Acad. Sci. USA* 89:4457.

Batra, J.K., Chaudhary, V.K., Fitzgerald, D., and Pastan, I., 1990, TGF-alpha-anti-Tac(Fv)-PE40: A bifunctional toxin cytotoxic for cells with EGF or IL2 receptors, *Biochem. Biophys. Res. Comm.* 171:1.

Berzofsky, J.A., Hicks, G., Fedorko, J., and Minna, J., 1980, Properties of monoclonal antibodies specific for determinants on a protein antigen, myoglobin, *J. Biol. Chem.* 255:11188.

Better, M., Chang, C.P., Robinson, R.R., and Horwitz, A.H., 1988, *Escherichia coli* secretion of an active chimeric antibody fragment, *Science* 240:1041.

Chang, C.N., Landolfi, N.F., and Queen, C., 1991, Expression of antibody Fab domains on bacteriophage surfaces. Potential use of antibody selection, *J. Immunol.* 147:3610.

Chaudhary, V.K., Gallo, M.G., Fitzgerald, D.J., and Pastan, I., 1990, A recombinant single-chain immunotoxin composed of anti-Tac variable regions and a truncated diphtheria toxin, *Proc. Natl. Acad. Sci. USA* 87:9491.

Clackson, T., Hoogenboom, H.R., Griffiths, A.D., and Winter, G., 1991, Making antibody fragments using phage display libraries, *Nature* 352:624.

De Neve, M., De Loose, M., Jacons, A., Van Houdt, H., Kaluza, B., Weidle, U., Van Montagu, M., and Depicker, A., 1993, Assembly of an antibody and its derived antibody fragment in *Nicotiana* and *Arabidopsis*, *Transgenic Res.* 2:227.

Düring, K., 1988, Wundinduzierbare Expression und Sekretion von T4 Lysozym und monoklonalen Antikörpern in *Nicotiana tabacum,* PhD thesis, University of Köln.

Düring, K., Hippe, S., Kreuzaler, F., and Schell, J., 1990, Synthesis and self-assembly of a functional monoclonal antibody in transgenic Nicotiana tabacum, *Plant Mol. Biol.* 15:281.

Gram, H., Marconi, L.-A., Barbas III, C.F., Collet, T.A., Lerner, R.A., and Kang, A.S., 1992, *In vitro* selection and affinity maturation of antibodies from a naive combinatorial immunoglobulin library, *Proc. Natl. Acad. Sci. USA* 89:3576.

Hawkins, R.E., Russell, S.J., and Winter, G., 1992, Selection of phage antibodies by binding affinity. Mimicking affinity maturation, *J. Mol. Biol.* 226:889.

Hermes, J.D., Parekh, S.M., Blacklow, S.C., Pullen, J.K., and Pease, L.R., 1989, A reliable method for random mutagenesis: the generation of mutant libraries using spiked oligodeoxyribonucleotide primers, *Gene* 84:143.

Hiatt, A., Cafferkey, R., and Bowdish, K., 1989, Production of antibodies in transgenic plants, *Nature* 342:76.

Horton, R.M. , Hunt, H.D., Ho, S.N., Pullen, J.K., and Pease, L.R., 1989, Engineering hybrid genes without the use of restriction enzymes: gene splicing by overlap extension, *Gene* 77:61.

Hollander, Z., and Katchalski-Katzir, E., 1986, Use of monoclonal antibodies to detect conformational alterations in lactate dehydrogenase isoenzyme 5 on heat denaturation and on adsorption to polystyrene plates, *Mol. Immunol.* 23:927.

Hoogenboom, H.R., Marks, J.D., Griffiths, A.D., and Winter, G., 1992, Building antibodies from their genes, *Immunol. Rev.* 130:41.

Hoogenboom, H.R., and Winter, G., 1992, Bypassing hybridomas: human antibodies from synthetic repertoires of germ-line VH-gene segments rearranged *in vitro*, *J. Mol. Biol.* 227:381.

Huse, W.D., Sastry, L., Iverson, S.A., Kang, A.S., Alting, M.M., Burton, D.R., Benkovic, S.J. and Lerner, R.A., 1989, Generation of a large combinatorial library of the immunoglobulin in phage lambda, *Science* 246:1275.

Hussey, R.S., 1989, Monoclonal antibodies to secretory granules in esophageal glands of *Meloidogyne* species, *J. Nematol.* 21:392.

Jones, P.T., Dear, P.H., Foote, J., Neuberger, M.S., and Winter, G., 1986, Replacing the complementarity-determining regions in a human antibody with those from a mouse, *Nature* 321:522.

Jones, P, Ambler, D.J., and Robinson, M.P., 1988, The application of monoclonal antibodies to the diagnosis of plant pathogens and pests, *in*: Brighton Crop Protection Conference - Pest and Diseases - 1988, vol. 2. BCPC Registered Office, Thornthon Heath, Surrey, UK.
Kipps, T.J., 1985, Switching the isotype of monoclonal antibodies, *in*: "Hybridoma technology in the biosciences and medicine". Springer T.A., ed., New York: Plenum Press.
Knappik, A., Krebber, C., and Plückthun, A., 1993, The effect of folding catalysts on the *in vivo* folding process of different antibody fragments expressed in *Escherichia coli*, *Bio/Technology* 11:77.
Köhler, G., and Milstein, C., 1975, Continuous cultures of fused cells secreting antibody of predefined specificity, *Nature* 256:495.
Marasco, W.A., Haseltine, W.A., and Chen, S.Y. 1993 Design, intracellular expression , and activity of a human anti-human immunodeficiency virus type-1 gp120 single chain antibody, *Proc. Natl. Acad. Sci. USA* 90:7889.
Marks, J.D., Griffiths, A.D., Malmqvist, M., Clackson, T., Bye, J.M., and Winter, G., 1992, By-passing immunization: building high affinity human antibodies by chain shuffling, *Bio/Technology* 10:779.
Marks, J.D., Hoogenboom, H.R., Bonnert, T.P., McCafferty, J., Griffiths, A.D., and Winter, G., 1991, By-passing immunization; Human antibodies from V-gene libraries displayed on phage, *J. Mol. Biol.* 222:581.
McCafferty, J., Griffiths, A.D., Winter, G., and Chiswell, D.J., 1990, Phage antibodies: filamentous phage displaying antibody variable domains, *Nature* 348:552.
Metzger, D.W., Ch'ng, L. -K., Miller, A., and Sercarz, E.E., 1984, The expressed lysozyme specific B-cell repertoire. I. Heterogeneity in the monoclonal anti-hen egg white lysozyme specificity repertoire, and its difference from the *in situ* repertoire, *Eur. J. Immunol.* 14:87.
Milstein, C., and Cuello, A.C., 1984, Hybrid hybridomas and the production of bi-specific monoclonal antibodies, *Immunol. Today* 5:299.
Orlandi, R., Gussow, D.H., Jones, P.T., and Winter, G., 1989, Cloning immunoglobulin variable domains for expression by the polymerase chain reaction, *Proc. Natl. Acad. Sci. USA* 86:3833.
Owen, M., Gandecha, A., Cockburn, B., and Whitelam, G. 1992, Synthesis of a functional anti-phytochrome single-chain F_v protein in transgenic tobacco, *Bio/Technology* 10:790.
Schots, A., Hermsen, T., Schouten S., Gommers, F.J., and Egberts, E., 1989, Serological differentiation of the potato-cyst nematodes *Globodera pallida* and *G. rostochiensis*. II. Preparation and characterization of species specific monoclonal antibodies, *Hybridoma* 8:401.
Schots, A., De Boer, J., Schouten, A., Roosien, J., Zilverentant, J.F., Pomp, H., Bouwman-Smits, L., Overmars, H., Gommers, F.J., Visser, B., Stiekema, W.J. and Bakker, J., 1992, 'Plantibodies': a flexible approach to endow plants with new properties, *Neth. J. Pl. Pathol.* 98, supplement 2:183.
Skerra, A., and Plückthun, A., 1988, Assembly of a functional immunoglobulin Fv fragment in *Escherichia coli*, *Science* 240:1038.
Söderlind, E., Simonsson Lagerkvist, A.C., Dueñas, M., Malmborg, A.-C., Ayala, M., Danielsson, L., and Borrebaeck, C.A.K., 1993, Chaperonin assisted phage display of antibody fragments on filamentous bacteriophages, *Bio/Technology* 11:503.
Tang, Y., Hicks, J.B., and Hilvert, D. 1991 *In vivo* catalysis of a metabolically essential reaction by an antibody, *Proc. Natl. Acad. Sci. USA* 82:1074.
Winter, G. and Milstein, C., 1991, Man-made antibodies, *Nature* 349:293.
Waterhouse, P., Griffiths, A.D., Johnson, K.S., and Winter, G., 1993, Combinatorial infection and *in vivo* recombination: a strategy for making large phage antibody repertoires, *Nucl. Ac. Res.* 21:2265.
Wu, X.-C., Ng, S.-C., Near, R.I., and Wong, S.-L., 1993, Efficient production of a functional single-chain antibody via an engineered *Bacillus subtilis* expression-secretion system, *Bio/technology* 11:71.

DISCUSSION

BIRD asked if any of the anti-nematode Abs cross-react with the host. SCHOTS answered no. BAZZICALUPO noted that the immune response (Abs plus cell mediated immunity) fails to clear parasites of animals. SCHOTS pointed out that this is not always the case. For example, expression of gut Abs by sheep can clear parasitic nematodes, and expression of anti-tick Abs in cattle is an effective control measure. ATKINSON wondered whether it is worth doing a bio-assay to test antibodies prior to making the transgenic plants. SCHOTS answered that making a transgenic plant is the only bio-assay possible. JONES asked what nematode molecules should be targeted to make Abs. SCHOTS suggested that stylet exudate proteins would be one appropriate target. JONES also asked what promoter would be good to drive plantibody expression.

SCHOTS reported that they have tried the 35S promoter, but that a feeding site-specific promoter might be better. HUSSEY commented that the 35S promoter is down regulated in giant cells, but OPPERMAN said that he has not found this to be the case. HUSSEY asked about the location of the expressed plantibody. SCHOTS said that the normal murine secretion signals were used, and that expression could be up to 2% of the total plant protein. He further noted that if a complete antibody is expressed, artifactual Fab(2) fragments tend to form. If murine secretion signals are used it has to be expected that the antibodies are secreted in the intercellular spaces. WILLIAMSON wondered that, in light of comments made by BAKKER about the high degree of divergence between *Globodera* spp., why has it been so difficult to find Abs that distinguish them. SCHOTS answered that although a high degree of divergence exists the majority of the proteins are homologues and are therefore almost identical. BAILLIE, BIRD and JONES were concerned that intracellular expression of Abs might be cyto-toxic. SCHOTS answered that this has indeed sometimes been reported for bacteria. For eukariotic organisms, i.e. mammalian cells, yeast and plants, cyto-toxicity was not observed.

REPETITIVE DNA IN PLANT-PARASITIC NEMATODES: USE FOR INTERSPECIFIC AND INTRASPECIFIC IDENTIFICATION

Pierre Abad

I.N.R.A.
Laboratoire de Biologie des Invertébrés
B.P. 2078
06606 Antibes Cedex, France

INTRODUCTION

Plant parasitic nematodes are serious damaging pests in many economically important crops, affecting both yield and quality of harvests. *Meloidogyne*, *Globodera* and *Heterodera* genera have the most cosmopolitan distribution and destructive effects on virtually all crop plants (Hyman and Powers, 1991). Accurate and reliable identification of plant parasitic nematodes is necessary for many aspects of their control and management. Therefore, attempts have been made to distinguish between nematode species and pathotypes. However, difficulties have arisen with respect to the precise identity of some isolates of these nematodes and thus, although nematode identification is extremely important, it is nevertheless a difficult task.

In some cases, the use of morphological characters is insufficient for identification because definitive characteristics sometimes overlap. This unreliable identification often leads to confusion between particularly damaging nematodes and other relatively harmless but morphologically similar species.

To overcome this problem of identification, biochemical techniques have been applied and proteins, carbohydrates and lipids have all been used with varying degrees of success to characterize nematode species, host races and pathotypes (Hussey, 1979). Biochemical approaches mostly include the separation of proteins (soluble proteins and allozymes) by one-dimensional polyacrylamide gel electrophoresis (1D-PAGE), protein fractionation using isoelectric focusing (IEF), and electrophoretic resolution in two dimensions using IEF followed by PAGE (2D-gel electrophoresis). However, proteins, like morphological characters, tend to be highly conserved between closely related taxa and their expression is subject to environmental and developmental factors. Thus, it may be difficult to draw conclusions.

Similar problems of phenotypic variation are encountered using serological techniques, with the additional difficulty that cross-reactivity may occur between different antibodies and antigens. The use of monoclonal antibodies is potentially a more powerful immunological tool in that it overcomes or minimizes the above problems and therefore this technique shows much promise.

However, one of the major drawbacks of the electrophoretic and serological techniques is that they analyse variations of the protein fraction. Such compounds represent the products of genetic expression and constitute only small fraction of the nematode genomic variation. Furthermore, the "non-coding" regions of the genome are not accessed by such techniques. Direct analysis of the nematode genome avoids this problem. Nucleic acid analysis should provide the ultimate resolution in identification since analysis of genomic variations of both "coding" and "non-coding" sequences of the genome are possible.

Advances in Molecular Plant Nematology
Edited by F. Lamberti *et al.*, Plenum Press, New York, 1994

GENOME ORGANIZATION

The eukaryotic genome consists of both unique and repeated sequences. The repetitive component is a complex class made up of repeated genes, transposable elements and diverse, short repeated sequences whose origin and function are unknown. Repeated sequences and unique sequences are intermingled throughout the DNA. Analysis of the nematode genome provides insights into the presence of the large amount of apparently non-coding DNA, the relationship between the DNA sequence and the karyotype, the origin and the role of short, interspersed repetitive DNA and transposable elements, and the mechanisms of genomic transmission and evolution. Molecular studies of nematode genomes have included a general characterization of the properties of the repetitive components. This has generally been done by reassociation kinetics where two or three genomic components can be analyzed. The quickly and rather quickly reannealing components represent highly and moderately repetitive sequences respectively, whereas the slowly reannealing components represent unique sequences in the haploid genome. The proportion of the repetitive component can fluctuate from one nematode species to another. In *P. equorum*, 85% of the genomic components are satellite DNA sequences while unique and middle repetitive DNA sequences represent only 15% of the genome (Moritz and Roth, 1976). Conversely in *C. elegans*, repetitive DNA sequences represent only 17% of the genomic component and unique sequences represent 83% (Sulston and Brenner, 1974). In *Meloidogyne*, the repetitive DNA fraction represents 20% of the genome (Paeblo and Triantaphyllou, 1989).

In this review, I will focus on the use of repetitive sequences as molecular tools for plant-parasitic nematode identification at both inter- and intraspecific levels. I shall mainly consider the work done on nematode genera of importance to worldwide agriculture such as *Meloidogyne*, *Globodera*, *Heterodera*, but I shall also consider *Bursaphelenchus* a nematode currently under study in our laboratory in Antibes and which causes one of the major conifer diseases in the world, particularly in Japan.

NATURE OF REPETITIVE DNA SEQUENCES

Studies of a large number of organisms have revealed first, that several classes of repetitive DNA exist in eukaryotic genomes and second, that this repetitive DNA can be divided into two classes according to its genomic distribution and its reiteration in the genome.

Concerning the genomic distribution, the first class is made up of long and short sequences dispersed throughout the genome as unlinked copies (Jelinek and Schmid, 1982). These sequences consist of functional genes or gene-related sequences; the latter are generally the most abundant and not transcribed. There are approximately 70 genes for 18S and 28S, and 110 genes for 5S RNA in the haploid genome of *C. elegans* (Sulston and Brenner, 1974; Files and Hirsh, 1981). In *Meloidogyne*, the same copy number has been found (Vahidi et al., 1991; Piotte, 1993). Furthermore, many protein-coding genes are members of dispersed repetitive gene families such as collagen, histone, myosin heavy chain, major sperm protein, actin, and so forth. In addition, there are some functional multiple copies of several transposable elements. The second class consists of relatively short, repeated units arranged in large blocks of tandemly reiterated arrays referred to as satellite DNA (Skinner, 1977; Beridze, 1986). These repetitive DNA sequences can vary in copy number from 10 to over 10^5 per haploid genome, and in proportion from less than 1% to more than 66% of the genome (Skinner, 1977).

Concerning the reiteration copies of repetitive sequences, it is generally assumed that moderate repetitive DNA is composed of gene families with multiple copies from two to some hundreds e.g., ribosomal genes, histone genes, collagen genes and some transposable element families. The highly repetitive DNA is composed of sequence families with repetition frequency higher than 1, 000 copies e.g., satellite DNA and some transposable element families.

In the eukaryotic genome, the fastest reannealing portion of the repetitive component in the reassociation kinetics consists of both highly reiterated sequences arranged in tandem array (Beridze, 1986) and sequences arranged as inverted repeats which reanneal in an intramolecular "fold-back" reaction (Sulston and Brenner, 1974). *C. elegans* seems lacking in satellite DNA, but contains up to 2400 regions with 300bp-long fold-back structures. Such short, interspersed repetitive DNA makes up 10% of the genome (Emmons et al., 1980). Conversely, the genome of *P. equorum* contains up to 85% of satellite DNA which is largely eliminated during the process of chromatin diminution (Boveri, 1887). This phenomenon is also found in *A. lumbricoides* and in a number of other nematode species (Tobler, 1986). Molecular analysis carried out in *P. equorum*, *A. lumbricoides* and *A. suum* indicated that the bulk of the DNA lost from the somatic cell consists of highly reiterated satellite sequences (Tobler, 1986) but transposable elements and ribosomal copies are also lost(Aeby et al., 1986; Etter et al., 1991).

However, since most of the repetitive sequence members are not transcribed, they are not under selection pressure and thus nucleotide sequence variations accumulate much more rapidly than unique transcribed sequences in evolutionary terms. Futhermore, repetitive DNA sequences are largely more sensitive as they will have proportionally more target sequences per genome than those directed at DNA sequences present in low copy number. Therefore, the abundance and the variability of repetitive DNA can be used to demonstrate sequence variations in closely related taxa such as species and populations.

RESTRICTION FRAGMENT LENGTH POLYMORPHISMS

Digestion of genomic DNA with restriction enzymes generates a unique set of different sized DNA restriction fragments dependent upon the base sequence of the genome. Nucleotide substitutions, insertions or deletions that create or destroy restriction sites modify the restriction profile and therefore generate restriction fragment length polymorphism (RFLPs).

The first approach is the examination of the size distribution of restriction fragments in ethidium bromide-stained agarose gels. The observed pattern results in a homogeneous smear that constitutes the background where distinct bands representing the multiple copies of repetitive DNA sequences (e.g. ribosomal genes, histone genes, transposable elements, satellite DNA) are present and RFLPs between such bands can be used as diagnostic characters. RFLPs have been used successfully to separate the two closely related potato cyst nematodes (PCN), *G. rostochiensis* and *G. pallida* (Burrows and Boffey, 1986; Dejong et al., 1989). The observation of the large majority of repetitive DNA bands is in striking contrast to the morphological similarity between *G. rostochiensis* and *G. pallida* . These large differences suggest that the two species are considerably more distinct than is evident from their conserved morphology. These data corroborate the large differences obtained with two-dimensional gel electrophoresis of proteins (Bakker and Bouwman-Smits, 1988). Other plant-parasitic nematodes have been distinguished in this way, e.g. *Meloidogyne* (Curran et al., 1985; Curran et al., 1986); *Heterodera glycines* (Kalinski and Huettel, 1988).

In the second approach, the digested genomic DNA is transferred onto a nitrocellulose filter (Southern, 1975) which is then subsequently hybridized with radiolabeled total genomic DNA. Using this approach, Bolla et al. (1988) have shown great genomic differences in repetitive DNA sequences between two closely related *Bursaphelenchus* species; *B. xylophilus*, and *B. mucronatus* . Furthermore, minute genomic differences are also apparent between two *B. xylophilus* pathotypes.

If the Southern blot technique is used, sequences with a moderate copy number can be detected. This involves hybridization of radiolabeled cloned DNA probes to genomic DNA and thus a wide variety of cloned probes has been used to detect RFLPs. In the first step, evolutionarily conserved sequences mainly from *C. elegans* have been used to identify plant-parasitic nematodes. One of the best representative examples is the rDNA cluster probe from *C. elegans* which reveals not only interspecific variation, but also intraspecific variation in plant-parasitic nematodes. This probe is able to differentiate race A and race B of *M. hapla* (Curran and Webster, 1987). In the second step, conserved sequences mainly from *C. elegans* have been used to clone homologous sequences in plant-parasitic nematodes. From partial rDNA *C. elegans* probes, Webster et al. (1990) have cloned a fragment containing the 18S and 28S fractions from *B. xylophilus* and *B. mucronatus*. RFLPs obtained with these two *Bursaphelenchus* probes demonstrate genotypic differences between isolates and clearly separate infraspecific groups of European, Asian, and North American origin (Webster et al., 1990).

DNA probes used to detect RFLPs do not have to be from known repetitive sequences. The only required property for these unknown repetitive sequences is that they hybridize with restriction genomic digest in order to produce distinct banding patterns. Due to the repetitive nature of the probes, the autoradiograms exhibit extensive RFLPs both between and within plant-parasitic nematode species.

We have chosen this approach to aim of analyze the relationships in both *Bursaphelenchus* and *Meloidogyne* genera. As a first step, partial genomic libraries were carried out and hybridized with total genomic DNA probes. Among clones of variable repetitivity (defined by the hybridization signal intensity), those with middle intensity, which putatively represented moderately repeated sequences in the genome, were isolated at random and were amplified in order to probe genomic DNA of different isolates from *Bursaphelenchus* and *Meloidogyne*. Tarès et al. (1992) have identified the precise relationships between isolates belonging to the pinewood nematodes. With these probes, the existence of a *B. xylophilus* group and a *B. mucronatus* group within these pinewood nematodes is confirmed as is the existence of three geographical subgroups in the *B. xylophilus* species: the American, Canadian and Japanese groups with a close relationship between the USA and Japanese isolates. This supports the hypothesis that the *B. xylophilus* isolate, which was recently introduced into Japan, originated from the USA. Furthermore, Tarès et al. (1992) have clarified the taxonomic status of some atypic isolates of pinewood nematodes. A similar approach has been used to analyze the phylogenetic relationships between amphimictic and parthenogenetic nematodes of the genus *Meloidogyne* (Castagnone-Sereno et al., 1991; Piotte et al., 1992; Castagnone-Sereno et al., 1993). These studies are very interesting since the species concept for parthenogenetic organisms cannot be based upon the reproductive criterion used for amphimictic forms. Using anonymous moderately repetitive sequences (Fig. 1), Castagnone-Sereno et al. (1993) described the evolutionary relationships between populations and species of *Meloidogyne* and a dendrogram has been established (Fig. 2). This clearly shows the early split of *M.hapla* and

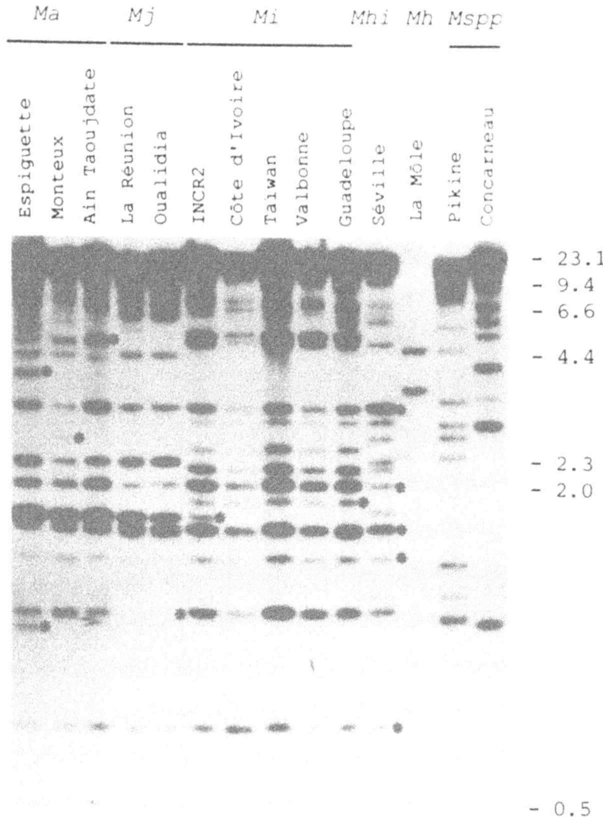

Figure 1. Hybridization of Southern blot of the *Bam*HI-digested DNAs of 14 *Meloidogyne* populations with an anonymous moderately repetitive sequence (pMiK13). Molecular weight markers are indicated in kilobases. *Meloidogyne* species are reported as : Ma, *M. arenaria* ; Mj, *M. javanica* ; Mi, *M. incognita* ; Mhi, *M. hispanica* ; Mh, *M. hapla* and Mspp undescribed species (from Piotte et al., 1992).

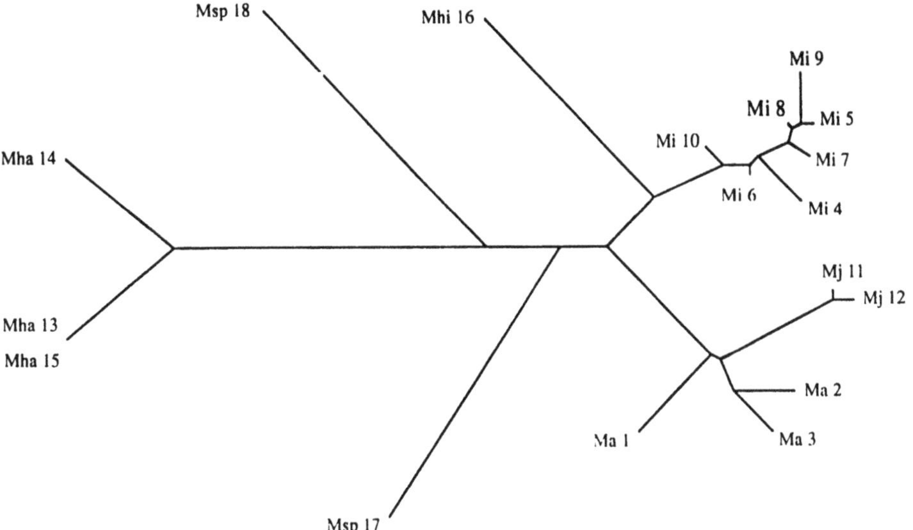

Figure 2. Phylogenetic tree for 18 *Meloidogyne* populations inferred from FITCH analysis performed on a distance matrix. The distance matrix was defined according to the analysis of RFLPs data obtained after hybrization of the Bam HI-digested DNAs of 18 *Meloidogyne* populations with three anonymous moderately repetitive sequences (4.3, 13.15 and 18.10). Similarity coefficients were calculated for all pairwise combinations of RFLP groups using the formula: $F = 2Nxy/(Nx+Ny)$, where Nx = number of bands in genotype x, Ny = number of bands in genotype y and Nxy = number of bands shared by genotypes x and y. The genetic distance was deduced from F values according to the formula: $d = 1-F$. This topology of tree was chosen among the 536 ones examined by the program. *Meloidogyne* species are reported as: Ma, *M. arenaria*; Mj, *M. javanica*; Mi, *M. incognita*; Mhi, *M. hispanica*; Mh, *M. hapla* and Mspp undescribed species. Numbers are for population codes (from Castagnone-Sereno et al., 1993).

the mitotically parthogenetic species cluster (*M. incognita, M. arenaria, M. javanica, M. hispanica*), and confirms the amphimictic ancestral mode of reproduction of root-knot nematodes. Moreover in this cluster, the existence of larger polymorphisms between species than within species confirms that each of them has to be considered as a true species, as was previously demonstrated by electrophoretic analysis of proteins and isoenzymes (Dalmasso and Bergé, 1983). In addition, Castagnone-Sereno et al. (1993) found intraspecific polymorphism, especially among *M. incognita* and *M. arenaria* isolates. This result is very interesting since intraspecific variation at soluble protein and enzymatic levels is generally very low in *Meloidogyne* species (Dalmasso and Bergé, 1983). The homogeneity in this genus was recently confirmed by RAPD-PCR analysis which showed no differences between populations of *M. incognita* and *M. arenaria* (Cenis, 1993). The intraspecific polymorphism found in this study may be correlated with the repetitive nature of the probe, taking into account their potential variability which is higher for non-coding regions than for expressed genes.

SPECIES-SPECIFIC SEQUENCES

There are two approaches to the use of DNA probes in species identification. One of these approaches has been described above. It results in probes that hybridize with the DNA of various species of plant-parasitic nematodes with generations of RFLPs. The alternative approach is to generate probes that are specific for DNA of a given species and which will not hybridize to DNA from other species. In such cases, hybridization of plant-parasitic nematode DNA will identify the organism for which the probe is specific, while ruling out other species.

Species-specific probes are quite important tools since they can be used in positive or negative assays for nematode species, races or pathotypes. As for RFLPs, species-specific DNA does not have to be from known repetitive sequences.

Unknown Species-Specific DNA

Burrows and Perry (1988) have used an unknown species-specific DNA approach to differentiate populations of *G. pallida* and *G. rostochiensis* on dot-blot assays. In this technique, the DNA uncut by endonucleases from the tested plant-parasitic nematodes was loaded on the nitrocellulose filter as a dot and was subsequently hybridized with the radiolabeled probe. Using genomic libraries constructed from the plant-parasitic nematode of interest (*G. pallida*), they screened relatively small libraries (< 500 clones) and yielded species-specific fragments. In dot-blot assay coupled with species-specific DNA probes, they isolated two DNA fragments that differentiate *G. pallida* from *G. rostochiensis* (Burrows and Perry, 1988). As shown in Fig. 3, these two probes hybridized with all *G. pallida* populations tested but not with *G. rostochiensis*. These probes readily detected 80 ng of purified potato cyst nematode DNA, which represented less DNA than from a second stage juvenile (Burrows and Perry, 1988). However, radiolabeled probes are dangerous, expensive and unstable since specific radioactivity decreases within a few weeks, and they need frequent relabeling. This is particularly true for ^{32}P labeled probes. In order to prevent the disavantages of the radiolabeled probes, Burrows (1988) successfully used a biotinyl-labeled probe to identify *G. pallida*. In spite of its lower sensitivity, this procedure is a very quick and safe means of detection which can be used as a simple diagnostic kit for routine identification.

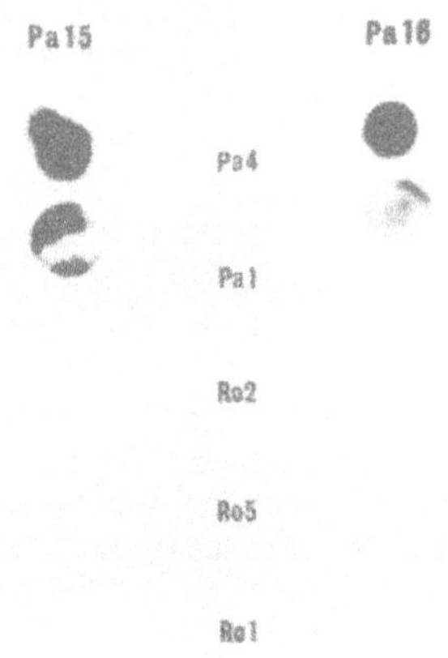

Figure 3. Dot blot autoradiograph showing hybridization of radio labeled chimeric plasmids from Pa 15 and Pa 16 to *G. pallida* DNA Pa 4 and Pa 1) but not to three populations of *G. rostochiensis* DNA (Ro 1, Ro 5 and Ro 1) (from Burrows and Perry, 1988).

In the same way, by differential screening of nematode DNA, Stratford et al. (1992) developed DNA probes derived from highly abundant sequences that are diagnostic for either *G. rostochiensis* and *G. pallida*. These clones provide sensitive and reliable probes for direct identification of individual potato cyst nematode using a simple dot blot procedure. In addition, Stratford et al. (1992) examined the genomic organization of these clones and showed that some of them exist in long tandem arrays of 200 copies or more.

Ribosomal DNA

Ribosomal DNA genes have been useful in classification at all taxonomic levels. They are arranged in tandem arrays structure where NTS (non transcribed spacers) link the 18S and 28S coding regions. Since NTS are non transcribed they are not sensitive to the selection pressure and thus accumulate sequence variations much more rapidly than transcribed sequences. Therefore non-coding spacers may provide information on specific and subspecific grouping. Webster et al. (1990) have used rDNA to differentiate populations of the pine wood nematode species complex. By using restriction and gene maps, they identified ribosomal repeats for both *B. xylophilus* and *B. mucronatus*, and subsequently by the *Exo*III deletion technique, they identified the restriction fragment containing the NTS region. The NTS DNA probe from *B. xylophilus* reacted positively to *B. xylophilus* isolates and negatively to *B. mucronatus* isolates. Conversely, NTS DNA probe from *B. mucronatus* was positive with *B. mucronatus*

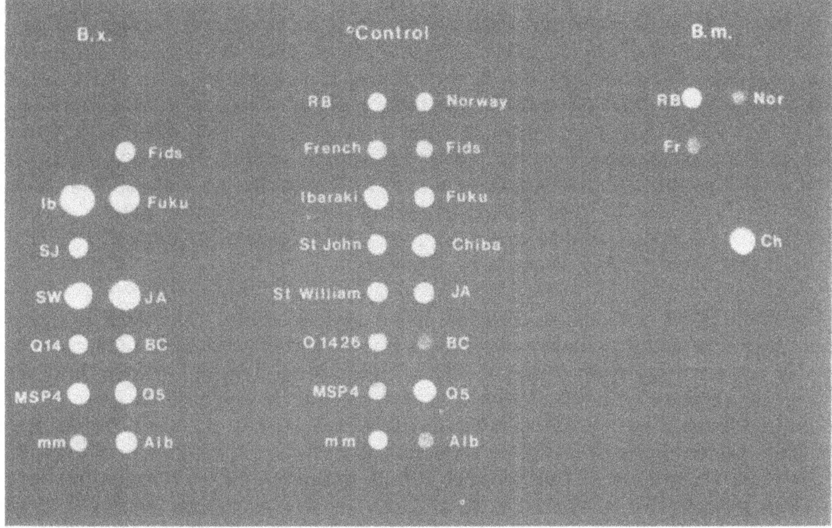

Figure 4. DNA from 16 isolates was spotted in triplicate onto nitrocellulose filter to be tested against NTS probes from *Bursaphelenchus xylophilus* (MSP-4) and *B. mucronatus* (RB), and a control probe (pBm3). The isolates are: RB (Japan); French; Ibaraki (Japan); St John (Canada); St William (Canada); Q1426 (Canada); MSP-4 (USA); mm (Canada); Norway; Fids (Canada); Fukushima (Japan); Chiba (USA); BxJA (USA); British Columbia (Canada); Q52A (Canada); Alb (Canada) (from Webster et al., 1990).

isolates and negative with *B. xylophilus* isolates. These two probes enabled the unequivocal segregation of sixteen isolates into two groups, either the *B. xylophilus* group or the *B. mucronatus* group by reciprocal dot-blot tests (Fig. 4). Furthermore, a close examination of the intensity of each isolate indicated that the *B. mucronatus* group comprises at least two subgroups. The position of the French specimen is clarified and is shown to have a much closer affinity to *B. mucronatus* than to *B. xylophilus*. As well, the French specimens appear to have a closer affinity with those from Norway than they do with the Japanese isolates. However, the degree of sensitivity of the dot-blot technique as used in this study is insufficient to provide definitive answers about subspecific relationships of a wide range of isolates.

Satellite DNA

The satellite DNA is localized mainly in the constitutive heterochromatin in an apparently genetically inert and compact part of chromosomes at both centromeric and telomeric regions of the chromosomes (John and Miklos, 1979; Brutlag, 1980; Miklos, 1985). This class of repetitive DNA can vary in copy number from 10 to over 10^5 per haploid genome. These copies are arranged in tandem arrays and can constitute a large proportion of the genomic sequences (Skinner, 1977). Satellite DNA has been found in the genome of almost all metazoan species. It is widely represented in plant genomes (Pruitt and Meyerowitz, 1986). Among vertebrates, it is represented from fishes (Datta et al., 1988) and amphibians (Vignali et al., 1991) to mammals (Brutlag, 1980; Novak, 1984; Fowler et al., 1989). In the invertebrates, available information is mainly from crustaceans (Skinner, 1977; Fowler and Skinner, 1985), echinoderms (Sainz et al., 1989), molluscs (Ruiz-Lara et al., 1992) and some insects (Brutlag and Peacock, 1979; Davis and Wyatt, 1989). From nematodes, information is scarce and concerns *A. lumbricoides* (Tobler et al., 1972; Moritz and Roth, 1976; Müller et al., 1982) and free-living nematodes such as *C. elegans* (Uitterlinden et al., 1989) and *Panagrellus redivivus* (de Chastonay et al., 1990). Despite extensive attemps to assign a role for these sequences in genome structure, no function has yet been established for satellite DNA although some hypotheses have been advanced which are largely devoid of experimental support (Bostock, 1980). In fact, the wide sequence variation amongst even related species and the lack of transcription suggest that any function they may have does not depend on retention of a particular sequence. Nevertheless, due to their ability to exhibit variation within and between subspecies, they offer a good prospect for population or individual identification (Forejt, 1973; MacKay et al., 1978) and they could provide a valuable tool for disclosing evolutionary relationships between species (Beridze, 1986). Generally, satellite DNA sequences have been described as species specific (Bachmann et al., 1989). Therefore, we isolated and characterized the satellite DNA fraction in from *Bursaphelenchus* and *Meloidogyne* genera in order to test its usefulness for taxon discrimination at both inter- and intraspecific levels (Tarès et al., 1993; Tarès et al., 1994; Piotte, 1993; Piotte et al., 1994).

Satellite DNA Structure. Satellite DNA can be easily identified in agarose gels following digestion with restriction enzymes, some of which generate a monomer unit. If all repeats in a tandem array were identical, DNA digestion to completion with these restriction enzymes would generate in Southern blot a very simple banding pattern which would include a major band constitued by the monomer unit. However, during the evolution of a genome, satellite DNA turnover has occurred by sequence amplification, unequal crossing-over, point mutation and gene conversion. Furthermore, a high frequency of point mutations is acting on satellite DNA blocks causing creation or destruction of some restriction enzyme sites which are visualized in Southern blot by a typical ladder hybridization. In this ladder pattern, regular periodic spaces between the oligomeric bands are observed (Fig. 5).

Satellite DNA in *Bursaphelenchus*. Isolation and identification of a *Msp*I repeated element from *B. xylophilus* was done by digestion of total genomic DNA from *B. xylophilus* with different restriction enzymes. Electrophoresis of the enzymatic digests on agarose gels containing EtdBr revealed the presence of a strong band of approximately 160 bp in *Msp*I and in *Alu*I digests (Fig. 6 A). The *Msp*I fragment used as a probe in Southern blot revealed the presence of a ladder of multimer of 160 bp repeats in the *Msp*I, but also in *Dra*I and *Alu*I digests of genomic DNA (Fig. 6B).

The primary structure of the satellite was analyzed. It showed an unambiguous consensus sequence with an average size of the repeats of 160 bp long (Fig. 7). The *B. xylophilus Msp*I satellite could be compared to the satellite found in *Drosophila melanogaster* and in *Tenebrio molitor* where no particular subgroups of satellite sequences exist and where variations are 3.6% and 2% respectively (Hsieh and Brutlag, 1979; Davis and Wyatt, 1989). However, the degree of variation seemed to increase at particular sites along the repeat lengths and some hot spot sites of mutations were found. The satellite appears to make up 30% of the total genomic DNA and therefore represents a large part of the *B. xylophilus* genome compared to satellite and minisatellite classes found in other nematodes. Satellite sequences make up

Satellite DNA composed of tandemly repeat unit (R) reiterates thousands times in tandem arrays and restricted by X enzymes.

repeat unit

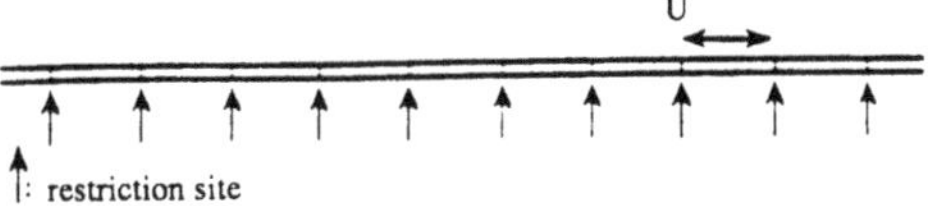

During evolution, point mutations were accumulated in satellite sequence with disappeance of some X restriction sites.

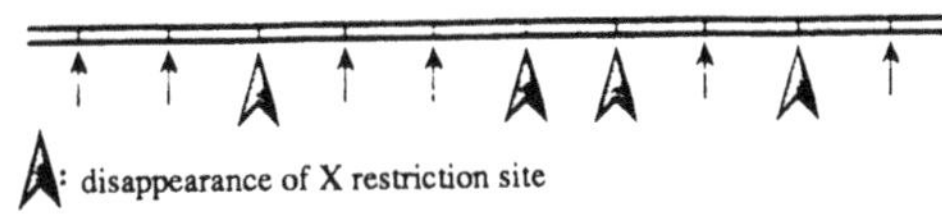

The digestion of satellite DNA with X enzyme generates multiple oligomeres fragments of the repeat unit R.

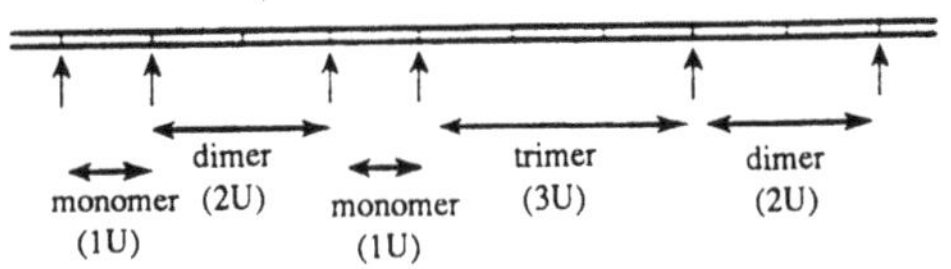

Electrophoresis on agarose gel allows the separation of the different oligomers of the repeat. A typical ladder pattern is thus obtained after hybridization of the corresponding Southern blot with the labelled satellite DNA probe.

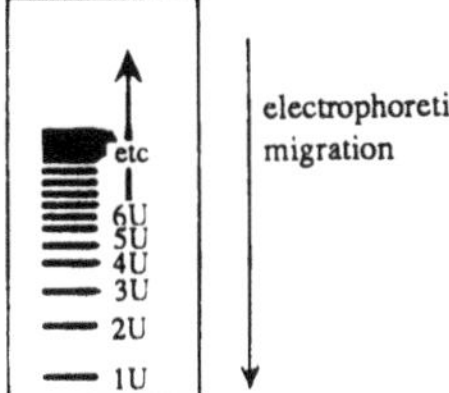

Figure 5. Genomic organization of satellite DNA.

approximately 17% in *P. redivivus* (de Chastonay et al., 1990) and about 20% in *A. lumbricoides* genomes (Tobler et al., 1972). In *C. elegans*, satellite as well as minisatellite sequences represent a very low percentage of the genome (La Volpe et al., 1988; Uitterlinden et al., 1989).

The *Msp*I satellite sequence is species specific since it hybridizes only with genomic DNA from the *B. xylophilus* isolates. Furthermore, the *Msp*I satellite DNA probe is effective directly on squashed nematodes spotted onto a filter using a very simple procedure with no need to extract DNA. The use of this *Msp*I satellite DNA as a species-specific probe leaves no doubt as to the identity of the nematode (Fig. 8). Therefore, in the pinewood nematode species complex, the identification of the plant pathogenic species *B. xylophilus* may be carried out both rapidly and reliably. This step may be crucial for the success of a direct detection of *B. xylophilus* in wood samples and will prove to be of great use in ecological and population studies.

The abundance, the variability and the species-specificity of this satellite family are used to demonstrate its usefulness in fingerprinting since it generates RFLPs between the *B. xylophilus* isolates. In Southern blotting experiments, we show that the American and Japanese isolates hybridize more strongly to the *Msp*I probe than do the Canadian isolates. Therefore, Japanese isolates of *B. xylophilus* seem to be closer to the USA ones than to the Canadian isolates. This result is in agreement with those previously obtained with other probes (Harmey and Harmey, 1993; Tarès et al., 1994). This observation supports the hypothesis that a *B. xylophilus* isolate reached Japan from North America (De Guiran and Bruguier, 1989), probably from the United States.

Polymorphisms in the hybridization ladder pattern are generated by the *Msp*I satellite DNA probe on genomic DNA of six *B. xylophilus* isolates digested to completion with *Alu*I, *Dra*I, *Hae*III and *Hin*dIII. The most important polymorphism, involving mainly qualitative

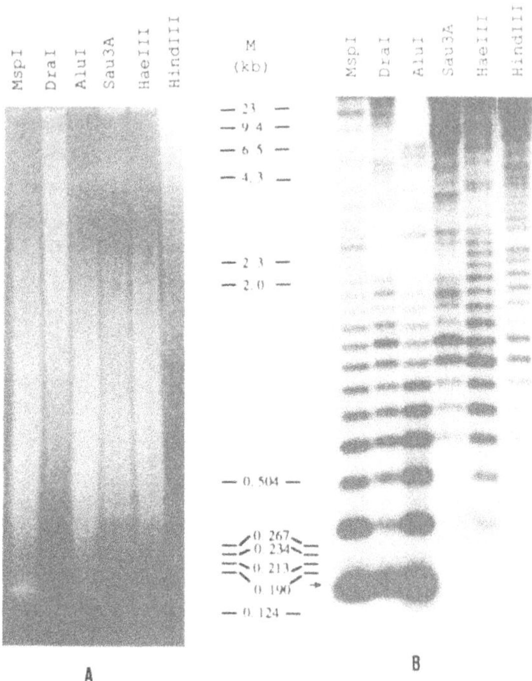

Figure 6. A : Restriction analysis of total *B. xylophilus* strain J10. DNA samples were digested with restriction enzymes listed on the top of the individual lanes. Restriction digests were electrophoresed on 1% agarose gel containing ethidium bromide. B : Autoradiograph of the gel in A after transfer to a nylon membrane and hybridization with 160-bp *Msp*I repeat fragments (mixed) of *B. xylophilus* DNA isolated from agarose gel and ^{32}P-labeled. The arrow indicates the position of *Msp*I monomer repeat of about 160bp (from Tarès et al., 1993).

hybridization differences, is observed in the ladder from the undecamer upward e.g. between 1.6 and 9 kb of the hybridization ladder pattern. For example in the *Dra*I digestion, the absence of signal for the oligomer n=11 (indicated by A) in the ladder hybridization-pattern differentiates the Canadian isolate Bc from the other tested isolates. In addition, extensive differences are observed in the different hybridization ladder patterns obtained with other restriction enzymes between of the tested *B. xylophilus* isolates (Fig. 9).

Satellite DNA in *Meloidogyne*. Satellite DNA from *M. hapla* "La Môle" was isolated by *Sty*I digestion-to-completion of its DNA. In this case, a strong band that corresponds to a highly repeated fragment of approximately 170 bp is visualized in agarose gel electrophoresis (Fig. 10). The *Sty*I satellite DNA has been sequenced (Fig. 11). The average size of the monomers, deduced from the consensus sequence is 169 bp. *M. hapla* satellite DNA is A+T rich (68%) with clusters of As and Ts. Consensus sequence harbors several restriction sites and a restriction polymorphism exists between the repeat units. The sites scattered throughout the units could be due to the accumulation of mutations during the evolutionary history of these sequences. All variations between monomers are point substitutions and deletions. The average divergence from the consensus sequence from this *Sty*I satellite family is estimated at 3%. These data on the *Meloidogyne* satellite may be compared with some observations on other satellite DNAs. This *Sty*I satellite DNA family from *Meloidogyne* seems to be closer to the satellite families observed in *Drosophila melanogaster* and*Tenebrio molitor*, where no particular subgroups exist and where divergence values in the sequenced satellite monomers from the consensus are 3.6% and 2% (Hsieh and Brutlag, 1979; Davis and Wyatt, 1989) than to other classes of satellite DNA found in rats (Sealy et al., 1981) and in *A. lumbricoides* (Müller et al., 1982) where satellites can be divided into variant classes or in bovine satellite (Pech et al. 1979) where the overall variation level is 12%.

```
                                            2                                                                       1               1'  2'
                                          ---->                                                                   ---->          ----> <----
C.s.          10         20         30         40         50         60         70         80         90        100        110        120        130        140        150        160
(bp)  CCGGTGTCTA GTATAATATC AGAGTTTTTC GCCCAGTGTC GGCGTAATTT GATTCCAAAA AAGCTGAAAC TTGCCATGCT AAAATCTCAG GCGATTAGCT TTCGAATGGT GTATGTCTTG TCAATTCACT CCGTCATCAC TAATTCACTT TTTTTTAAAA
      MspI                                                           AluI                             DdeI        AluI  TaqI                                                          DraI

160   .......... .......... .......... .......... .......... .......... .......... .......... .......... .......... .......... .......... .......... .......... .......... ..........
160   .......... .......... .......... .......... ....C..... .......... .......... .......... .......... .......... .......... .......... .......... ....G..... .......... ..........
160   .......... A......... .......... .......... ....C..... ...G...... G......... .......... .......... .......... ......C... .......... .......... .T........ .......... ......T...
159   .....A.... .......... .......... .......... .......... .......... .......... .......... .......... .......... .......... .......... .......... .T........ .......... ..........
160   .......... .......... .......... .......... .......... .......... .......C.. .......... .......... ..A...C... .......... .......... .......... .......... .......... ..........
160   ....C..... .......... .......... .......... .......... .......... .......... .......... .......... .......... .......... .......... .......... .......... .......... ......T...
160   .......... .......... .......... .......... ......G... ...C...... .......... ........G. .......T.. .G.G..C... .......... ......A... .......... ..T....... .......... ......C...
160   .......... .........A .......... .......... ....C..... .......... .......... .......... .......... .......... .......... .......... ...T...... .......... .......... ..........
157   ......C... .......... .....G.... ...TC.G.GG C..--..... .......... .......... .....G.... .......T.C .ACG..C... .......... ......A... .......... ...A...... ........C. .....-....
158   .......... .......... .....G.... .........G ...TAGG... .......... .T........ .......... .......... .......... .......... .......... .......... .........- .......... ....-C....
160   .......... .......... .......... .......... ......G... .......... .......... .......... .......... .......... .......... .......... .......... ..T....... .......... ..........
160   .......... .......... ....A..... ..A......G ......G... .......... .......... .......... ......AT.. .A.G..C... .......... .......... .......... .......... ........C. .....C....
160   .......... .......... .......... .......... ....C..... ....G..... G......... .......... .......... ..A....... ....G..A.. .......... .......... .T........ .......... ..........
```

Figure 7. Consensus sequence (C.s.) and variant positions in the repeat units. The variant nucleotides (nt) in 13 sequenced repeat-unit monomers are shown in 5' to 3' orientation. The C.s. from the sequences of the 13 cloned DNA monomers is shown at the top of the table. The nt assignments were decided on the basis of nt occurence of 50% or higher at each position in the sequences ; nt differences from the C.s. are listed : dots indicate identical nt, and dashes specify deleted nt. Restriction sites are indicated below the C.s. The repeat-unit C.s. exhibited a high A+T content of 62%. The perfect direct repeat of 8 bp (1 and 1') and the inverted repeat of 7 bp are indicated by arrows above the sequence (from Tarès et al., 1993).

Individuals	*B. xylophilus* J10	*B. xylophilus* Bc	*B. mucronatus*	*C. elegans*
1				
5				
10				
20				

Figure 8. Direct hybridization of the ^{32}P-labeled *B. xylophilus* J10 satellite DNA monomer on different numbers of squashed nematodes belonging to *B. xylophilus* J10, *B. xylophilus* Bc, *B. mucronatus* and *C. elegans* (from Tarès et al., 1994).

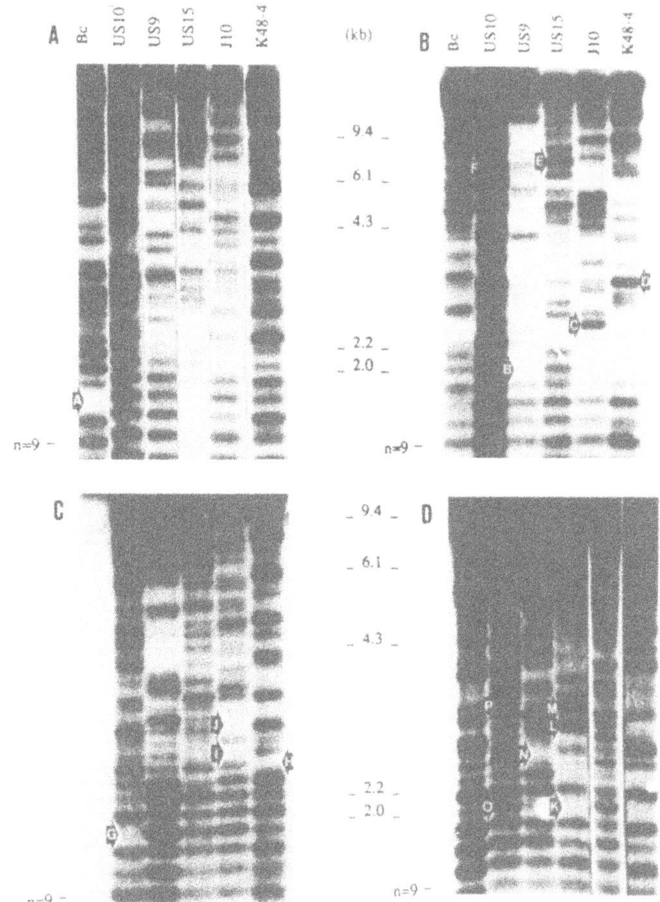

Figure 9. Autoradiogram of genomic Southern blot of six *B. xylophilus* isolates digested by four restriction enzymes and hybridized with the *B. xylophilus* J10 satellite DNA monomer. (A) : *Dra*I, (B) : *Alu*I, (C) : *Hae*III, (D) : *Hin*dIII. The agarose gel was Southern blotted and hybridized with *B. xylophilus* ^{32}P labeled satellite DNA. Molecular weight markers are shown (kb). The abbreviation n=9 indicated the position of the nonamer in the hybridization ladder. The lettered arows indicate polymorphisms characterizing each of the tested isolates (from Tarès et al., 1994).

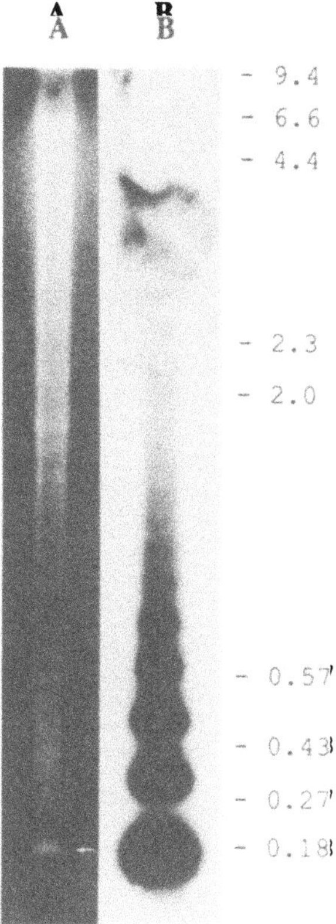

Figure 10. Restriction analysis of total genomic DNA from *M. hapla*. A: The genomic DNA was digested to completion with *Sty*I and fractionated by electrophoresis through 1% agarose gels containing ethidium bromide. A 170-bp fragment is detected by transillumination of agarose gel with UV (marked by the arrow). B: The agarose gels were Southern blotted and hybridized with the 170 bp ^{32}P-labeled cloned monomer (from Piotte, 1993).

```
                                                   XbaI
             10         20         30         40         50         60         70         80         90        100        110        120
Cons CTTGGAATTA GAAATAATTG TTTATATGAG TTCCTTGTAA AGCAACCTCT AGATGCCACC AGATATAAAA AATTGTTGCA CAGTCAAAAC TGGAAATGGT CTTCATTATT GAATATCTTA
1    .......... ......T... .....**... G......... .......... .......... .......... .......... .......... .......... .......... ..G.......
2    .......... .....T.... .......... G......... ......T... .......... .......... .......... A......... .......... .......... ..........
3    .......... .......... ...**..... .......... .......... .......... .......... .......... .......... .......... .......... .......A..
4    .......... .......... .......... .......... .......... .......... .......... .......... .......... .......... .......... .......A..
5    .......... .......... .......... .......... .......... .......... .......... .......... A......... .......... .......... .......A..
6    .......... .......... A.A....... .....A.... ......T... .........A .......... .......... .......... .......... .......... ..TG......
7    .......... .......... ..A*...... .......... .......... .......... .......... .......... .......... .......... .......... ..*G......
8    .......... ......T... ...*...... .......... .......... .......... .......... .......... .......... .......... .......... ..*A..A...
9    .......... .......... ...*...... .......... ......T... .......... .......... .......... .......... .......... .......... ..G.......

       BamI
            130        140        150        160       169
Cons CAGAATGCTG TATCTATCCT GAAAAAATCC TGTGTCGAAC TTAATTTTC
1    T......... .......... .......... ......A... .........
2    .......... .......... .......... .......... .........
3    .......AAA A......... .......... .......... ....AG...
4    .......... .......... .......... .......... ....A....
5    .......... .......... .......... .......... .........
6    T...G..... .......... .......... .........T .........
7    T......... .......... .......... .......... .........
8    .......... .......... .......... .........A .........
9    T......... .......... .......... .......... .........
```

Figure 11. Nucleotide sequences of cloned satellite monomers from *M. hapla*. Consensus sequences (Cons) have been deduced from the alignment of nine variants of *Sty*I satellite DNA. Base assignments were decided on the basis of 50% occurrence or higher at each position in the sequences. For each clone, only the non-homologous nt are shown. Asterisks appearing in some variant positions of the cloned sequences indicate deletions (from Piotte et al., 1994).

The *Sty*I satellite from *M. hapla* appears to make up 5% of the *Meloidogyne* genome. Assuming that the genome size of *Meloidogyne* is about 51 000 kb (Pableo and Triantaphyllou, 1989), this satellite fraction corresponds to approximately 15 000 copies per haploid genome.

Southern blot experiments, using *Sty*I satellite DNA of *M. hapla* as probe indicate that this satellite is species-specific since it gives a signal only with the *M. hapla* isolates. The species-specificity and the reiteration of *M. hapla* satellite sequence are indicative of its high detection power when used as probe. In the cases where unreliable identification lead to confusion between some particularly damaging nematodes, this satellite DNA may represent a potential field tool. This is so for distinguishing between *M. hapla* and *M. chitwoodi*, which are sympatric and morphologically similar.

We have tried to use this technique to differentiate *M.hapla* from *M.chitwoodi* (Fig. 12). In squashed nematodes experiments, the Dutch populations of *M.hapla* tested give a strong signal while Dutch populations of *M.chitwoodi* never hybridize with *M.hapla* satellite DNA. The intensity of the signal increases with the quantity of material (one female<five females<egg mass) and the root fragment around the female, in the case of the gall, does not prevent the hybridization. As positive control, *M.hapla* "La Môle" hybridizes with its own satellite DNA and as negative control, *M.incognita* "Taïwan" does not hybridize at all with this probe. The hybridization of healthy plant material is always negative.

	1 femelle	5 femelles	1 masse d'oeufs	1 galle
M. incognita Taïwan				
M. hapla La Môle				
M. chitwoodi Netherlands 1				
M. chitwoodi Netherlands 2				
M. hapla Netherlands 1				
M. hapla Netherlands 2				

Figure 12 : Squashed nematodes experiments. Squashed *M. hapla*, *M. incognita* and *M. chitwoodi* are hybridized with ^{32}P-labeled *M. hapla* satellite DNA. For each population, one female, five females, one egg mass and one gall were squashed on a nylon membrane (from Piotte, 1993).

The principal advantage of this procedure is to avoid DNA extraction. Because of the high power of sequence detection of satellite DNA due to its representation in the *M. hapla* genome, simple squashes of nematodes, even in the root tissues, allowed effective hybridization with the satellite DNA probe. This satellite DNA could be developed as a specific non-radioactive probe. Such tools should provide a rapid, inexpensive and user-friendly method of *Meloidogyne* species identification.

At the intraspecific level, these satellite probes clearly allow one to distinguish between geographic populations. We easily distinguished between *M. hapla* populations using *M. hapla* satellite DNA as probe. "La Môle", "Angleterre" and "Frontignan" gave very different patterns when digested by *Alu*I and *Msp*I (Fig. 13). Population specific bands can be noticed whose sizes are, e.g., in *Alu*I digestion patterns, 4.3 kb for "La Môle", 6.2 kb for "Frontignan", 6.4 kb for "Angleterre", and in *Msp*I digestion patterns, 4.1 kb for "La Môle", 6.4 kb for "Frontignan" and 18 kb for "Angleterre". This clear polymorphism is a direct consequence of the rapid evolution of this DNA fraction (Brutlag, 1980). "La Môle" and "Frontignan" profiles were quite different but were closer to each other than they were to "Angleterre" profiles. Two

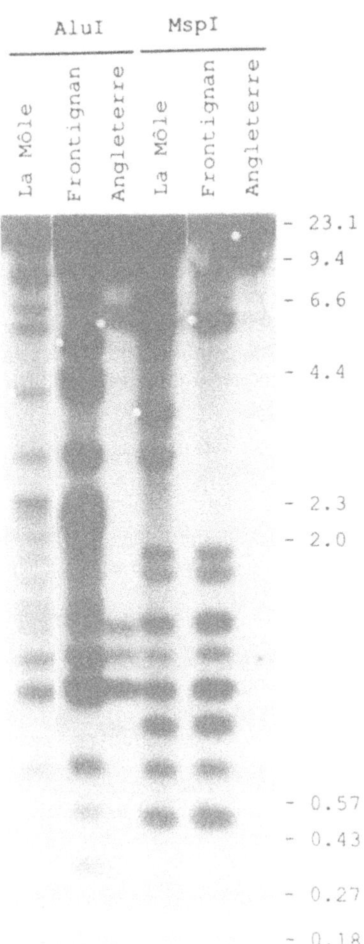

Figure 13. Autoradiogram of genomic Southern blot of three *M. hapla* populations digested to completion by *Alu*I and *Msp*I and hybridized *M. hapla* satellite DNA. Lane 1, La Môle ; lane 2, Frontignan ; lane 3, Angleterre. Numbers refer to the sizes of molecular markers in kilobases. The agarose gel was Southern blotted and hybridized with *M. hapla* satellite DNA ^{32}P-labeled (from Piotte, 1993).

hypotheses have to be developed. Either this result is linked to geographical separation of tested isolates, or a relationship exists between this result and the mode of reproduction of these populations. "Angleterre" is a parthenogenetic mitotic population while "La Môle" and "Frontignan" are parthenogenetic meiotic. These modes of reproduction certainly correspond to different mechanisms and speeds of evolution that could explain the result obtained.

CONCLUSION

During the last decade, molecular techniques have provided new and precise tools for nematode identification. Repetitive DNA sequences represent one of the best molecular markers since most of the repetitive sequence members lack transcription and are not under selection pressure. With these sequences used as probes, separations of sibling species complexes and race pathotypes are now possible. The resolution power of repetitive DNA sequences may be correlated with their potential variability which is higher for non-coding regions than for expressed genes. Furthermore, repetitive DNA sequences are largely more sensitive as they will have proportionately more target sequences per genome than those directed at DNA sequences present in low copy number. Repetitive DNA sequences have been shown to be effective in techniques such as restriction fragment length polymorphism analysis and species-specific hybridization probes for direct analysis of nematode DNA. This review illustrates the applications of satellite DNA as a taxonomic diagnostic marker in the *Bursaphelenchus* and *Meloidogyne* genera. In these cases, isolated satellites were species-specific and therefore allowed to direct probing of squashed nematodes. Since the resolution of such squash blot hybridization is very high and since the experimental procedure is very easy and not time consuming, it should be possible to introduce it to field work without the need for a well-equiped laboratory. Another application of satellite DNA in taxonomy is its use for systematic analyses at the species-population boundary which might give interesting insights into their relationships. The examples given above illustrate that satellite DNAs can serve as good tools for phylogenetic analyses. This does not mean that satellite DNA analysis guarantees good solutions to all taxonomic problems of closely related populations or species but the previous examples of satellite DNA from *Bursaphelenchus* and *Meloigogyne* support the fact that it is worthwhile to apply satellite DNA techniques if there is a need for a rapidly evolving taxonomic marker.

REFERENCES

Aeby, P., Spicher, A., De Chastonay, Y., Müller, F., and Tobler, H., 1986, Structure and genomic organization of provirus-like elements partially eliminated from the somatic genome of *Ascaris lumbricoides*, *EMBO J.* 5: 3353.

Bachmann, L., Raab, M., and Sperlich, D., 1989, Satellite DNA and speciation. A species specific satellite DNA of *Drosophila guanche*, *Z. Zool. Syst. Evol.-Forsch* 27: 84.

Bakker, J., and Bouwman-Smits, L., 1988, Contrasting rates of protein and morphological evolution in cyst nematode species, *Phytopathology* 78: 900.

Beridze, T.G., 1986, Satellite DNAs. Springer-Verlag, Berlin.

Bolla, R.I., Weaver, C., and Winter, R.E.K., 1988, Genomic differences among pathotypes of *Bursaphelenchus xylophilus*, *J. Nematol.* 20: 309.

Bostock, C., 1980, A function for satellite DNA?, *Trends Biochem. Sci.* 5: 117.

Boveri, T., 1887, Ueber Differenzierung der zellkerne Während der Furchung des Eies von *Ascaris megacephala*, *Anat. Anz.* 2: 688.

Brutlag, D.L., 1980, Molecular arrangement and evolution of heterochomatic DNA, *Annu. Rev. Genet.* 14: 121.

Brutlag, D.L., and Peacock, W.J., 1979, DNA sequences of the 1.672 g cm^{-3} satellite of *Drosophila melanogaster*, *J. Mol. Biol.* 135: 565.

Burrows, P.R., 1988, The differentiation of *Globodera pallida* from *G. rostochiensis* using species-specific DNA probes, *Nematologica* (Abstract. European Society of Nematologists Meeting, Uppsala, Sweden, August 1988) 34: 260.

Burrows, P.R., and Boffey, S.A., 1986, A technique for the extraction and restriction endonuclease digestion of total DNA from *Globodera rostochiensis* and *G. pallida* second stage juveniles, *Revue Nematol.* 9:199.

Burrows, P.R., and Perry, R.N., 1988, Two cloned DNA fragments which differentiate *Globodera pallida* from *G. rostochiensis*, *Revue Nématol.* 11: 441.

Castagnone-Sereno, P., Piotte, C., Abad, P., Bongiovanni, M., and Dalmasso, A., 1991, Isolation of a repeated DNA probe showing polymorphism among *Meloidogyne incognita* populations, *J. Nematol.* 23: 316.

Castagnone-Sereno, P., Piotte, C., Uijthof, J., Abad, P., Wajnberg, E., Vanlerberghe-Masutti, F., Bongiovanni, M., and Dalmasso, A., 1993, Phylogenetic relationships between amphimictic and parthenogenetic nematodes of the genus *Meloidogyne* as inferred from repetitive DNA analysis, *Heredity* 70: 195.

Cenis, J.L., 1993, Identification of four major *Meloidogyne* spp. by random amplified polymorphic DNA (RAPD-PCR), *Phytopathology* 83: 76.

Curran, J., Baillie, D.L., and Webster, J.M., 1985, Use of genomic DNA restriction fragment length differences to identify nematodes species, *Parasitology* 90: 137.

Curran, J., Mc Clure, M.A., and Webster, J.M., 1986, Genotypic differenciation of *Meloidogyne* populations by detection of restriction fragment length polymorphism in total DNA, *J. Nematol.* 18: 83.

Curran, J., and Webster, J.M., 1987, Identification of nematodes using restriction fragment length difference on species specific probes, *Can. J. Plant Pathol.* 9: 162.

Dalmasso, A., and Bergé, J.B., 1983, Enzyme polymorphism and the concept of parthenogenetic species, exemplified by *Meloidogyne*, In : "Concepts in nematodes systematics", A.R., Stone, Platt, H.M., and Khalil, L.F., eds., Academic Press, London and New York.

Datta, U., Dutta, P., and Mandal, R.K., 1988, Cloning and characterization of a highly repetitive fish nucleotide sequence, *Gene* 62: 331.

Davis, C.A., and Wyatt, G.R., 1989, Distribution and sequence homogeneity of an abundant satellite DNA in the beetle, *Tenebrio molitor*, *Nucleic Acid Res.* 17: 5579.

De Chastonay, Y., Müller, F., and Tobler, H., 1990, Two highly reiterated nucleotide sequences in the low C-value genome of *Panagrellus redivivus*, *Gene* 93: 199.

De Guiran, G., and Bruguier, N., 1989, Hybridization and phylogeny of the pine wood nematodes (*Bursaphelenchus* spp.), *Nematologica* 35: 321.
Dejong, A.J., Bakker, J., Roos, M., and Gommers, F.J., 1989, Repetitive DNA and hybridization patterns demonstrate extensive variability between the sibling species *Globodera rostochiensis* and *G. pallida, Parasitology* 99: 133.
Emmons, S.W., Rosenzweig, B., and Hirsch, D., 1980, Arrangement of repeated sequences in the DNA of the nematode *Caenorhabditis elegans, J. Mol. Biol.* 144: 481.
Etter, A., Aboutanos, M., Tobler, H., and Müller, F., 1991, Eliminated chromatin of Ascaris contains a gene that encodes a putative ribosomal protein, *Proc. Natl. Acad. Sci.* USA 88: 1593.
Files, J.G., and Hirsh, D., 1981, Ribosomal DNA of *Caenorhabditis elegans, J. Mol. Biol.* 149: 223.
Forejt, J., 1973, Centrometric heterochromatin polymorphism in the house mouse, *Chromosoma* 43 : 187.
Fowler, R.F., and Skinner, D.M., 1985, Cryptic satellites rich in inverted repeats comprise 30% of the genome of a hermit crab, *J. Biol. Chem.* 260: 1296.
Fowler, J.C.S., Skinner, J.D., Burgoyne, L.A., and Drinkwater, R.D., 1989, Satellite DNA and higher-primate phylogeny, *Mol. Biol. Evol.* 6: 553.
Harmey, J., and Harmey, M.A., 1993, Detection and identification of *Bursaphelenchus* species with DNA fingerprinting and Polymerase Chain Reaction, *J. Nematol.* 25: 406.
Hsieh, T.S., and Brutlag, D., 1979, Sequence and sequence variation within 1.688 g/cm3 satellite DNA of *Drosophila melanogaster, J. Mol. Biol.* 135: 465.
Hussey, R.S., 1979, Biochemical systematics of nematodes--a review, *Helminth. Abstr. Ser. B Plant Nematol.* 48: 141.
Hyman, B.C., and Powers, T.O., 1991, Integration of molecular data with systematic of plant parasitic nematodes, *Ann. Rev. Phytopathol.* 29: 98.
Jelinek, W.R., and Schmid, C.W., 1982, Repetitive sequences in eukaryotic DNA and their expression, *Ann. Rev. Biochem.* 51: 813.
John, B., and Miklos, G.L.G., 1979, Functionnal aspects of satellite DNA and heterochomatin, *Int. Rev. Cytol.* 58: 1.
Kalinski, A., and Huettel, R.N., 1988, DNA restriction fragment length polymorphism in races of the soybean cyst nematode, *Heterodera glycines, J. Nematol.* 20: 532.
La Volpe, A., Ciaramella, M., and Bazzicapulo, P., 1988, Structure, evolution and properties of a novel repetitive DNA family in *Caenorhabditis elegans, Nucleic Acid Res.* 16: 8213.
MacKay, R.D.G., Bobrow, M., and Cooke, H.J., 1978, The identification of a repeated DNA sequence involved in the karyotype polymorphism of the human Y chromosome, *Cytogenet. Cell. Genet.* 21: 19.
Miklos, G.L.G., 1985, Sequencing and manipulating repeated DNA, In : "Molecular Evolutionnary Genetics", R.J., Mc Intyre, ed., Plenum, New York.
Moritz, K.B., and Roth, G.E., 1976, Complexity of germline and somatic DNA in *Ascaris, Nature* 259: 55.
Müller, F., Walker, P., Aeby, P., Neuhaus, H., Felder, H., Back, E., and Tobler, H., 1982, Nucleotide sequence of satellite DNA contained in the eliminated genome of *Ascaris lumbricoides, Nucleic Acid Res.* 10: 7493.
Novak, U., 1984, Structure and properties of a highly repetitive DNA sequence in sheep, *Nucleic Acids Res.* 12: 2343.
Pableo, E.C., and Triantaphyllou, A.C., 1989, DNA complexity of the root-knot nematode (*Meloidogyne* spp.) genome, *J. Nematol.* 21: 260.
Pech, M., Streeck, R.E., and Zachau, H.G., 1979, Patchwork structure of a bovine satellite DNA, *Cell.* 19: 765.
Piotte C., 1993, Organisation du génome chez les nématodes phytoparasites du genre *Meloidogyne* : étude de la fraction répétée et utilisation en typage moléculaire et phylogénie, Thèse de l'Université de Paris 6, France.
Piotte, C., Castagnone-Sereno, C., Bongiovanni, M., Dalmasso, A., and Abad, P., 1994, Cloning and characterization of two satellite DNAs in the low C-value genome of the nematode *Meloidogyne* spp, *Gene (in Press).*
Piotte, C., Castagnone-Sereno, P., Uijthof, J., Abad, P., Bongiovanni, M., and Dalmasso, A., 1992, Molecular characterization of species and populations of *Meloidogyne* from various geographic origins with repeated-DNA homologous probes, *Fundam. Appl. Nematol.* 15: 271.
Pruitt, R.E., and Meyerowitz, E.M., 1986, Characterization of the genome of *Arabidopsis thaliana, J. Mol. Biol.* 187: 169.
Ruiz-Lara, S., Prats, E., Sainz, J., and Cornudella, L., 1992, Cloning and characterization of a highly concerved satellite DNA from the mollusc *Mytilus edulis, Gene* 117: 237.
Sainz, J., Azorin, F., Cornudella, L., 1989, Detection and molecular cloning of highly repeated DNA in the sea cucumber sperm,*Gene* 80: 57.
Sealy, L., Hartley, J., Donelson, J., Chalkley, R., Hutchinson, N., and Hamkalo, B., 1981, Characterization of a highly repetitive sequence DNA family in rat, *J. Mol. Biol.* 145: 291.
Skinner, D.M., 1977, Satellite DNAs, *Bioscience* 27: 790.
Southern, E.M., 1975, Detection of specific sequence among DNA fragments separated by gel electrophoresis, *J. Mol. Biol.* 90: 503.
Stratford, R., Shields, R., Goldsbrough, A., and Fleming, C., 1992, Analysis of repetitive DNA sequences from potato cyst nematode and their use as diagnostic probes, *Phytopathology* 82: 881.

Sulston, J.E., and Brenner, S., 1974, The DNA of *Caenorhabditis elegans*, *Genetics* 77: 95.
Tarès, S., Abad, P., Bruguier, N., and De Guiran, G., 1992, Identification and evidence for relationships among geographical isolates of *Bursaphelenchus* spp. (Pinewood nematode) using homologous DNA probes, *Heredity* 68: 157.
Tarès, S., Lemontey, J.M., De Guiran, G., and Abad, P., 1993, Cloning and characterization of a highly conserved satellite DNA sequence specific for the phytoparasitic nematode *Bursaphelenchus xylophilus*, *Gene* 129: 269.
Tarès, S., Lemontey, J.M., De Guiran, G., and Abad, P., 1994, Use of species-specific satellite DNA from *Bursaphelenchus xylophilus* as a diagnostic probe, *Phytopathology (in Press).*
Tobler, H., 1986, The differentiation of germ and somatic cell lines in nematodes. Results and problems in cell differentiation, In "Germ line-soma differentiation ", W. Hennig, ed., Heidelberg, Springer-Verlag.
Tobler, H., Smith, K.D., and Ursprung, H., 1972, Molecular aspects of chromatin elimination in *Ascaris lumbricoides*, *Dev. Biol.* 27: 190.
Uitterlinden, A.G., Slagboom E.P., Johnson T.E., and Vijg, J., 1989, The *Caenorhabditis elegans* genome contains monomorphic minisatellites and simple sequences, *Nucleic Acid Res.* 17: 9527.
Vahidi, H., Purac, A., Le Blanc, J.M., and Honda, B.M., 1991, Characterization of potentially functional 5S rRNA-encoding genes within ribosomal DNA repeats of the nematode *Meloidogyne arenaria*, *Gene* 108: 281.
Vignali, R., Rijli, F.M., Batistoni, R., Fratta, D., Cremisi, F., and Barsacchi, G., 1991, Two dispersed highly repeated DNA families of *Triturus vulgaris meridionalis* (Amphibia, Urodela) are widely conserved among Salamandridae, *Chromosoma* 100: 87.
Webster, J.M., Anderson, R.V., Baillie, D.L., Beckenbach, K., Curran, J., and Rutherford, T.A., 1990, DNA probes for differentiating isolates of the pinewood nematode species complex, *Revue Nematol.* 13: 255.

PART III

HOST-PARASITE INTERACTIONS: THE PLANT

ARABIDOPSIS THALIANA AS A MODEL HOST PLANT TO STUDY MOLECULAR INTERACTIONS WITH ROOT-KNOT AND CYST NEMATODES

Andreas Niebel, Nathalie Barthels, Janice de Almeida-Engler, Mansour Karimi, Isabelle Vercauteren, Marc Van Montagu, and Godelieve Gheysen

Laboratorium voor Genetica
Universiteit Gent
K.L. Ledeganckstraat 35
B-9000 Gent, Belgium

INTRODUCTION

Arabidopsis thaliana is a small cruciferous plant that is considered as being a "botanical *Drosophila*" (Whyte, 1946) for a number of reasons recently reviewed by Meyerowitz (1989). Thanks to its small size and ability to grow well at room temperature both *in vitro* and in the greenhouse, it is easy and cheap to maintain large populations of *Arabidopsis* plants under laboratory conditions. *Arabidopsis* has a short generation time: approximately 3 months from seed to seed. It is extremely prolific and its seeds are very small (about 50,000 seeds fit in a 1.5-ml Eppendorf tube). These features make *A. thaliana* highly suited for classical genetic approaches.

Several linkage maps of the *A. thaliana* genome consisting of visible and/or molecular markers have been produced. A high density map containing 252 randomly amplified polymorphic DNA markers (RAPD) and 60 restriction fragment length polymorphism markers (RFLP) has been constructed (Reiter et al., 1992). An integrated map comprising 125 classical genetic markers and 306 RFLP markers has also been generated recently (Hauge et al., 1993). These maps provide valuable tools for gene mapping experiments and for map-based gene cloning approaches.

Arabidopsis thaliana is also highly suited for large-scale mutagenesis experiments. The embryos contain only two cells that will give rise to the entire germ line (Li and Rédei, 1969). Mutations in one of these cells will thus lead to the formation of only limited chimeras (two sectors). Using point mutation-inducing agents such as ethane methylsulfonate (EMS), or deletion-inducing agents such as fast neutrons, large collections of mutants in numerous biosynthetic or developmental pathways have been obtained. Additionally, genetic transformation and regeneration into whole plants are well established in *Arabidopsis* (Valvekens et al., 1988) and have also led to the generation of molecularly-tagged mutants. The advantage of such tagged mutants being that the isolation of the corresponding genes is much more straightforward. *Agrobacterium tumefaciens*-mediated

Advances in Molecular Plant Nematology
Edited by F. Lamberti *et al.*, Plenum Press, New York, 1994

transformation can lead to insertion mutants tagged by the T-DNA (Walden et al., 1991) and a number of genes have already been cloned by this approach. Recently, transposon gene tagging has also been shown to work in *Arabidopsis*. Heterologous transposon systems from maize have been successfully used to clone a male sterility gene and a putative isopentenyltransferase gene (Aarts et al., 1993; Bancroft et al., 1993). An endogenous transposon system has also been shown to function (Tsay et al., 1993).

The haploid genome of *A. thaliana* consists of five chromosomes only and is the smallest known among flowering plants (approximately 10^8 base pairs). This is equivalent to the *Caenorhabditis elegans* genome, 15 times smaller than the tobacco genome and approximately 50 times smaller than the genome of widespread monocotyledonous crops such as maize or wheat (Arumuganathan and Earle, 1991).

Numerous cDNA and genomic libraries have been made in *A. thaliana*. To date, approximately one-third of the *Arabidopsis* genome has been cloned so far into yeast artificial chromosomes (YAC), which are being placed on the linkage maps and are very useful for the cloning of mapped genes. Finally, a world-wide genome sequencing project has been started. Despite the serious competition from the rice sequencing project, *A. thaliana* will probably be the first higher plant to have its genome entirely sequenced. These features make *A. thaliana* an ideal model system, more suited than any other plant for molecular biological approaches and in particular for gene cloning.

Although initially thought to be a "pathogen-less" plant, *A. thaliana* appeared to be a suitable host for most types of plant pathogens: bacteria (Simpson and Johnson, 1990), fungi (Koch and Slusarenko, 1990), viruses (Ishikawa et al., 1991), and even nematodes (Sijmons et al., 1991). Nowadays, a large number of plant pathologists, particularly those interested in cloning plant resistance genes, has been attracted by the many advantages of using *Arabidopsis* as a model host. In 1992 the *Arabidopsis* Pathogen Network (ARAPANET), a network stimulating fruitful exchange of information, techniques, and material, between plant pathologists working with the same plant, was created.

GENETIC APPROACHES

Initial Approaches

While Sijmons et al. (1991) were establishing culture and inoculation conditions *in vitro* to show that *A. thaliana* was a good host for several plant-parasitic nematodes, we were performing similar experiments under greenhouse conditions. Studying root parasites, it was important to have quick access to the underground part of the plants. Therefore, we started growing and infecting *A. thaliana* on sand, supplemented with a nutrient solution. This allowed us to easily score for nematode infection after rinsing the roots in water. This hydroponic inoculation system was particularly suited for the study of root-knot nematode infections. When inoculated early after germination (7 to 14 days), with large inocula (1,000 L2 larvae/plant), infection rates of up to 130 egg-laying-females (ELF)/plant could be obtained on the relatively small root system of *A. thaliana*. In the case of cyst nematode infection, the washing step had to be performed carefully in order not to remove the females protruding from the roots.

Under greenhouse conditions, we could show that *A. thaliana* was susceptible to two species of root-knot nematodes, *Meloidogyne incognita* and *M. javanica*, and also to two cyst nematodes known to infect cruciferous plants, *Heterodera schachtii* and *H. trifolii*. Other cyst nematodes that infect only a restricted number of solanaceous plants (*Globodera pallida* and *G. rostochiensis*) did not give any sign of infection even when large inocula were used. The migratory parasite *Ditylenchus destructor* was also tested without success.

Ecotype Screenings

Arabidopsis thaliana ecotypes, bearing monogenic resistance traits to certain fungal or bacterial isolates but not to others have been isolated (Whalen et al., 1991). Several bacterial avirulence genes have already been characterized and cloned (Dong et al., 1991). On the plant side, the mapping of the genes involved in the resistance and their isolation by map-based cloning is well advanced.

In an attempt to identify nematode-resistant ecotypes, we performed an initial screening with *M. incognita* on 10 ecotypes from various geographic origins. Although some differences in reproduction rates (ELF/plant) have been observed, no fully resistant ecotype could be identified (Table 1). Similarly, Sijmons et al. (1991) performed a more extensive ecotype screen *in vitro* with *H. schachtii*. About 100 ecotypes have already been screened, without success (F. Grundler, pers. comm.). This is surprising, especially since previously a number of bacteria- and fungus-resistant ecotypes had been easily found. It is of course possible that insufficient numbers of ecotypes have been tested so far. On the other hand, *A. thaliana* might be a poor host for sedentary nematodes under natural conditions, presumably because of its short life cycle, allowing only limited reproduction of those nematodes. Consequently co-evolution phenomena leading to gene-for-gene relationships might not have taken place. These unpromising results from the ecotype screenings led us to investigate other possible genetic approaches.

Table 1. Ecotype screening.

Ecotype	Origin	Egg-laying females
Llagostera (Ll-o)	Spain	11.4 ± 8.9
Chisobra (Chs-o)	Russia	29.0 ± 21.0
Landsberg erecta (La-o)	Poland	30.5 ± 14.5
Tossa de mar (Ts-1)	Spain	43.8 ± 35.0
Bensheim (Be-0)	Germany	44.0 ± 19.8
Columbia (Col-o)	USA (Mo)	45.0 ± 30.2
Zürich (Zü-o)	Switzerland	51.0 ± 8.8
Rschew (Rsch-o)	Russia	57.4 ± 8.7
Wassilewskija (Ws-o)	Russia	90.8 ± 47.3

Mean value of 5 plants analyzed 5 weeks after inoculation with 1,000 L2 larvae of *Meloidogyne incognita.*

Mutant Screenings

As no naturally occurring resistances had been found, we have screened for resistance to *M. incognita* in mutagenized *Arabidopsis* plants. The aim was to identify plant genes important or essential for each stage of the complex interaction between root-knot nematodes and plants. A non-lethal mutation in one of these genes should lead to reduction or even absence of infection symptoms.

In our laboratory, a chemically mutagenized (using EMS) *A. thaliana* ecotype Columbia seed stock had been generated. Since this ecotype was a good host for *M. incognita*, we chose to use it for our mutant screen. Figure 1 describes the screening scheme we used. In two successive inoculation experiments, almost 5,000 mutagenized M_2 seeds were inoculated and checked for nematode infection and reproduction 5 weeks after inoculation. Following comparison with control plants, 17 plants showing no sign of infection and 13 showing reduced infection were isolated and put back in soil to set seed. Viable offsprings were obtained and tested for 18 of these plants. Finally, 5 mutants with altered responses to *M. incognita* were retained and called AMi (acronym for *Arabidopsis/Meloidogyne incognita*).

AMi 1 Mutant. A high level of resistance was observed with AMi 1. When ten mutant M_3 plants were infected with *M. incognita*, only one plant developed a small swelling, while control plants had an average of 8.5 ELF/plant. There is no apparent phenotypic effect of the AMi mutation: stems, leaves, and floral organs look normal. Upon closer observation, the root system of this mutant appears slightly shorter and more branched than that of the wild-type plants.

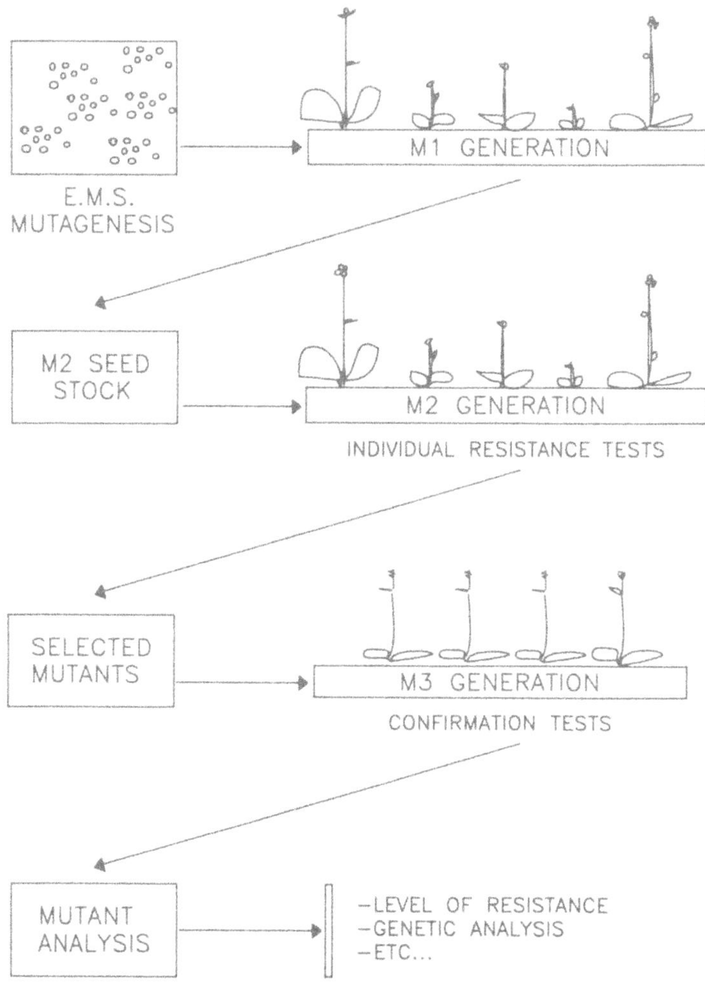

Figure 1. Screening scheme for the isolation of *Meloidogyne incognita*-resistant *Arabidopsis thaliana* mutants.

AMi 2 Mutant. Thirty-two M_4 plants have been infected (because of low fertility only few viable M_3 seeds had been obtained). A 4-fold (statistically significant) reduction in ELF/plant was observed in the mutant as compared with control plants. AMi 2 is slightly affected in several organs. It is generally a little smaller and weaker than control plants, and its fertility is low. Some parts (sometimes up to 100%) of the root system of AMi 2 have a strange phenotype. The side roots remain very short and have a swollen, bulbous appearance. This phenotype was not always observed *in vitro*.

AMi 3 Mutant. Ten M_3 plants have been infected with *M. incognita*. A 4.5-fold (statistically significant) reduction in ELF/plant was observed in the mutants as compared to control plants. While the aerial organs of this mutant looked normal, the root system was slightly affected in its overall structure. It was shorter and more branched giving it a "bushy" appearance.

AMi 4 Mutant. About 16% of the offspring (after self-fertilization) of this mutant had an aberrant phenotype. These plants showed a strong reduction in growth and barely produced seeds. Their root system was abnormally compact and branched, with very short side roots. Infections were hardly ever observed in these abnormal plants while the other mutant offspring showed infection rates comparable to those of control plants.

AMi 5 Mutant. About 33% of the offspring (after self-fertilization) of this mutant, presented an even more extreme phenotype. These plants were hardly viable, only developed a few leaves and almost no roots. Flowers have been obtained in some cases, but they were male sterile. Hardly any infection was ever found in these mutants.

Due to the pronounced abnormality of their root system and the complex segregation of the mutant phenotype in self-fertilizations, no further work was done on AMi 4 and AMi 5.

Genetic Analysis of the Mutants

Crosses between AMi mutants and wild-type *A. thaliana* cv. Columbia have been made (using the mutants as males). The F_1 plants obtained from these crosses were self-fertilized and their progeny analyzed for segregation of the resistant phenotype.

For AMi 3, 42 out of 172 F_2 plants analyzed (24.4%) had a bushy root phenotype. This root phenotype appears to be associated with resistance. Eleven of the 42 segregating plants have been inoculated and were less infected than control plants. These results suggest that AMi 3 is a monogenic recessive mutation. No clear results are available yet for AMi 1 and AMi 2.

In vitro Tests

To analyze more precisely at which stage of the interaction the AMi mutations are acting, it is essential to infect and observe the mutant plants *in vitro*. Therefore, the resistance levels to *Meloidogyne* of the AMi mutants had to be tested *in vitro*. Much to our surprise, the *in vitro* resistance tests gave rather diverging results from the sand tests (Table 2). This type of difference between resistance levels *in vitro* and in soil had already been observed in the past for other systems. The mustard cultivar Maxi, for example, is highly resistant to *H. schachtii* in the field and sensitive *in vitro* (F. Grundler, pers. comm.). Only for AMi 2, the resistance was also observed *in vitro*.

Table 2. Resistance tests.

	Galls/mutant plant			Galls/control plant		
	AMi 1	AMi 2	AMi 3	AMi 1	AMi 2	AMi 3
Tests in sand	0.1 ± 0.3	2.1 ± 1.6	3.2 ± 4.1	8.4 ± 5.2	8.4 ± 4.6	18.3 ± 6.7
In vitro tests*	5.5 ± 5.3	0.3 ± 0.9	2.0 ± 1.6	7.0 ± 6.1	6.4 ± 4.5	4.2 ± 3.2

* Mean value of 10 plants infected with 30 L2/plant.

We then decided to check if penetration of larvae in the roots of AMi mutants was the same as in control roots. For this purpose we infected the different plants *in vitro* with the same inoculum. After 24 hours, the roots were fixed and stained with acid fuchsine (a dye that stains *Meloidogyne* nematodes within the roots) (Daykin and Hussey, 1985). The larvae present inside the roots were counted. As seen in Table 3, the numbers are lower for all three mutants tested. However, they are only statistically significant for AMi 2. These results suggest that partial inhibition of larval penetration in the roots of AMi 2 (possibly because of the modified root structure of these mutants) might explain the reduced infection symptoms.

Table 3. Penetration test.

Mutant	Larvae/mutant root system	Larvae/control root system
AMi 1	5.5 ± 5.3	7.0 ± 6.1
AMi 2	0.3 ± 0.9	6.4 ± 4.5
AMi 3	2.0 ± 1.6	4.2 ± 3.2

Mean values obtained with 10 plants; number of larvae present in roots 24 hours after inoculation.

AMi 6, AMi 7, AMi 8, AMi 9, AMi 10, AMi 11, and AMi 12 Mutants

As screening for mutants is easier *in vitro* we performed additional screenings under these conditions. Approximately 10,000 mutagenized plants were tested and 30 selected. Out of these 30 candidates tested in the M_3 generation, 7 have been retained as being affected in their response to *M. incognita* and need to be further analyzed.

Conclusion

Mutagenized seeds (15,000) have been screened for resistance to *M. incognita* and 11 partially resistant mutants have been isolated. The further analysis and characterization of these mutants should provide interesting information about this complex plant-parasite interaction.

It is possible that most plant genes, necessary for the proper development of the parasite thus leading to full resistance if mutated, are also essential to the plant. Nevertheless, this mutagenesis approach should lead to the isolation of interesting resistant mutants, that will allow a "genetic dissection" of the different steps of the interaction between *M. incognita* and plants. Higher numbers of mutagenized plants will have to be screened. It also may be useful to vary the source of mutagenizing agent to get different types of mutations. Ideally, we hope to find tagged mutants (see below).

MOLECULAR APPROACHES

One of the most common techniques used to isolate plant genes induced by pathogens, is the differential screening of cDNA libraries constructed from infected material. We have used this technique to isolate potato genes induced by *G. pallida* (Niebel et al., 1993b) and tomato genes induced by *M. incognita* (Van der Eycken et al., 1992). As yet we have not used this approach on *A. thaliana*.

Promoter Tagging

Recently, we have started another approach to identify nematode-responsive promoters. Teeri et al. (1986) have demonstrated that a T-DNA gene fusion vector can be used effectively in plants to identify and characterize regulatory elements. Data from two laboratories (Koncz et al., 1989; Herman et al., 1990) indicate that during *A. tumefaciens*-mediated transformation T-DNAs integrate preferentially in sequences that can be transcribed. These observations were used to design a promoter trapping approach in *Arabidopsis*. Constructs containing a promoterless β-glucuronidase (GUS) sequence close to the T-DNA border together with a selectable marker were introduced into *A. thaliana* via root transformation (Valvekens, 1989). The subsequent screening for β-glucuronidase (*gus*) expression allows the identification of lines containing a fusion between the *gus* gene and 5' regulatory plant sequences (Kertbundit et al., 1991). Inducible promoters can be identified by screening the transformed lines for *gus* expression following various stimuli. Genomic sequences flanking the inserted T-DNA (likely to contain the desired regulatory elements) can be amplified by an inverse polymerase chain reaction and, subsequently isolated (Lindsey et al., 1993).

This approach is particularly elegant for the isolation of nematode-responsive promoters in plants. Indeed, using promoter trapping, it is possible to screen for transgenic lines expressing *gus* at any stage of the interaction (penetration, migration, induction or maintenance of feeding site, reproduction). Additionally, *gus* expression can be assayed in the entire infected or non-infected plant. A temporal and spatial expression analysis can thus be obtained rapidly and easily. In contrast, differential screenings of cDNA libraries give a picture of the molecular events after nematode infection only in a given organ and at a given time point.

So far, our main effort has been to optimize all aspects of the *A. thaliana* transformation protocol in order to efficiently produce a large collection of transgenic lines. Until now, 1,500 T_1 lines have been obtained. We have started to screen these lines for *gus* expression after root-knot and cyst nematode infection.

Approximately 30 lines have been infected with *H. schachtii* and 7 showed staining in syncytia, one week after inoculation. The GUS staining was not restricted to syncytia. One line expressed the *gus* gene only in veins of young leaves and in syncytia. About 40 other lines have been infected with *M. incognita* and for two of these, GUS staining was found in galls. One of them, Arm1 (*Arabidopsis/Meloidogyne* 1), expresses *gus* in giant cells and, at a lower level, in initiating side roots. Arm1 is also the only line for which staining was found after both root-knot and cyst nematode infection.

Conclusion. These preliminary results indicate that it is possible to use this approach to identify promoters that are up-regulated in nurse cells. The high frequency of lines expressing *gus* in nurse cells suggests even that the main part of the work will be to select the most interesting lines among all the positive candidates identified. These lines will then be further analyzed and the corresponding regulatory sequences will be isolated.

Another application of the promoter tagging approach lies in the possibility of screening for mutants among the lines expressing *gus* in nematode-feeding sites. The T_1

lines obtained are heterozygous for the T-DNA. After self-fertilization, the effect of the homozygous mutation on nematode infection can then be studied on 25% of the progeny. This might prove to be a useful source of tagged nematode-resistant *A. thaliana* mutants.

Induction by Nematodes of the Plant Cell Cycle in Nurse Cells

An important point that will be further developed by Grundler et al. (this volume), is the simplicity and translucent character of the *A. thaliana* root. All the observation described (see below) have been made directly on whole roots. This would not have been possible in any of the usual hosts of *H. schachtii* or *M. incognita*.

Giant cells and syncytia induced by root-knot and cyst nematodes, respectively, are functionally similar organs. Both are essentially efficient "food factories", upon which the nematodes rely entirely for feeding. They also share several structural features: they are multinucleate their nuclei are hypertrophied, the metabolic activity of these cells is very high, and their cell walls develop ingrowths (typical of transfer cells), thought to enhance solute transport from the xylem towards the nurse cell. Despite these functional and structural similarities, giant cells and syncytia differ completely in the way they are formed. Giant cells have been described as developing by repeated mitosis without cytokinesis and subsequent cell expansion. Syncytia develop by fusion of cells upon cell wall breakdown and cell expansion. No sign of mitosis has been described in that case. Given these facts, we were interested to study the cell cycle in nurse cells from a molecular point of view.

Cdc2 is a protein kinase originally isolated from yeast that is a key regulator of the cell cycle. In *A. thaliana*, the gene encoding a functional homologue of Cdc2 (*cdc2a*) has been recently isolated (Ferreira et al., 1991). To follow *cdc2a* expression after nematode infection, we used transgenic *A. thaliana* plants transformed with a chimeric construct containing the promoter of *cdc2a* fused to the *gus* gene. In uninfected roots, GUS staining was visible in the meristematic zone of root tips, emerging lateral roots, and in the pericycle and vascular parenchyma of young roots (Hemerly et al., 1993).

***cdc2a* Expression After *Meloidogyne incognita* Infection.** No significant effects on *gus* expression could be seen during penetration and migration of the larvae, except for an inhibition effect in the meristematic zones disturbed by migrating larvae. As soon as the larvae became sedentary and induced a feeding site, very intense staining was observed around the head of feeding nematodes and in the neighboring tissues that had started to form a typical gall. The GUS staining became even stronger as the gall developed further. This high expression level was observed for about 3 days, after which the intensity of the GUS staining started to fade away, and to become practically absent in later stages.

***cdc2a* Expression After *Heterodera schachtii* Infection.** *Heterodera* also infects older root parts where *cdc2* expression is absent. This was very useful for the observation of the very early phases of syncytium induction. GUS staining was thus observed in the initial syncytial cell, less than an hour after syncytium induction by *Heterodera*. Very intense GUS staining was then observed for approximately 3 days throughout the syncytium. Then again the intensity of the staining faded away and no staining was detectable in syncytia at the adult stage of the parasite.

Conclusions. Cell divisions have already been described extensively both in giant cells induced by *Meloidogyne* species and in neighboring gall cells. Our results on *cdc2* expression confirm these morphological data at the molecular level. The expression of *cdc2* in syncytia induced by *H. schachtii* was unexpected since cell division has never been described in these structures (metaphase chromosomes have never been reported). A possibility is that *Heterodera* only induces a partial mitosis that is blocked in an early phase. The strong and transient induction of a key regulator of the cell cycle after root-knot and cyst nematode infection, is, however, clearly correlated with the initial steps of nurse cell formation. This suggests that the early control by the nematode of the plant cell cycle in developing nurse cells could play a key role in the complex redifferentiation process observed.

Acknowledgments

The authors thank Wout Boerjan for providing the mutagenized seed stock, Adriana Hemerly and Paulo Ferreira for P*cdc2-gus* plants and extensive discussions, Liz Corben and Tom Gerats for critical reading, Martine De Cock for help with the manuscript, and Vera Vermaercke for artwork. This work was supported by grants from the Belgian Programme on Interuniversity Poles of Attraction (Prime Minister's Office, Science Policy Programming, #38) and the Vlaams Actieprogramma Biotechnologie (No. 073). A.N., N.B., and M.K. are indebted to the European Community, the Instituut ter Aanmoediging van het Wetenschappelijk Onderzoek in Nijverheid en Landbouw, and the Iranian Ministry of Higher Education for a predoctoral fellowship, respectively; and J.d.A.-E. to the Conselho Nacional de Pesquisa for a postdoctoral fellowship. G.G. is a Senior Research Assistant of the National Fund for Scientific Research (Belgium).

REFERENCES

Aarts, M.G.M., Dirkse, W.G., Stiekema, W.J., and Pereira, A., 1993, Transposon tagging of a male sterility gene in *Arabidopsis*, *Nature* 363:715.

Arumuganathan, K., and Earle, E.D., 1991, Nuclear DNA content of some important plant species, *Plant Mol. Biol. Reporter* 9:208.

Bancroft, I., Jones, J.D.G., and Dean, C., 1993, Heterologous transposon tagging of the DRL1 locus in Arabidopsis, *Plant Cell* 5:631.

Daykin, M.E., and Hussey, R.S., 1985, Staining and histopathological techniques in nematology, *in* "An Advanced Treatise on Meloidogyne", Barker, K.R., Carter, C.C., and Sasser, J.N. eds. Department of Plant Pathology and US Agency for International Development, Raleigh.

Dong, X., Mindrinos, M., Davis, K.R., and Ausubel, F.M., 1991, Induction of *Arabidopsis* defense genes by virulent and avirulent *Pseudomonas syringae* strains and by a cloned avirulence gene, *Plant Cell* 3:61.

Ferreira, P.C.G., Hemerly, A.S., Villarroel, R., Van Montagu, M., and Inzé, D., 1991, The *Arabidopsis* functional homolog of the $p34^{cdc2}$ protein kinase, *Plant Cell* 3:531.

Hauge, B.M., Hanley, S.M., Cartinhour, S., Cherry, J.M., Goodman, H.M., Koornneef, M., Stam, P., Chang, C., Kempin, S., Medrano, L., and Meyerowitz, E.M., 1993, An integrated genetic/RFLP map of the *Arabidopsis thaliana* genome, *Plant J.* 3:745.

Hemerly, A.S., Ferreira, P.C.G., de Almeida Engler, J., Engler, G., Van Montagu, M., and Inzé, D., 1993, *cdc2* expression in *Arabidopsis thaliana* is linked with competence for cell division, *Plant Cell* 5:1711.

Herman, L., Jacobs, A., Van Montagu, M., and Depicker, A., 1990, Plant chromosome/marker gene fusion assay for study of normal and truncated T-DNA integration events. *Mol. Gen. Genet.* 224:248.

Ishikawa, M., Obata, F., Kumagai, T., and Ohno, T., 1991, Isolation of mutants of *Arabidopsis thaliana* in which accumulation of tobacco mosaic virus coat protein is reduced to low levels, *Mol. Gen. Genet.* 230:33.

Kertbundit, S., De Greve, H., Deboeck, F., Van Montagu, M., and Hernalsteens, J.-P., 1991, *In vivo* random β-glucuronidase gene fusions in *Arabidopsis thaliana*, *Proc. Natl. Acad. Sci. USA* 88:5212.

Koch, E., and Slusarenko, A., 1990, *Arabidopsis* is susceptible to infection by a downy mildew fungus, *Plant Cell* 2:437.

Koncz, C., Martini, N., Mayerhofer, R., Koncz-Kalman, Z., Körber, H., Redei, G.P., and Schell, J., 1989, High-frequency T-DNA-mediated gene tagging in plants, *Proc. Natl. Acad. Sci. USA* 86:8467.

Li, S.L., and Rédei, G.P., 1969, Estimation of mutation rate in autogamous diploids, *Radiation Bot.* 9:125.

Lindsey, K., Wei, W., Clarke, M.C., McArdle, H.F., Rooke, L.M., and Topping, J.F., 1993, Tagging genomic sequences that direct transgene expression by activation of a promoter trap in plants, *Transgenic Res.* 2:33.

Meyerowitz, E.M., 1989, *Arabidopsis*, a useful weed, *Cell* 56:263.

Niebel, A., De Almeida, J., Van Montagu, M., and Gheysen, G., 1993a, Identification of root knot nematode-resistant mutants in *Arabidopsis thaliana*, *in* "Mechanisms of Plant Defense Responses", (Development in Plant Pathology, Vol. 2), Fritig, B., and Legrand, M. eds. Kluwer Academic Publishers, Dordrecht.

Niebel, A., Van der Eycken, W., Van Montagu, M., and Gheysen, G., 1993b, Isolation and characterization of two potato genes induced during cyst nematode infection. *Arch. Int. Physiol. Biochim. Biophys.* 101, B18.

Reiter, R.S., Williams, J.G.K., Feldmann, K.A., Rafalski, J.A., Tingey, S.V., and Scolnik, P.A., 1992, Global and local genome mapping in *Arabidopsis thaliana* by using recombinant inbred lines and random amplified polymorphic DNAs, *Proc. Natl. Acad. Sci. USA* 89:1477.

Sijmons, P.C., Grundler, F.M.W., von Mende, N., Burrows, P.R., and Wyss, U., 1991, *Arabidopsis thaliana* as a new model host for plant-parasitic nematodes, *Plant J.* 1:245.

Simpson, R.B., and Johnson, L.J., 1990, *Arabidopsis thaliana* as a host for *Xanthomonas campestris* pv. *campestris*, *Mol. Plant-Microbe Interactions* 3:233.

Teeri, T., Herrera-Estrella, L., Depicker, A., Van Montagu, M., and Palva, E.T., 1986, Identification of plant promoters *in situ* by T-DNA-mediated transcriptional fusions to the *npt*-II gene, *EMBO J.* 5:1755.

Tsay, Y.-F., Frank, M.J., Page, T., Dean, C., and Crawford, N.M., 1993, Identification of a mobile endogenous transposon in *Arabidopsis thaliana*, *Science* 260:342.

Valvekens, D., Van Montagu, M., and Van Lijsebettens, M., 1988, *Agrobacterium tumefaciens*-mediated transformation of *Arabidopsis* root explants using kanamycin selection, *Proc. Natl. Acad. Sci. USA* 85:5536.

Van der Eycken, W., Niebel, A., Inzé, D., Van Montagu, M., and Gheysen, G., 1992, Molecular study of root knot induction by the nematode *Meloidogyne incognita*. *Med. Fac. Landbouww. Univ. Gent* 57/3a:895.

Walden, R., Hayashi, H., and Schell, J., 1991, T-DNA as a gene tag, *Plant J.* 1:281.

Whalen, M.C., Innes, R.W., Bent, A.F., and Staskawicz, B.J., 1991, Identification of *Pseudomonas syringae* pathogens of *Arabidopsis* and a bacterial locus determining avirulence on both *Arabidopsis* and soybean, *Plant Cell* 3:49.

Whyte, R.O., 1946, "Crop Production and Environment," Faber & Faber, London.

ARABIDOPSIS THALIANA AND *HETERODERA SCHACHTII*: A VERSATILE MODEL TO CHARACTERIZE THE INTERACTION BETWEEN HOST PLANTS AND CYST NEMATODES

Florian M.W. Grundler[1], Annette Böckenhoff[1],
Kay-Peter Schmidt[1], Miroslaw Sobczak[1],
Wladislaw Golinowski[2] and Urs Wyss[1]

[1]Institut für Phytopathologie
Universität Kiel
Hermann-Rodewald-Straße 9
24118 Kiel, Germany

[2]Katedra Botaniki
SGGW Warszawa
Rakowiecka 26/30
02528 Warsaw, Poland

INTRODUCTION

Arabidopsis thaliana was shown to be a good host plant for a number of plant parasitic nematodes by Sijmons et al. (1992). Several specific properties make this plant an ideal target to study and analyze the interaction with cyst and root-knot nematodes (Wyss and Grundler, 1992a; Sijmons et al., 1994). The thin and translucent roots provide excellent possibilities to observe both, nematode behavior and plant responses by *in vivo* microscopy (Wyss and Grundler, 1992b; Wyss et al., 1992). In consequence, exactly timed samples for fixation and subsequent ultrastructural examination can be taken (Golinowski and Grundler, 1992; Grundler et al., 1994). The syncytia induced by female juveniles of *Heterodera schachtii* are usually covered by only a few cell layers. They are therefore amenable to manipulations with microcapillaries. Substances can be injected into the syncytial feeding cells with the aid of specially adapted microinjection equipment to study physiological and biochemical features associated with nematode nutrient uptake (Böckenhoff and Grundler, 1994). In addition a great number of *H. schachtii* individuals and syncytia can be obtained to prepare extracts for a biochemical and molecular analysis and also for the production of antibodies (Grundler et. al, 1993). Transgenic *A. thaliana* plants, carrying a GUS reporter gene, enable the observation of plant gene regulation in nematode feeding sites (Goddijn et al., 1993). The application of several complementary approaches may finally elucidate the complex chain of events in the interactions between the plant and *H. schachtii*.

ROOT ANATOMY

The anatomical structure of the root of *A. thaliana* is the simplest known in dicotyledonous plants. The vascular cylinder just behind the tip of young roots is covered by three single cell layers, the epidermis, cortex and endodermis (Fig. 1a). The outer layer of the vascular cylinder is formed by a single pericycle layer. The phloem elements, consisting of proto- and metaphloem, are located opposite each other at two poles. The protoxylem elements differentiate in the other two poles of the cylinder. The metaxylem elements - in main roots usually three, in lateral roots three to five - differentiate in a row between the protoxylem poles, thus separating the vascular cylinder into two halves. The conductive elements are embedded within procambial cells, which later give rise to the differentiation of secondary phloem and xylem elements. During secondary growth the epidermis, cortex and endodermis degenerate, while new covering tissue is formed from pericycle and procambial cells.

NEMATODE INVASION

The infective juveniles (J2) of *H. schachtii* are able to invade all parts of the root, and penetrate also the stem and leaves of *A. thaliana* in agar cultures (Sijmons et al., 1991). Epidermal cells are ruptured by stylet thrusts when the juveniles enter the young roots. In most cases they move immediately towards the vascular cylinder, which they reach by a destructive intracellular migration. In some cases migration through the cortical layer is extended to a long necrotic path parallel to the vascular cylinder, while in others only the anterior end of the body is pushed into the root tissue and retracted again, leaving necrotic spots within the outer root tissue.

FEEDING BEHAVIOR

Within the vascular cylinder the J2 orientate themselves towards one of the two xylem poles. The destructive behavior changes into a subtle exploration and finally one of the procambial cells, located adjacent to a xylem element, is selected for the induction of the syncytium (Golinowski et al., in prep.). The cell is perforated by careful stylet thrusts and the stylet tip stays protruded for a period of several hours (Wyss, 1992). Feeding does not occur during this "preparation period", which obviously prepares both nematode and plant cell for their future functions. Subsequent feeding of the juveniles is characterized by a repetition of a behavioral pattern composed of three different phases. During phase one nutrients are withdrawn from the syncytial cell via the inserted stylet by continuous pumping of the metacorpal bulb. During the second phase the stylet is retracted and reinserted, while in the third phase secretions from the dorsal glands are released through the stylet orifice. These distinct feeding phases are maintained throughout the entire development (Wyss and Zunke, 1986; Wyss, 1992).

SYNCYTIA OF FEMALES

The anatomical changes in the plant during syncytium differentiation described by Golinowski et al. (in prep.) are as follows. A few hours after the induction, the initial syncytial cell fuses with neighboring procambial cells by partial dissolution of the cell walls. Precursor cells of xylem elements are primarily integrated into the syncytium. Thus, the syncytium expands along the differentiated xylem elements from the nematode's head in both directions. While integrated cells hypertrophy, the neighboring

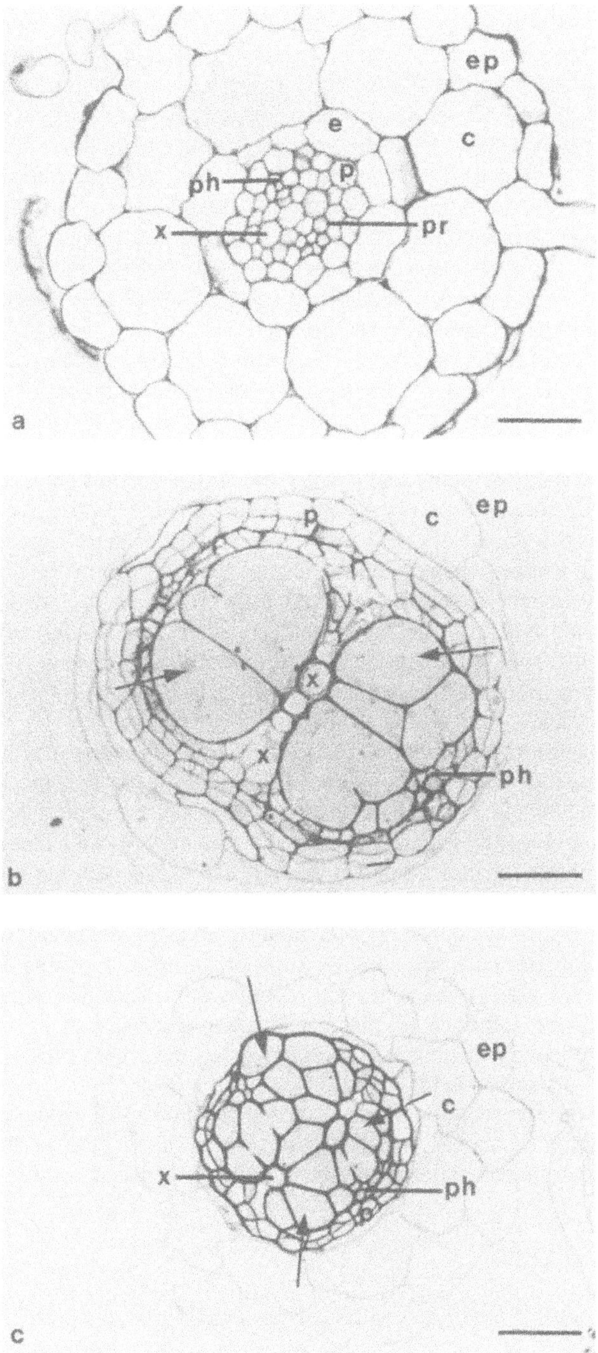

Fig.1: (a) Cross section through a control root of *A. thaliana* behind the root tip. (b) Cross section through a syncytium of a fourth stage female juvenile. (c) Cross section through a syncytium of a third stage male juvenile at the third stage. ep = epidermis, c = cortex, e = endodermis, p = pericycle, pr = procambium, ph = phloem, x = xylem, arrows = syncytial cells, bar = 20 μm.

and surrounding procambial and pericyclic cells undergo divisions (Fig. 1b). With the expansion of the syncytium, division activity is permanently induced in the distal regions. New cells arising from the divisions are partially included into the syncytium. At the same time the endodermis degenerates, while the cortex and epidermis may burst due to the expansion of the syncytium. At the distal parts of the syncytium only a few fused and slightly hypertrophied cells form syncytium extensions, which protrude finger-like into the vascular cylinder. Also in these regions procambial and pericyclic tissue show a distinct hyperplastic response. The differentiated metaxylem elements at the two poles are pushed apart by the hypertrophying syncytial cells but become embedded between the syncytial cells, thus forming two supporting axes, along which the syncytium differentiates. The development of primary xylem is suppressed by syncytium differentiation but in later stages secondary xylem is formed in the tissue around the syncytium. The phloem poles are also pushed apart but the phloem elements proliferate as an early response to syncytium differentiation.

The ultrastructural events during syncytium differentiation have also been studied by Golinowski et al. (in prep.). As an early response to the nematode's activity the cytoplasm of the initial syncytial cell becomes condensed, the tonoplast disintegrates and the vacuole is reorganized into a high number of small vesicles. The number of mitochondria, dictyosomes and plastids rises and rough endoplasmic reticulum becomes more and more elaborate. Osmiophilic granules within the plasmalemma are relatively abundant. During the third stage (J3) many free ribosomes appear and the nuclei in the syncytium assume a lobed shape. The number of vesicles within the dense cytoplasm is now considerably lower. Towards the fourth stage (J4) a characteristic proliferation of smooth endoplasmic reticulum becomes clearly evident and tubular structures and lipid bodies are observed, which remain unchanged during the adult stage. After the death of the female the syncytium disintegrates, the cytoplasm becomes translucent and is filled with vesicles.

During syncytium differentiation distinct cell wall processes play an important role. Before integration into the syncytium, the walls of the cells become bent. In the bent regions the plasmalemma forms invaginations, so-called paramural bodies, into which fibrillar material and vesicles are integrated. At sites of cell wall dissolution paramural bodies are often observed, indicating that they are associated with the process of cell wall dissolution. Becoming detached from the cell wall and the plasmalemma, the paramural bodies, now designated as multivesicular bodies, undergo degradation. While the cell walls between syncytial cells are dissolved at certain sites, those at the syncytial/non-syncytial border are always thickened and contain only very few plasmodesmata. Cell wall ingrowths at the interface of the metaxylem elements, which are typical for the feeding cells of sedentary nematodes, develop only in the region close to the nematode's head from the third stage onwards.

In spite of numerous observations, many questions still have to be resolved by detailed investigations, mainly in connection with the very early events during the induction of the initial syncytial cell and the physiology of nematode feeding.

SYNCYTIA OF MALES

As the sex of *H. schachtii* juveniles can only be determined at the end of the second stage, it is difficult to decide whether the site of syncytium induction in male juveniles differs from that of female J2. Nevertheless, a number of features, characteristic for syncytia of male juveniles, have been observed by Sobczak et al. (in prep.) in those cases in which male development was enhanced by specific growth conditions. Generally the syncytia of male juveniles remain much smaller than those of females at a corresponding stage. The reduced size results from a less extended expansion along the central cylinder and also from a considerably reduced hypertrophy of the syncytial cells. Close to the nematode's head hypertrophied pericyclic cells are integrated into the syncytium. The expansion of the syncytium then normally proceeds along the procambial cells along the vascular cylinder. However, fusion of cells by dissolution of cell walls occurs only close to the juvenile's head. In many cases groups of cells in the vicinity of the differentiating syncytium are triggered to an hyperplastic response without regular differentiation, thus forming a callus-like tissue. Remote from the feeding site, the syncytial cells are recognizable by their slight hypertrophy and the condensed cytoplasm, which, however, does not stain as intensively as in syncytia of females. The syncytial cells quite often are not fused and are scattered over the whole vascular cylinder. In principle the ultrastructural features are very similar to those found in female syncytia. The lower degree of cytoplasm condensation and the less elaborate system of cell wall ingrowth are the most pronounced differences.

The anatomical and ultrastructural results show that *A. thaliana* is a good host for *H. schachtii*, in which functional feeding sites with typical features are differentiated. A number of biological properties render this excellent model plant more suitable for detailed anatomical and cytological studies than hosts of economic importance, which are much more difficult to handle.

MICROINJECTION INTO SYNCYTIA

Injection experiments were performed to study the range of molecular weights and sizes of substances which can be taken up by the nematode from its syncytium (Böckenhoff and Grundler, 1994). With the aid of specially adapted microinjection equipment it is possible to inject substances *in situ* into the syncytia of J4 and adult females. For this purpose a commercially available pneumatic microinjector (Eppendorf, Hamburg, Germany) was used to inject the fluorescent dye Lucifer Yellow CH through glass microcapillaries with a tip diameter of about 0.3 μm. After exploratory tests, the internal pressure of root cells was measured in collaboration with Prof. Zimmermann at the Institut für Biotechnologie, Universität Würzburg. Within procambial cells of the vascular cylinder pressures of about 4 bars were recorded, whereas inside syncytia the pressures were 9 bars and more. Therefore the equipment was adapted to admit injections at pressures of up to 10 bars. After successful injections, the dye (0.457 kDa) spread quickly within the syncytial cytoplasm and was subsequently taken up by the feeding nematodes as indicated by their fluorescing digestive tracts.

In order to determine the range of molecular sizes of substances taken up by the nematodes, fluorochrome-labeled dextrans of different sizes were injected into the syncytia. Dextrans of 3, 10, 20, 40 and 70 kDa were used as probes. The nematodes were able to ingest dextrans up to a weight of 20 kDa, while the probes of 40 kDa and 70 kDa were not taken up. Apparently an exclusion limit exists between 20 kDa and 40 kDa.

In transport studies the effective size of a substance rather than its molecular weight must be regarded as a reliable parameter. The hydration sizes of globular molecules in aqueous solutions are described by the Stokes radius. Although dextran molecules in solution have a certain degree of flexibility, their hydrodynamic properties are well approximated by those of hard spheres. The Stokes radius of a 20 kDa dextran is about 3.2 nm and that of a 40 kDa dextran about 4.4 nm. It can therefore be assumed that for the injected dextrans a molecular sieve in the range between 3.2 and 4.4 nm prevents the uptake of larger molecules.

The fluorescent dye Lucifer Yellow, which is not membrane permeable, has often been used to study the symplastic contact between plant cells (e.g. Fisher, 1988). After injection into syncytia the dye remained within the syncytial cytoplasm and was not

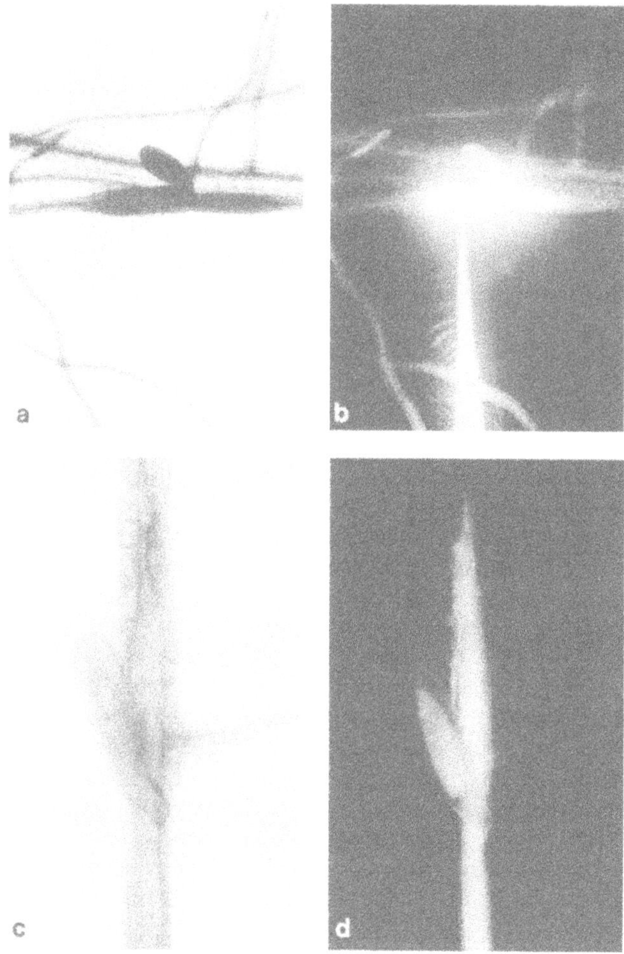

Fig. 2: Microinjection of the fluorescent dye Lucifer Yellow CH into syncytia of fourth stage female juveniles. (a) Light micrograph. (b) The same specimen in a fluorescence micrograph during dye injection with a microcapillary. (c) Light micrograph of an early fourth stage female juvenile after dye injection. (d) The same specimen in a fluorescence micrograph view, showing ingested fluorescent dye in the digestive tract.

transported into adjacent cells. This indicates an absence of functional plasmodesmata within the syncytial cell wall. On the basis of the observations we conclude that the plasmalemma of the syncytium must be perforated by the nematodes' stylet and that the feeding tubes, through which the nematodes are supposed to withdraw the nutrients, are not entirely covered with a membrane.

Many questions arise from the fact that the nematodes feed from cells with an enhanced internal pressure by using a mechanism that acts as a molecular sieve. The ultrastructural observations support the data from microinjection. The abundance of organelles within a condensed cytoplasm indicates a high metabolic activity, which may lead to an enhanced internal osmotic pressure. The thickened cell walls enclosing the syncytium may act as a physical antagonist against the high pressure. Higher levels of proteins and an enhanced proportion of free amino acids were often found in syncytia (Krauthausen and Wyss, 1982; Betka et al., 1991; Grundler et al., 1991). The experiments with dextrans do not lead to general conclusions on the size and conformation of proteins and other substances, which may serve the nematodes as nutrients. Only a minor proportion of these molecules may have a globular shape and it is possible that they are predigested before uptake and are thus altered in conformation. Additional parameters, like the charge of the molecules, may play a role in their uptake. Current experiments with injections of proteins have begun, which may give further information (Böckenhoff and Grundler, unpublished).

CHARACTERIZATION OF SYNCYTIAL PROTEINS

The pronounced structural changes of the root during the differentiation of syncytia are also manifested at the physiological and biochemical level. The shift in the protein level has already been mentioned above.

There are several ways to determine the nature and molecular background of proteins with altered expression patterns in syncytia. With the help of differentially screened cDNA libraries of infected and uninfected root tissue, the expression of proteins can be studied at the transcriptional level. Separation and purification of proteins by electrophoretic methods, on the other hand, enable microsequencing, production of

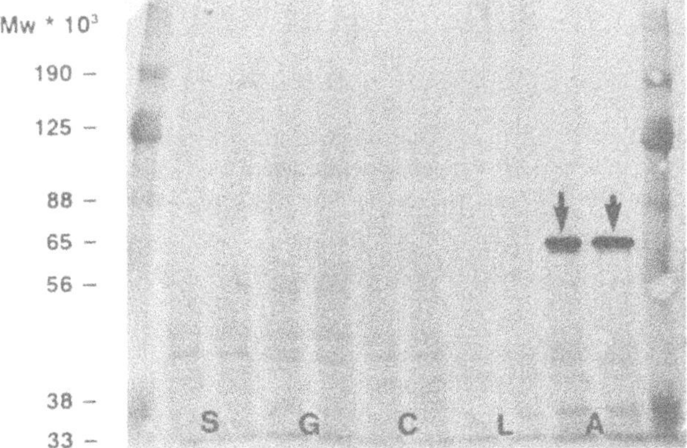

Fig. 3: Western blot of a SDS-PAGE of root protein extracts of uninfected *Sinapsis alba* (S), *Glycine max* (G), *Cucumis sativa* (C), *Lycopersicon esculentum* (L) and *Arabidopis thaliana* (A). The extracts were arranged in double tracks. The monoclonal antibody binds specifically to the *A. thaliana* protein (arrow).

degenerated oligonucleotides and PCR-amplification of the desired DNA sequences. Another approach is the production of antibodies against syncytial proteins and a subsequent differential screening. Candidates can then be used to screen already existing expression libraries of *A. thaliana* roots.

With all mentioned approaches the problem exists that it must be determined in which way the synthesis of a single protein is changed at a generally elevated protein level of infected tissue. Considering a four fold rise of the protein level in syncytia, proteins which occur at a disproportional level are of special interest. As GUS expression studies show, some promoters are downregulated in syncytia even to a non-detectable level (Goddijns et al., 1993). We assume that a *de novo* production of a syncytium-specific protein most likely does not occur.

We use the last two of the mentioned approaches. Monoclonal antibodies against syncytial proteins were raised in collaboration with K. Davies and P. Burrows, Rothamsted Experimental Station, U.K., and screened with ELISA techniques against extracts of syncytia, uninfected root tissue, tops of *A. thaliana* plants and female nematodes. Four of more than two hundred clones secreted antibodies against one protein, specifically produced in roots of *A. thaliana* (Fig. 3) and expressed in syncytia at an altered level. The monoclonal antibodies are currently used to screen a cDNA expression library of *A. thaliana* roots in mammalian COS cells, which was established by W. Kammerloher, Institut für Biochemie, Universität München, Germany (Kammerloher and Schäffner, 1993).

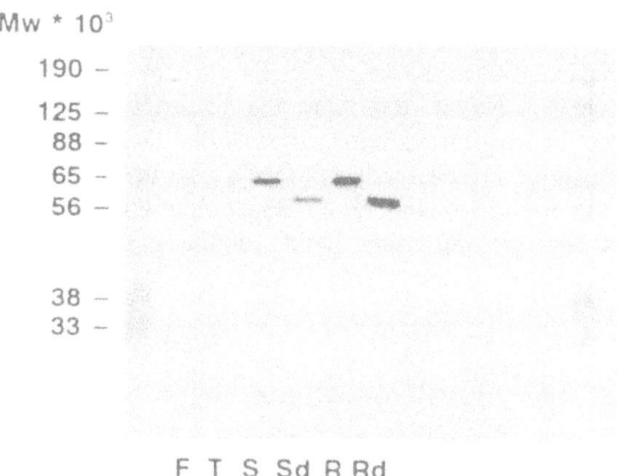

Fig. 4: Western blot of a SDS-PAGE of a protein extract of *A. thaliana* roots (R) and tops (T), syncytia (S) and females (F) of *H. schachtii*. Syncytium and root extracts were also treated with N-glycosidase F. The probes indicate that the antibody binds to a root-specific glycoprotein at an epitope on the peptid chain. d = extract digested with N-glycosidase F.

In the second approach the proteins of a root extract were separated by 2-D electrophoresis revealing about 700 distinct spots. Differences between uninfected and infected root material are currently being studied. The protein recognized by the four antibodies could be collected and purified to a high degree. This approach provides sufficient purified protein to allow the determination of amino acid sequences of polypeptide fragments of the protein. On the basis of these sequences, degenerate oligonucleotides will be synthesized in order to use suitable constructs as PCR primers to amplify genomic sequences.

Treatment with N- glucosidase F proved that the protein is a glycoprotein (Fig.4). As the digested protein retained its immunogenity, the epitope must be located on its polypeptide chains. The complete protein is 63 kDa, the deglycosidated protein fragment is 58.5 kDa. The final characterization of the protein will clarify its biochemical properties and its physiological function. In this way it becomes possible to determine proteins, which are under the direct or indirect influence of the nematode. Immunological detection on ultrathin sections and microinjections of the labeled specific antibody and protein into syncytia will complete these studies. We are aware of the fact that there are many proteins in a syncytium with altered expression patterns. However, the characterization of a few with different patterns may help to find some general features, in order to clarify the processes responsible for a syncytium-specific regulation of protein synthesis.

REFERENCES

Betka, M, Grundler, F. and Wyss, U., 1991, Influence of changes in the nurse cell system (syncytium) on the development of the cyst nematode *Heterodera schachtii*: Single amino acids, *Phytopathology* 81:75.

Böckenhoff, A. and Grundler, F.M.W., 1994, Studies on the nutrient uptake by the beet cyst nematode *Heterodera schachtii* by *in situ* microinjection of fluorescent probes into the feeding structures in *Arabidopsis thaliana*, *Parasitology*, in press.

Fisher, O.G., 1988, Movement of lucifer yellow in leaves of *Coleus blumei* Benth., *Plant, Cell and Environment*, 11:639.

Goddijn, O., J., M., Lindsey, K., Van der Lee F., M., Klap, J., C. and Sijmons, P. C., 1993, Differential gene expression in nematode induced feeding structures of transgenic harbouring promoter-gusA fusion constructs. *Plant J.*, 4:863.

Golinowski, W. and Grundler, F.M.W., 1992, The structure of feeding sites (syncytia) of *Heterodera schachtii* in roots, stems and leaves of *Arabidopsis thaliana*, *Nematologica* 38:413.

Grundler, F., Betka, M. and Wyss, U., 1991, Influence of changes in the nurse cell system (syncytium) on sex determination and development of the cyst nematode *Heterodera schachtii*: Total amounts of proteins and amino acids, *Phytopathology* 81:70.

Grundler, F.M.W., Böckenhoff, A., Golinowski, W., Münch, A., Schmidt, K.-P., Sobczak, M., Davies, K., Burrows, P. and Wyss, U., 1993, Approaches to the analysis of the host-parasite interaction between *A. thaliana* and gall- and cyst- forming plant parasitic nematodes, in: "Abstracts of the International Conference in *Arabidopsis* Research", R. Hangarter, Scholl, R., Davies, K., Feldmann, K., eds., The Ohio State University, Columbus.

Grundler, F.M.W, Golinowski, W. and Wyss, U., 1994, Sedentary endoparasitic nematodes, in: "*Arabidopsis*, an Atlas of Morphology and Development", J. Bowman, ed., Springer, New York.

Kammerloher, W. and Schäffner, A. R., 1993, PIP - an *A. thaliana* specific plasma membrane MIP homologue cloned by expression in mammalian cells, in: "Abstracts of the International Conference in *Arabidopsis* Research", R. Hangarter, Scholl, R., Davies, K., Feldmann, K., eds., The Ohio State University, Columbus.

Krauthausen, J. and Wyss, U., 1982, Influence of the cyst nematode *Heterodera schachtii* on relative changes in the pattern of free amino acids at feeding sites, *Physiol. Pl. Path.* 21:425

Sijmons, P.C., Grundler, F.M.W, von Mende, N., Burrows, P.R., and Wyss, U., 1992, *Arabidopsis thaliana* as a new model host for plant-parasitic nematodes, *Plant J.* 1:245.

Sijmons, P.C., von Mende, N. and Grundler, F.M.W, 1994, Plant parasitic nematodes, in: "*Arabidopsis*", C. Somerville and Meyerowitz, E., eds., Cold Spring Harbour, New York.

Wyss, U., 1992, Observations on the feeding behaviour of *Heterodera schachtii* throughout development, including events during moulting, *Fundam. appl. Nématol.* 15:75.

Wyss, U. and Grundler, F.M.W., 1992a, Seminar: *Heterodera schachtii* and *Arabidopsis thaliana*, a model host-parasite interaction. *Nematologica* 38:488.

Wyss, U. and Grundler, F.M.W., 1992b, Feeding behaviour of sedentary plant parasitic nematodes, *Neth. J. Pl. Pathol.* 98: Supplement 2:165.

Wyss, U., Grundler, F.M.W. and Münch, A., 1992, The parasitic behaviour of second stage juveniles of *Meloidogyne incognita* in roots of *Arabidopsis thaliana*, *Nematologica* 38:98.

Wyss, U. and Zunke, U., 1986, Observations on the behaviour of second stage juveniles of *Heterodera schachtii* inside host roots, *Rev. Nématol.* 9:153.

MOLECULAR AND CELLULAR DISSECTION OF GIANT CELL FUNCTION

David McK. Bird and Mark A. Wilson

Department of Nematology and Graduate Genetic Group
University of California-Riverside
Riverside CA 92521 USA

INTRODUCTION

Plant parasitic nematodes significantly affect the productive capability of the world's farmlands. Sasser and Freckman (1987) have calculated that they reduce the yield of the world's forty major food staples and cash crops by an average of 12.3%, with the losses being substantially higher for some commodities (e.g., 20.6% for tomato). The US proportion, in monetary terms, is in excess of $5.8 billion annually. As a group, the root-knot nematodes (*Meloidogyne* spp.) are the major contributors to these losses.

Root-knot nematodes are sedentary endo-parasites that form complex associations with their host plant. Worms hatch as second-stage (L2) larvae and invade the root in the zone of elongation. They migrate intercellularly, first to the root apex and then to the developing vascular cylinder where permanent feeding sites are established (Wyss et al., 1992). In response to repeated stimulation from the parasite, cells in the root stele are directed to initiate DNA synthesis and nuclear division (but not cell division), and also to assume a new differentiated fate termed a giant cell.[†]

Since their first characterization by Cornu (1879), giant cells induced by various members of the genus *Meloidogyne* have been subjected to considerable scrutiny. These multinucleate cells become large and avacuolate, and undergo extensive remodeling of their cell wall (Bird, 1961). They are metabolically active (Bird, 1971) and serve as the obligate nutritive source for the developing nematode, in which the locomotive musculature ultimately atrophies. Giant cell formation, coupled with limited proliferation of nearby pericycle and cortical cells results in the characteristic root-knot gall.

Light and EM studies (reviewed by Endo, 1987) have shown that the parasite's hollow, retractable feeding stylet traverses the plant cell wall through a structure of unknown origin, termed a feeding plug, that remains after stylet withdrawal. The plasmalemma is pushed away from the wall by the stylet, the tip of which appears to be closely associated in the host cytoplasm with an electron dense feeding tube surrounded by rough endoplasmic reticulum. Feeding

[†] Many nematode species induce specific modifications in the host cells from which they feed. Although morphologically distinct, these cells can collectively be termed "nurse cells."

tubes are evident in the host cell after stylet withdrawal (Mankau and Linford, 1960). Recent work on the ring nematode, *Criconemella xenoplax*, (Hussey et al., 1992a) has shown that the stylet orifice is open to the host cytoplasm, with the plasmalemma apparently connected to the stylet. Callose is deposited between the stylet and the invaginated membrane (Hussey et al., 1992b) and is the only tangible defense response of a sensitive host.

Little is known of the molecular mechanisms underlying the induction and maintenance of giant cells and concomitant feeding by *Meloidogyne*. Giant cells are not transformed *per se*; their formation and maintenance require repeated stimulation from the nematode (Bird, 1962). The nature of the stimulus is unknown, but it has been widely speculated (see reviews by Hussey, 1989; Hussey et al., 1994) that the inductive agent for giant cell formation is a component of the stylet secretion, perhaps originating in the dorsal and/or subventral pharyngeal glands. The subventral glands are presumably also the source of material copiously secreted from the L2's stylet during its intracellular migratory phase (Wyss et al., 1992), and it seems likely that the dorsal gland provides digestive enzymes for the parasite. Stylet secretions may also contribute to the feeding plug and/or feeding tube; it is not known whether these are of host or parasite origin. Similarly, the host target for the presumed nematode ligand(s) is arcane, as is the manner in which this signal is transduced to elicit giant cell differentiation.

FEEDING SITE SELECTION

Of the many questions one might pose of the *Meloidogyne*-host interaction, none is seemingly more straightforward than asking "what host cell types can be giant cell progenitors?" Many studies have pointed to the vascular cylinder, but the literature is often vague as to the actual cell type although this is, perhaps, not too surprising as infection by the nematode induces the formation of a new cell type. Although there are claims of giant cells arising from pericycle, cortex and epidermal cells, careful studies (e.g., Krusberg and Nielsen, 1958; Jones and Payne, 1978) implicate the vascular parenchyma as the preferred feeding site. Krusberg and Nielsen (1958) further noted that "in root tips, giant cells were ... initiated in xylem parenchyma adjacent to xylem elements." These workers also noted that some giant cells formed from phloem parenchyma. Whether or not there is a biological difference between phloem and xylem initials (i.e., are the vascular parenchyma cells determined?) is unclear. Steeves and Sussex (1989) suggest that "it is reasonable to think that the events leading to the final differentiation of these cells do not occur much before the first visual evidence of differentiation."

It is worth reviewing the cellular events associated with normal metaxylem development (see Steeves and Sussex, 1989). Concomitant with differentiation of the xylem initial is an increase in cytoplasmic density, an increase in the number of organelles (particularly Golgi and endoplasmic reticulum) and reduction of the tonoplast. The cell begins to elongate and secondary cell wall is deposited, resulting in a variety of patterns of thickening. As is common in angiosperm (but not gymnosperm) development, both the number of nuclei and the ploidy level may increase (Nagl, 1978). Most nematologists would clearly recognize these features as being characteristic of giant cell formation. Indeed, during the first days of giant cell induction, giant cells can bear a striking resemblence to developing xylem [e.g., see Fig. 26 in Jones and Payne (1978)], as do the mononucleate nurse cells induced by the gymnosperm-parasite *Meloidodera floridensis* (Mundo-Ocampo et al., 1993).

Unpublished observations of the migratory and feeding-site selection behavior of *Heterodera schachtii* in *Arabidopsis thaliana*, by Golinowski et al.

(reported by Grundler et al., 1994), showed that the syncytium is initiated from a procambial cell. These workers further reported that cells subsequently incorporated into the developing syncytium also were from the procambium, and that syncytial formation suppressed development of primary xylem. This is consistent with the notion that the procambial cells recruited into the nurse cell were in fact xylem initials. Perhaps rather than "suppressing" xylem formation, syncytial formation depletes the supply of normal xylem precursors. It is significant that the final stages of metaxylem differentiation (in tracheid formation) involve cell wall dissolution, effectively forming what would be a syncytium if the cells remained alive. The outcome of precocious expression of this normally programmed cell-wall dissolution might well be the formation of a cell-type strongly resembling an *Heterodera*-induced syncytium.

Obviously, giant cells and syncytia are not metaxylem elements. Nevertheless, morphological evidence suggests that the normal route for their formation might involve large parts of the developmental pathway followed by differentiating metaxylem elements. If this is the case, it seems likely that the nematode cues used to initiate nurse cell formation might closely resemble normal plant effectors, and might work in concert with endogenous host signals (Bird, 1992). Although these nurse cell types appear generally to arise from metaxylem initials, it should be noted that this does not preclude their genesis being from some other cell type. Indeed, Jones and Northcote (1972) described the origin of potato cyst-nematode syncytia as being in the cortex, and Magnusson and Golinowski (1991) argued that the *Heterodera*-induced syncytia of *Brassica napus* were initiated in the pericycle, with xylem parenchyma cells subsequently recruited into the syncytium. Of course, what these results might show is that xylem initials might also lie within the cortex and pericycle. As we indicate below, *in situ* localization of giant cell or syncytial transcripts will shed light on these questions.

FEEDING SITE

At a developed feeding site mature giant cells and syncytia occupy much of the space normally occupied by the xylem (see Endo, 1987). Not surprisingly, as Dorhout et al. (1991) who used a split root assay, with one side only infected with the parasite demonstrated, heavy *M. incognita* infection directly affects the ability of roots to transport water. Similarly, whilst the growth of *M. incognita*-infected cotton is significantly suppressed compared with uninfected plants when soil water levels drop below field capacity, plants are asymptomatic when water is plentiful (O'Bannon and Reynolds, 1965) and in snap bean, tolerance to nematode infection appears to be directly related to the cultivar's ability to utilize water (Wilcox and Loria, 1986).

The location of these nurse cells raises the intriguing possibility that root-knot and cyst nematodes are xylem-feeders. Other parasites that feed directly from the xylem stream include various homopteran insects and 3,000 species of parasitic angiosperms (reviewed by Press and Whittaker, 1993). The former utilize a large cibarial pump in conjunction with a stout stylet directly to tap the xylem, and excrete large volumes fluid to concentrate dilute nutrients. In contrast, the parasitic plants interact *via* an haustorium. Interestingly, xylem parenchyma elements within the hyalin body of the haustorium (reviewed by Visser and Dörr, 1987) have many features in common with giant cells and syncytia. It is conceivable that giant cells and syncytia serve to integrate the nutrient-harvesting functions of a xylem-tapping haustorium.

Of course, it is possible that root-knot and cyst nematodes are in fact phloem-feeders. Because the giant cells associated with an individual nematode are linked *via* plasmodesmatal connections, it would be necessary for only one giant cell to contact the phloem. To test whether *M. incognita*-induced giant cells were in proximity with phloem elements, we used anti-phloem antibodies to challenge sections of *M. incognita*-infected and healthy tomato roots [antibody production and immunohistochemical details are presented elsewhere (Chen et al., submitted)]. As shown in Figure 1 (A and C), the antibodies are specific for phloem cells, and also cross-react with the parasite (panel D). Although these data are preliminary, the result shown in Figure 1D in which one of the giant cells appears to be in intimate contact with a sieve element, seems to be typical. Thus, the question of the source of nutrients for the parasite remains unresolved.

GIANT-CELL GENE EXPRESSION

Although we have argued that giant cells share features with other plant cells, notably developing metaxylem cells, they are a novel cell type and presumably arise by a pattern of gene expression different from that in other plant cells. It previously has been speculated that gene expression in giant cells might include genes normally expressed at different developmental times or in different cell types (Bird, 1992).

Figure 1. Immunohistochemical localization of phloem elements. Plant tissue was fixed, embedded in paraffin and 10 μm sections challenged with pre-immune serum (A and B) or anti-phloem antibodies (C and D); signals were detected using an alkaline phosphatase-conjugated second antibody. Challenging longitudinal sections of uninfected roots (A and C) confirmed that the antibodies were phloem cell-specific. In transverse sections of galls (B and D), one of the giant cells (asterisk) is adjacent to one of the phloem cell quadrants (unlabeled arrows). G: giant cell; N: nematode; p: phloem cell; SP: sieve plate; X: xylem.

Surprisingly little is known about gene expression in healthy root cells. In a survey of randomly chosen, cloned root mRNAs, Evans et al. (1988) were unable to identify any as being root-specific. Using a differential screen, Conkling et al. (1990) isolated genes encoding four, moderately to abundantly-expressed, root-specific transcripts from tobacco. The expression of one of these, *TobRB7*, is strongly up-regulated in giant cells induced in tobacco by *M. incognita* (Opperman et al., 1994).

Gurr et al. (1991) used a differential screening approach to identify a potato gene with expression "correlated with events in the immediate vicinity of the pathogen" (potato cyst nematode, *Globodera rostochiensis*), but the nature of this gene was not revealed. In a preliminary report, Niebel et al. (1993c) describe using a similar strategy which resulted in identification of two genes with expression up-regulated following *G. rostochiensis* infection. One appears to encode a potato catalase able to respond generally to pathogen attack; the other gene could not be identified. Thus, despite what obviously represented considerable efforts, this approach has been disappointing.

In contrast, strategies that incorporate a hybridization-competition (subtraction) step as part of a library screen, or employ the direct cloning of subtracted cDNA, have proven effective in identifying transcripts with differential expression profiles, including developmentally regulated mRNAs from *Dictyostelium discoidium* (Mangiarotti et al., 1981) and gastrulation-specific sequences from *Xenopus laevis* (Sargent and Dawid, 1985).

However, classical subtractive approaches, whether they be used to construct an enriched library from which individual clones may be characterized, or to produce probe for the differential screening of a standard library, have a number of limitations. Because of their requirement for large amounts of starting mRNA (typically tens of micrograms of poly-A (+) from the driver source and >1 μg from the target source), these methods alone are not suited for isolation of giant cell-specific transcripts. Furthermore, the clones that result from either approach represent moderately to abundantly expressed transcripts; rare transcripts tend to be under-represented. Finally, the differential enrichment that the subtractive steps are designed to achieve also results in enrichment of aberrant clones, and contamination of subtracted cDNA with double stranded material after hydroxylapatite chromatography can lead to a subtracted library with substantial representation of clones present in both driver and target. Differential screening is then required to confirm subtracted clones.

Recently, a number of techniques that address some of the shortcomings of the subtractive approach has been developed. By exploiting the polymerase chain reaction (PCR) to amplify small amounts of cDNA, libraries in the order of 10^6 primary recombinants have recently been produced starting with milligram amounts of tissue (Belyavsky et al., 1989; Domec et al., 1990; Welsh et al., 1990). PCR amplification has also been used to produce cDNA as driver for subtraction (Lebeau et al., 1991). To ensure that cDNAs representing transcripts expressed at low abundance are represented requires that enough cDNA remain after subtraction for efficient cloning. By performing the cloning event before the subtraction the requirement to work with potentially very small quantities of cDNA can be avoided. Several researchers (Duguid et al., 1988; Rubenstein et al., 1990; Swaroop et al., 1991) have developed methods to produce cDNA libraries in phagemid vectors, which may be easily and efficiently transformed following subtractive hybridzation.

Library Construction

To identify genes either uniquely expressed in giant cells (compared with uninfected root cells) or with elevated expression levels, we elected to construct a subtracted cDNA bank from *M. incognita*-induced, tomato giant cells, and designed a strategy (outlined in Figure 2) based on PCR-amplification and phagemid cloning (Wilson et al., 1994). A factor that we considered crucial to the success of this approach was the biological material with which we started. Nematodes were maintained on host roots grown in axenic organ culture. This facilitated the isolation of giant cells free from the parasite and with only a minimal amount of surrounding gall tissue. Because the developmental stage of the nematode could be determined, giant cells of equivalent maturation could be harvested. Furthermore, it was possible to harvest giant cells very rapidly and with a minimum of manipulation, thereby limiting the opportunity for general host wound responses to be expressed.

Target	Driver
Poly-A$^+$ RNA captured on para-magnetic beads from giant cell total RNA	Poly-A$^+$ RNA captured on para-magnetic beads from healthy root total RNA
↓	↓
ds-cDNA synthesized on beads by RTase/RNase H/DNA pol I method	ds-cDNA synthesized on beads by RTase/RNase H/DNA pol I method
↓	↓
Not I adapter added to non-bead end (≡ 5'-end of mRNA)	*Not* I adapter added to non-bead end (≡ 5'-end of mRNA)
↓	↓
2nd strand cDNA eluted and extended with oligo-dT(15)-anchored *Sfi* I primer	2nd strand cDNA eluted and extended with oligo-dT(15)-anchored *Sfi* I primer
↓	↓
Limited PCR with *Not* I/*Sfi* I primers and size fractionation	Limited PCR with *Not* I/*Sfi* I primers and size fractionation
↓	↓
Not I/*Sfi* I cut PCR product cloned into phagemid vector, ssDNA recovered	Biotin-16-dUTP incorporated by PCR
↓	↓

Target and driver annealed together to Cot_{500}

↓

Biotinylated DNA removed on streptavidin-coated para-magnetic beads

↓

DNA transformed and giant cell cDNA clones recovered

Figure 2. Flow chart outlining strategy used to construct a subtracted bank of cDNAs representing transcripts up-regulated in *M. incognita*-induced, tomato giant cells (Wilson et al., 1994).

Fifty one milligrams of essentially pure giant cells were accumulated by dissection. Based on comparative ethidium bromide fluorescence, the yield of total RNA from this sample was estimated to be 11 μg. This represents a yield per gram fresh weight of tissue of approximately 4-fold higher than we obtained from cultured whole roots, and possibly is a reflection of increased transcriptional activity in giant cells. Interestingly, Grundler et al. (1994) have reported that the total amount of protein in *H. schachtii*-induced syncytia is elevated 4-fold above uninfected cells, and speculate, perhaps because an apparent down-regulation in *H. schachtii*-induced syncytia of promoterless *gus*A constructs randomly integrated in to the *A. thaliana* genome has been observed (Goddijn et al., 1993), that this might result from the specific amplification of a small number of proteins. Although we can not preclude this in our system, the number of recombinants we isolated that appear to correspond to distinct transcripts up-regulated in *M. incognita*-induced tomato giant cells (see below) is not consistent with this interpretation.

A library was constructed from the giant cell RNA as outlined in Figure 2; 287 insert-containing clones were recovered (Wilson et al., 1994). This corresponds to an overall enrichment of 4,860-fold of giant cell sequences over normal root sequences present in the cDNA library and although this figure fails to take in to account various other cloning losses, it appears to be substantially higher than is typically obtained by subtraction. Duguid et al. (1988), for example, achieved only an 100-fold enrichment of scrapie-modulated RNAs.

Library Analysis

Prior to analyzing individual clones, a variety of tests was performed to assess the overall quality of the subtracted library. In particular, we were interested in: 1) confirming that the clones were of plant and not nematode origin, 2) determining the complexity of our bank, and 3) identifying those clones that were derived from high copy number genes.

To determine the total number clones that defined highly repeated genomic sequences, replica filters of the library were probed with nick-translated tomato total DNA. Forty two of the 287 clones gave a high signal. Four of these clones were sequenced; three were derived from 25S rRNA and one from 16S rRNA. Further to confirm that clones not detected in this assay were indeed derived from low copy number genes, genomic Southern blots have subsequently been performed using nearly all the clones as probes. Most appear to be low copy or unique genes. Partial sequence analysis of some clones has confirmed the presence of restriction sites implicated by genomic Southern blots. These results also confirm that, with two exceptions (*M. incognita* sequences), the clones are of plant origin. Seven randomly selected clones were used as probes to hybridize back to the entire library. One clone detected two recombinants; the remaining six probes detected only the clones from which they were constructed. This result suggests that the complexity of the subtracted library might be high.

Expression profiles. Dot blots of RNA samples isolated from tomato tissues (galls and whole mature roots from tissue culture, mature leaves from green-house grown plants, and cotyledons, hypocotyls and roots from one-week-old seedlings) were hybridized with probes synthesized from giant cell cDNA clones. The presence of equal amounts of target RNA in each dot was confirmed by probing with ribosomal sequences. Results from 44 experiments are summarized in Table 1. A "+" indicates that a hybridization signal was detected, but gives no clue as to transcript abundance (exposure times varied significantly between the different filters despite the specific activities and amount of probe used in each experiment being essentially the same). Not all

Table 1. DNA sequence and RNA dot blot analysis of giant cell cDNA clones.

Clone DB#	Accession No.[2]	Identity[3]	Tissue source of RNA[1]					
			Gall	Driver	Leaf	Hyp.	Apex	Root
103	L23860	16 kD E_2 enzyme	–	–	–	+	+	–
104	Unsequencable clone		+	–	–	+	+	+
113	L24003	Pioneer	–	–	–	–	–	+
114	L24004	Pioneer	+	–	+	+	+	–
115	L24005	Pioneer	+	–	+	+	+	–
117	L24006	Heptamer repeat	–	–	–	+	+	+
118	L24013	Pioneer	–	–	–	–	–	–
125	L24007	Pioneer	–	–	–	+	+	–
137	L24008	Pioneer	–	–	–	+	+	–
139	L24010	Pioneer	+	–	+	–	–	+
140	L24014	Pioneer	+	–	+	–	–	+
141	L24009	Pioneer	+	–	–	+	+	–
142	L24011	anti-Ef-3	–	–	–	–	+	–
161	L24015	Pioneer	–	–	–	–	–	–
163	L24017	16 kD E_2 enzyme	–	–	–	+	+	–
164	L24016	Pioneer	–	–	–	–	–	+
165	L24018	Pioneer	–	–	+	–	–	–
166	L24019	Pioneer	–	–	+	–	+	–
173	L26982	Pioneer	+	+	+	–	–	+
197	L24020	Pioneer	–	–	–	–	–	–
199	L26983	anti-Ala tRNA synthase	–	–	–	–	+	–
203	L24023	Pioneer	–	–	–	–	–	–
205	L24024	Pioneer	–	–	–	–	+	–
207	L24025	Pioneer	–	–	–	–	–	–
208	L24060	Pioneer	–	–	–	–	–	–
210	L24027	Pioneer	–	–	–	–	+	–
212	L24028	Pioneer	–	–	–	–	+	–
215	L24061	Zn-finger domain	–	–	–	–	+	–
216	L24030	Pioneer	–	–	–	–	–	–
217	L24031	Laminin B receptor	+	–	+	–	–	+
220	L24033	anti-PEP-carboxylase	–	–	–	–	–	–
221	L24054	Pioneer	–	–	+	–	–	–
222	L24055	Pioneer	–	–	–	–	–	–
223	L24056	Pioneer	–	–	–	–	–	–
224	L24057	Pioneer	–	–	–	–	–	–
226	L24058	Proton ATPase	–	–	–	+	–	–
239	L24059	eF IIe 5' UTR	–	–	–	+	+	–
240	L24062	Pioneer	–	–	–	+	+	–
244	L24064	Pioneer	–	–	+	–	+	–
263	L24065	Pioneer	–	–	–	+	+	+
265	L24067	anti-Tnt1-94 transposon	–	–	–	+	+	+
275	L24069	Pioneer	–	–	–	+	+	–
279	L24110	Pioneer	–	–	–	–	–	+
280	L24111	Myb DNA-binding site	–	–	–	+	+	–

[1]RNA samples from the tissues indicated (Driver: RNA from cultured, uninfected roots and used as driver for the subtractive cloning; Hyp: hypocotyls; Apex: cotyledons plus shoot apex; Root: primary root from young seedlings).
[2]GenBank accession number.
[3]Putative identity of partial cDNA clones based on DNA or deduced amino acid homology with sequences in GenBank or PIR. Clones with no meaningful homology are termed pioneers.

the probes (e.g., DB#118) gave a signal; presumably these cDNAs represent low abundance messages. Most probes also failed to detect transcripts in gall RNA despite the fact that only cDNA synthesized from giant cell mRNA was exposed to the vector (Figure 2) thereby ensuring that all clones in our bank must encode giant cell transcripts. Although this is partly due to giant cells representing only a small fraction of the total mass of the gall (it was not technically feasible to collect the numbers of giant cells required to isolate sufficient RNA for blot analysis), it further suggests that the clones in our bank do not encode abundant transcripts. Ultimately, the (elevated) presence in giant cells of transcripts defined by each clone in our bank will need to be confirmed; *in situ* analyses are currently in progress.

Only one clone (DB#173) produced a signal in the RNA from mature roots. This is important because mature root was the source of the RNA used as driver in construction of the subtractive library, and provided strong evidence that the subtraction was effective. Those probes that gave a signal hybridized to various subsets of the RNA samples. Eleven of the cDNA clones detected transcripts in seedling root (Table 1), and presumably encode functions associated with young, expanding roots. Three of these appeared to be root-specific. In a survey of randomly chosen, cloned root mRNAs, Evans et al. (1988) were unable to identify any as being root-specific, suggesting that this class of message might be a rare in healthy root. It will be interesting to determine the function of these transcripts in our bank. Although different clones reveal a range of expression patterns, many detected transcripts in RNA isolated from hypocotyls and/or cotyledon plus apex tissue, generally at levels much higher than in uninfected seedling root, and nine cDNA clones detected a signal in leaf RNA. Overall, these results give a picture of giant cells sharing transcripts with actively dividing and expanding tissues, and also non-root tissues.

DNA sequencing. Partial cDNA sequence of 44 cDNA clones is presented in Table 1 as GenBank accession numbers (Bilofsky and Burks, 1988). Reflective of the directional nature of the cloning, each sequence began with a poly-T tract (ranging in length from 4 to 75 residues), corresponding to the poly-A tail of the transcript. The presence of a long oligomeric tail rendered some clones (e.g., DB#104) unsequencable. Four independent rRNA clones from the same bank also were determined. The sequences corresponded exactly to those published for tomato (Kiss et al., 1989). This suggests also that the number of PCR-introduced artifacts in our bank might be low.

Sequences were compared to others in the public domain DNA and protein databanks using the BlastN and BlastX algorithms (Altschul et al., 1990). With some exceptions, only those database matches involving homology with the correct strand and in the correct part of the matching gene (i.e., the 3'-end) were considered, and a score of at least 100 was chosen as indicating a potentially valid homology. Many of the cDNA sequences failed these tests, and are listed as "Pioneers" in Table 1.

A database search using a partial sequence from the cDNA clone DB#249 as a query revealed that its inferred product shares homology with that of the *RB7-5A* gene from tobacco (Conkling et al., 1990). Significantly, this gene is strongly up-regulated in *M. incognita*-induced tobacco giant cells (Opperman et al., 1994), and the presence of its tomato homologue in our bank provides strong corraborative evidence as to the quality of our library. Also consistent with previous findings is the absence (in the clones analyzed to date) of sequences encoding wound response functions. Niebel et al. (1993b) showed that expression of an extensin is up regulated during invasion and at the onset of feeding by *M. incognita*, but that the transcripts for this gene declined as the pathogen developed (and are presumably absent in giant cells associated with adult female nematodes).

The identity of the DB#103 transcript was confirmed as encoding an E_2 enzyme, a key component of the protein ubiquitination pathway. Conjugation of proteins to the 76-residue ubiquitin polypeptide is an essential step in targeting them for extralysosomal degradation and is an important element of cellular regulation (reviewed by Finley and Chau, 1991). By playing a role in modulating the levels of key proteins (e.g., the yeast mating type regulator, MATα2 repressor, and *Xenopus laevis* cyclin and Mos kinase proteins), ubiquitination exerts an influence on DNA repair, response to stress, regulation of meiotic progression and cellular differentiation, including vascular differentiation in tobacco (Bachmair et al., 1990).

Detailed biochemical and genetic analyses have defined the three enzymatic steps in the ubiquitin pathway and subsequent proteolysis of the ubiquinated target protein (reviewed by Hershko and Ciechanover, 1992). Briefly, the α-carboxyl group of ubiquitin is activated by the formation of a thiol ester with the ubiquitin activating enzyme (E_1). Multiple E_1-encoding genes have been identified in higher plants (Hatfield and Vierstra, 1992). The transfer of the ubiquitin moiety from E_1 to the ultimate target is mediated by the ubiquitin-carrier proteins (E_2s or UBCs), generally in consort with an ubiquitin-protein ligase (E_3). Because of their obvious role in determining the target specificity, the E_2 enzymes have been subjected to extensive scrutiny. Studies in yeast have revealed that some UBCs are required to modulate specific cellular functions, whereas others play more general and overlapping roles (Seufert et al., 1990).

Genes encoding ubiquitin-carrier proteins have been isolated from a number of organisms including yeast (Seufert et al., 1990) and *Arabidopsis* (Sullivan and Vierstra, 1991; Girod et al., 1993), and the tertiary structure of the *A. thaliana* UBC-1 protein has been determined. Based on sequence homology, yeast and higher plants appear to share many E_2s, although there may be functional differences. For example, the genes encoding UBC8-UBC12 in *A. thaliana*, in contrast with their yeast homologs, are not induced by heat shock (Girod et al., 1993). It is possible that in higher plants different E_2s are expressed in a tissue and/or developmentally specific manner. Because its transcript is present at elevated levels in giant cells, the DB#103 cDNA, which encodes part of a putative E_2, might define such a gene.

The DB#163 cDNA is identical with the DB#103 sequence, but has an additional 19 residues immediately before the poly-A tail. Because the gene encoding these transcripts, *LeUBC10*, appears to be unique, the different 3'-ends might arise by differential RNA processing.

Computer translation of the DB#117 cDNA revealed a hypothetical protein with multiple contiguous repeats of a seven amino acid motif, shown aligned in Figure 3, and terminating with a stop codon 47 bp 5' from the poly-A tail. This motif is strikingly similar to the motif TSPSYSP, a structure diagnostic for the carboxy terminal domain (CTD) of the large subunit of RNA polymerase II. *Arabidopsis* has 40 copies of this repeat (Dietrich et al., 1990) and *C. elegans* has 42 (Bird and Riddle, 1989). Although not previously cloned from tomato, it seems likely that DB#117 encodes this gene. It is not surprising that an RNA polymerase II subunit might be up-regulated in giant cells.

The conservative substitution of lysine for serine has been observed (although not to this degree) at certain positions in the *Arabidopsis* and soybean proteins (Dietrich et al., 1990). The aspartic acid to serine also is conservative, but the tyrosine to threonine is novel. These changes certainly would alter the degree of phosphorylation and shape of the CTD and thus presumably its function. It is significant to note that *Arabidopsis* employs alternative splicing to generate multiple CTDs, and soybean actually has multiple genes encoding this protein. Southern blotting suggests that the DB#117 region is unique.

```
 K PSSDS
YK PSYDNS
YKKPSYDSG
YK PSYDNG
YKKPSYDSG
YK PSYDS
YK PSYDN
YK PSYDS
YK PSYDN
YK PSYDRL-stop.
```

TS PSYSP

Figure 3. Carboxyl terminus domain (CTD) of the deduced DB#117-gene product, aligned to show the heptamer repeat. The asterisk represents the carboxy terminus. In bold below is the canonical heptamer repeat for the CTD of the largest subunit of RNA polymerase II.

BlastX analysis revealed significant homologies between the inferred DB#280 product and different members of the *myb* gene family. Homology is highest with the DNA binding domain of the petunia *myb Ph3* gene product, although multiple alignment with Mybs from other plants and vertebrates revealed that this homology extends further. As nuclear transcription regulators, members of the Myb family play pivotal roles in the regulation of cellular proliferation (Calabretta and Nicolaides, 1992). Through interactions with other trans-activators (e.g., members of the Ets family) Myb proteins effect growth control and oncogenesis (Wasylyk et al., 1993). The *Arabidopsis* gene *GL1*, required for the initiation of trichome development, encodes a Myb-type protein (Marks et al., 1991). It is likely that other plant developmental processes are mediated by Myb; it is an intriguing possibility that giant cell formation is one of these processes, involving either transcriptional activation or repression. In *Arabidopsis* (Marks et al., 1991), maize and barley (Marocco et al., 1989), the *myb* genes exist in a multi gene family. The part of *myb* spanned by DB#280 appears unique in the tomato genome.

Homology between the DB#226 sequence and the 3'-end of the *pma4*-encoded isoform of a plasmalemma H^+-ATPase from *Nicotiana plumbaginifolia* (Moriau et al., 1993) was observed. The DB#226-gene appears to be different from those encoding two previously cloned tomato isoforms (Ewing et al., 1990). The high degree of homology between the 3' untranslated regions (UTR) of these genes (the tobacco stop codon, and presumably also the DB#226 terminator, is a further 60 residues 5') might indicate a biological role for this part of the sequence. Based on measurements of transmembrane potentials following various chemical treatments, it has been postulated that giant cells possess proton efflux pumps (Jones et al., 1975). Using an *in vivo* fluorescence assay Dorhout et al. (1992) demonstrated ATP-dependent acidulation of the walls of giant cells, but not of other cells at the infection site. It seems likely that the DB#226-gene product is responsible for this activity. A range of functions have been ascribed to proton-ATPases. Changes in intracellular pH can have global influences on physiological functions, including regulation of transcription and translation. Proton-ATPases clearly play a major role in cells undergoing intense solute transport, and also have been proposed to mediate acidification of the cell wall to facilitate cell expansion (Serrano, 1989). Cell expansion and solute flux clearly are features of giant cells.

Other sequences in the giant cell cDNA bank, including a number of the pioneer clones, encode recognizable structural motifs. For example, part of the DB#215 sequence specifies a zinc finger domain. Additional homologies have been observed in the UTRs of some genes. The 3'-UTR of DB#239 contains a $(GCC)_5$ element previously noted in the 5'-UTR of human general transcription factor TFIIE (Sumimoto et al., 1991); sequences such as these may play a role in translational regulation.

Four clones in the cDNA bank (DB#142, DB#199, DB#220, DB#265) have high degrees of homology with sequences in GenBank (Blast scores of 246, 499, 806 and 173 respectively) but to the anti-sense strand. These sequences terminate with a poly-A tail, suggesting that they do not represent artifactually cloned sequences. Significantly, when used as probes in the dot blot assay, DB#142, DB#199, and DB#265 detected transcripts in plant tissues (Table 1). The DB#265 transcript, which encodes a sequence with anti-sense homology to the tobacco retroviral-like transposon Tnt 1-94 (Grandbastien et al., 1989), and also the *Arabidopsis* copia-like element (Voytas and Ausubel 1989), appears to be particularly abundant in seedling root tissue. It is an intriguing possibility that these anti-sense transcripts play a role in giant cells, and perhaps also in normal plant cells. Further analysis of these and other genes with up-regulated expression in giant cells will likely prove interesting.

CONCLUSION

We have isolated a suite of genes with apparently up regulated expression in *M. incognita*-induced giant cells. Based on preliminary characterization it appears that some of these genes (e.g., a plasmalemma H^+-ATPase) encode functions likely to be directly responsible for the biochemical make-up of giant cells, whereas others may play regulatory roles in giant cell formation (e.g., a putative transcription factor and an apparently vascular parenchyma-specific E_2 enzyme). A major goal of our research will be to use our cloned sequences to dissect the molecular cascade, beginning with the nematode-encoded ligand(s), that leads to giant cell formation. We also are interested in using these clones as tools to study normal vascular differentiation.

Obviously, our sequences provide the basis from which nematode-resistant transgenic plants might be designed. In particular, these genes define a pool of promoters which presumably are up regulated following nematode infection. Additionally, because it is likely that many of these sequences are required for giant cell function, then their inactivation might be effective in curtailing giant cell development, and hence nematode development.

Our findings in tomato are not consistent with previous observations in *Arabidopsis*. Results obtained by Goddijn et al. (1993) suggest that the major transcriptional changes in this species following *M. incognita*-infection are down-regulation events, including down-regulation of well defined elements such as the CaMV35S promoter. In contrast, we have found the expression of many sequences to be up-regulated, and other workers (e.g., Opperman, pers. comm.) have observed that the same is true for the 35S promoter. Despite the fact that *M. incognita*-infections can be established and maintained on *Arabidopsis* in the laboratory, it seems likely that it is not a *bona fide* nematode host. A much cited supposed utility of *Arabidopsis* is the ability to perform genetic analyses. Indeed, Niebel et al. (1993a; 1994) have reported the isolation of root knot nematode-resistant mutants in *Arabidopsis*. Although these studies are likely to be very productive, until a comprehensive set of balancer chromosomes is developed, genetic analyses are effectively limited to the study of non-essential genes. Based on our preliminary results, many of the genes involved in the sensitive-host response to *M. incognita*-infection may indeed be essential. Consequently, we see value in studying both model and crop plants concurrently, and hope that information can be exchanged between these "camps."

REFERENCES

Altschul, S.F., Gish, W., Miller, W., Myers, E.W., and Lipman, D.J., 1990, Basic local alignment search tool, *J. Mol. Biol.* 215:403.

Bachmair, A., Becker, F., Masterson, R.V., and Schell, J., 1990, Perturbation of the ubiquitin system causes leaf curling, vascular tissue alterations and necrotic lesions in a higher plant, *EMBO J.* 9:4543.

Belyavsky, A., Vinogradova, T., and Rajewsky, K., 1989, PCR-based cDNA library construction: general cDNA libraries at the level of a few cells, *Nucl. Acids Res.* 17:2919.

Bilofsky, H.S., and Burks, C., 1988, The GenBank (R) sequence data bank, *Nucl. Acids Res.* 16:1861.

Bird, A.F., 1961, The ultrastructure and histochemistry of a nematode-induced giant cell, *J. Biophys. Biochem. Cytol.* 11:701.

Bird, A.F., 1962, The inducement of giant cells by *Meloidogyne javanica*, *Nematologica* 8:1.

Bird, A.F., 1971, Quantitative studies on the growth of syncytia induced in plants by root knot nematodes, *Int. J. Parasitol.* 2:157.

Bird, D.McK., 1992, Mechanisms of the *Meloidogyne*-host interaction, *in*: "Nematology: from Molecule to Ecosystem," F.J. Gommers, and Maas, P.W.Th., eds., ESN Inc., Dundee.

Bird, D.McK., and Riddle, D.L., 1989, Molecular cloning and sequencing of *ama-1*, the gene encoding the largest subunit of *Caenorhabditis elegans* RNA polymerase II, *Molec. Cell. Biol.* 9:4119.

Calabretta, B., and Nicolaides, N.C., 1992, *C-myb* and growth control. *Crit. Rev. Euk. Gene Expression* 2:225.

Conkling, M.A., Cheng, C.-L., Yamamoto, Y.T., and Goodman, H.M., 1990, Isolation of transcriptionally regulated root-specific genes from tobacco, *Plant Physiol.* 93:1203.

Cornu, M., 1879, Études sur le *Phylloxera vastatrix*, *Mém. Prés. Acad Sci. Paris.* 26:164.

Dietrich, M.A., Prenger, J.P., and Guillfoyle, T.J., 1990, Analysis of the genes encoding the largest subunit of RNA polymerase II in *Arabidopsis* and soybean, *Plant Mol. Biol.* 15:207.

Domec, C., Garbay, B., Fournier, M., and Bonnet, J., 1990, cDNA library construction from small amounts of unfractionated RNA: association of cDNA synthesis with polymerase chain reaction amplification, *Anal. Biochem.* 188:422.

Dorhout, R., Gommers, F.J., and Kollöffel, C., 1991, Water transport through tomato roots infected with *Meloidogyne incognita*. *Phytopathology* 81:379.

Dorhout, R., Kolloffel, C., and Gommers, F.J., 1992, Alteration of distribution of regions with high net proton extrusion in tomato roots infected with *Meloidogyne incognita*, *Physiol. Molec. Plant Pathol.* 40:153.

Duguid, J.R., Rowher, R.G., and Seed, B., 1988, Isolation of cDNAs of scrapie-modulated RNAs by subtractive hybridization of a cDNA library, *Proc. Natl. Acad. Sci. USA* 85:5738.

Endo, B.Y., 1987, Histopathology and ultrastructure of crops invaded by certain sedentary endoparasitic nematodes, *in*: "Vistas on Nematology," J.A. Veech, and Dickson, D.W., eds., Society of Nematologists Inc., Hyattsville.

Evans, I.M., Swinhoe, R., Gatehouse, L.N., Gatehouse, J.A., and Boulter, D., 1988, Distribution of root mRNA species in other vegatative organs of pea (*Pisum sativum* L.). *Mol. Gen. Genet.* 214:153.

Ewing, N.N., Wimmers, L.E., Meyer, D.J., Chetelat, R.T., and Bennett, A.B., 1990, Molecular cloning of tomato plasma membrane H^+-ATPase, *Plant Physiol.* 94:1847.

Finley, D., and Chau, V., 1991, Ubiquitination, *Annu. Rev. Cell Biol.* 7:25.

Girod, P.-A., Carpenter, T.B., Nocker, S., Sullivan, M.L., and Vierstra, R.D., 1993, Homologs of the essential ubiquitin conjugating enzymes UBC1, 4, and 5 in yeast are encoded by a multigene family in *Arabidopsis thaliana*, *Plant J.* 3:545.

Goddijn, O.J.M., Lindsey, K., van der Lee, F.M., Klap, J.C., and Sijmons, P.C., 1993, Differential gene expression in nematode-induced feeding structures of transgenic plants harbouring promoter-*gus*A fusion constructs, *Plant Cell* 4:863.

Grandbastien, M.A., Spielmann, A., and Caboche, M., 1989, Tnt-1, a mobile retroviral-like transposable element of tobacco, isolated by plant genetics, *Nature* 337:376

Grundler, F.M.W., Böckenhoff, A., Schmidt, K.-P., Sobczak, M., Golinowski, W, and Wyss, U., 1994, *Arabidopsis thaliana* and *Heterodera schachtii*: a versatile model to characterize the interaction between host plants and cyst nematodes, *in:* "NATO ARW: Advances in Molecular Plant Nematology," F. Lamberti, De Giorgi, C., and Bird, D.McK., eds., Plenum Press, New York.

Gurr, S.J., McPherson, M.J., Scollan, C., Atkinson, H.J., and Bowles, D.J., 1991, Gene expression in nematode-infected plant roots. *Mol. Gen. Genet.* 226:361.

Hatfield, P.M., and Vierstra, R.D., 1992, Multiple forms of ubiquitin-activating enzyme E_1 from wheat, *J. Biol. Chem.* 267:14799.

Hershko, A., and Ciechanover, A., 1992, The ubiquitin system for protein degradation, *Annu. Rev. Biochem.* 61:76.
Hussey, R.S., 1989, Disease-inducing secretions of plant-parasitic nematodes, *Annu. Rev. Phytopathol.* 27:123.
Hussey, R.S., Davis, E.L., and Ray, C., 1994, *Meloidogyne* stylet secretions, *in:* "NATO ARW: Advances in Molecular Plant Nematology," F. Lamberti, De Giorgi, C., and Bird, D.McK., eds., Plenum Press, New York.
Hussey, R.S., Mims, C.W., and Westcott, S.W., 1992a. Ultrastructure of root cortical cells parasitized by the ring nematode *Criconemella xenoplax*, *Protoplasma* 167:55.
Hussey, R.S., Mims, C.W., and Westcott, S.W., 1992b. Immunocytochemical localization of callose in root cortical cells parasitized by the ring nematode *Criconemella xenoplax*, *Protoplasma* 171:1.
Jones, M.G.K., and Northcote, D.H., 1972, Nematode-induced giant-cell formation in roots of *Impatiens balsamina*, *J. Nematol.* 10:70.
Jones, M.G.K., Novacky, A., and Dropkin, V.H., 1975, Transmembrane potentials of parenchyma cells and nematode-induced transfer cells, *Protoplasma* 85:15.
Jones, M.G.K., and Payne, H.L., 1978, Early stages of nematode-induced syncytium – a multinucleate transfer cell, *J. Cell Sci.* 10:789.
Kiss, T., Kiss, M., and Solomosy, F., 1989, Nucleotide sequence of a 25S rRNA gene from tomato, *Nucl. Acids Res.* 17:796.
Krusberg, L.R., and Nielsen, L.W., 1958, Pathogenesis of root-knot nematodes to the Porto (*sic*) Rico variety of sweetpotato, *Phytopathology* 48:30.
Lebeau, M., Alvarez-Bolado, G., Wahli, W., and Catsicas, S., 1991, PCR driven DNA-DNA competitive hybridization: a new method for sensitive differential cloning, *Nucl. Acids Res.* 19:4778.
Magnusson, C., and Golinowski, W., 1991, Ultrastructural relationships of the developing syncytium induced by *Heterodera schachtii* (Nematoda) in root tissue of rape, *Can. J. Bot.* 69:44.
Mangiarotti, G., Chung, S., Zuker, C., and Lodish, H.F., 1981, Selection and analysis of cloned developmentally-regulated *Dictyostelium discoidium* genes by hybridization-competition. *Nucl. Acids Res.* 9:947.
Mankau, R., and Linford, M.B., 1960, Host-parasite relationships of the clover cyst nematode *Heterodera trifolii* Goffart. Bulletin 667, University of Illinois, AES.
Marks, M.D., Esch, J., Herman, P., Sivakumaran, S., and Oppenheimer, D., 1991, A model for cell-type determination and differentiation in plants, *Symp. Soc. Expt. Biol.* 45:77.
Marocco, A., Wissenbach, M., Becker, D., Paz-Ares, J., Saedler, H., Salaminin, F., and Rohde, W., 1989, Multiple genes are transcribed in *Hordeum vulgare* and *Zea mays* that carry the DNA binding domain of the *myb* oncoproteins, *Molec. Gen. Genet.* 216:183.
Moriau, L., Bogaerts, P., Jonniaux, J.-L., and Boutry, M., 1993, Identification and characterization of a second plasma membrane H^+-ATPase gene subfamily in *Nicotiana plumbagininiflora*, *Plant Molec. Biol.* 21:955.
Mundo-Ocampo, M., Greene, M., Flaxman, M., and Baldwin, J.G., 1993, Morphology of nurse cell nuclei induced by *Meloidodera floridensis*: A computer graphics application, *J. Nematol.* 25:603.
Nagl, W., 1978, "Endopolyploidy and Polytene in Differentiation and Evolution," Elsevier/North Holland Biomedical, Amsterdam.
Niebel, A., de Almeida, J., Van Montagu, M, and Gheysen, G., 1993a, Identification of root knot nematode-resistant mutants in *Arabidopsis thaliana*, *in*: "Mechanisms of Plant Defense Responses," B. Fritig, and Legrand, M. eds., Kluwer Academic Publishers, Dordrecht.
Niebel, A., Barthels, N., de Almeida-Engler, J., Karimi, M., Vercauteren, I., Van Montagu, M, and Gheysen, G., 1994, Arabidopsis thaliana as a model host plant to study molecular interactions with root-knot and cyst nematodes, *in:* "NATO ARW: Advances in Molecular Plant Nematology," F. Lamberti, De Giorgi, C., and Bird, D.McK., eds., Plenum Press, New York.
Niebel, A., Engler, J.deA., Tiré, C., Engler, C., Van Montagu, M, and Gheysen, G., 1993b, Induction patterns of an extensin gene in tobacco upon nematode infection, *Plant Cell* 5:1697.
Niebel, A., Van der Eycken, W., Van Montagu, M, and Gheysen, G., 1993c, Isolation and characterization of two potato genes induced during cyst nematode infection, *Arch. Int. Physiol. Biophys.* 101:B18.
O'Bannon, J.H., and Reynolds, H.W., 1965, Water consumption and growth of root-knot-nematode-infected and uninfected cotton plants, *Soil Sci.* 99:251.
Opperman, C.H., Taylor, C.G., and Conkling, M.A., 1994, Root-knot nematode-directed expression of a plant root-specific gene, *Science* 263:221.

Press, M.C., and Whittaker, J.B., 1993, Exploitation of the xylem stream by parasitic organisms, *Phil. Trans. R. Soc. Lond.* B 341:101.
Rubenstein, J.L.R., Brice, A.E.J., Ciaranello, R.D., Denny, D., Porteus, M.H., and Usdin, T.B., 1990, Subtractive hybridization system using single-stranded phagemids with directional inserts, *Nucl. Acids Res.* 18:4833.
Sargent, T.D., and Dawid, I.B., 1985, Differential gene expression in the gastrula of *Xenopus laevis*, *Science* 222:135.
Sasser, J.N. and Freckman, D.W., 1987, A world perspective on nematology: the role of the society, *in*: "Vistas on Nematology," J.A. Veech, and Dickson, D.W., eds., Society of Nematologists Inc., Hyattsville.
Serrano, R., 1989, Structure and function of plasma membrane ATPase, *Ann. Rev. Plant Physiol.* 40:61.
Seufert, W., McGrath, J.P., and Jentsch, S., 1990, UBC1 encodes a novel member of an essential subfamily of yeast ubiquitin-conjugating enzymes involved in protein degradation, *EMBO J.* 9:4535.
Steeves, T.A., and Sussex, I.M., 1989, "Patterns in Plant Development," Cambridge University Press, Melbourne.
Sullivan, M.L., and Vierstra, R.D., 1991, Cloning of a 16-kDa ubiquitin carrier protein from wheat and *Arabidopsis thaliana*, *J. Biol. Chem.* 266:2387.
Sumimoto, H., Ohkuma, Y., Sinn, E., Kato, H., Shimasaki, S., Horikoshi, M., and Roeder, R.G., 1991, Conserved sequence motifs in the small subunit of human general transcription factor TFIIE, *Nature* 354:401.
Swaroop, A., Xu, J., Agarwal, N., and Weissman, S.M., 1991, A simple and efficient cDNA library subtraction procedure: isolation of human retina-specific cDNA clones, *Nucl. Acids Res.* 19:1954.
Visser, J., and Dörr, I., 1987, The haustorium, *in*: "Parasitic Weeds in Agriculture Vol. I: *Striga*," L.J. Musselman, ed., CRC Press, Boca Raton.
Voytas, D.F., and Ausubel, F.M., 1989, A copia-like element family in *Arabidopsis thaliana*, *Nature* 336:242.
Wasylyk, B., Hahn, S.L., and Giovane, A., 1993, The Ets family of transcription factors, *Eur. J. Biochem.* 21:7.
Welsh, J., Liu, J-P.X., and Efstatiadis, A., 1990, Cloning of PCR-amplified total cDNA: construction of a mouse oocyte cDNA library, *Genet. Anal. Techn. Appl.* 7:5.
Wilcox, D.A., and Loria R., 1986, Water relations, growth, and yield in two snap bean cultivars infected with root-knot nematode, *Meloidogyne hapla* (Chitwood), *J. Am. Soc. Hortic. Sci.* 11: 34.
Wilson, M.A., Bird, D.McK., and van der Knaap, E., 1994, A comprehensive subtractive cDNA cloning approach to identify nematode-induced transcripts in tomato, *Phytopathology* in press.
Wyss U., Grundler, F.M.W., and Münch, A., 1992, The parasitic behaviour of 2nd-stage juveniles of *Meloidogyne incognita* in roots of *Arabidopsis thaliana*, *Nematologica* 38:98.

NOVEL PLANT DEFENCES AGAINST NEMATODES

Howard J. Atkinson[1], Keith S. Blundy[2], Michael C. Clarke[1], Ekkehard Hansen[1], Glyn Harper[1], Vas Koritsas[1], Michael J. McPherson[1], David O'Reilly[2], Claire Scollan[1], S. Ruth Turnbull[1], and Peter E. Urwin[1]

[1]Centre for Plant Biochemistry and Biotechnology
University of Leeds
Leeds LS2 9JT UK

[2]Advanced Technologies (Cambridge) Ltd
210 Cambridge Science Park
Cambridge CB4 4WA
UK

INTRODUCTION

Nematodes are major pests of world agriculture (Nickle, 1991) and they have a global economic impact of >$100 billion per year (Sasser and Freckman, 1987). The most prevalent cause of economic loss in climates other than temperate ones are species of *Meloidogyne*, root-knot nematodes. Their wide host ranges ensure they are a major constraint on the development of agriculture in the tropics where they impose losses of 11-25% to a wide range of crops (Sasser, 1979). In contrast, cyst nematode (*Heterodera* and *Globodera* spp.) predominate in temperate agriculture and each species has a narrow host crop range. Species that are key pests of their respective crops include soybean (*Heterodera glycines)* sugar beet (*Heterodera schachtii*) and potato (*Globodera* spp.) cyst nematodes. For instance, potato cyst nematodes (*Globodera pallida* and *Globodera rostochiensis*) infest 40% of the UK national potato acreage. They are believed to impose an annual cost of £10-50 million on the industry in that country; the two *Globodera* spp. are also important in other potato growing areas.

Cultural control, resistant varieties and nematicides are often combined within a pest management strategy. Cultural control is widely practised but rotation is of limited value for nematodes with a host range as wide as that of *Meloidogyne* spp. It may also be against the economic interests of the specialist grower or those limited in their choice of alternative crops. Resistant cultivars have proven commercially successful for instance in the control of *Meloidogyne* on tomato and *Globodera rostochiensis* on potato. In both cases, virulent forms of nematodes occur that challenge these resistant cultivars. Natural genes for resistance have limited potential for some crops because suitable genes are either lacking or difficulties arise in breeding to commercial acceptability (Roberts, 1992).

Nematicides have an uncertain future and several have been withdrawn or restricted in use because of either toxicological hazard or environmental harm. Two fumigant nematicides (DBCP and EDB) were withdrawn from use in the USA in the 1980s. The continued use of DBCP on banana plantations elsewhere in the Americas has resulted in recent litigation brought in Texas by agricultural workers claiming impotence and sterility caused by exposure to the nematicide while at work (Anon 1993a). Its use on bananas for import to USA has recently been cancelled by the EPA. A more recent generation of nematicides based on oxime-carbamate and carbamate compounds are among the most environmentally damaging of agrochemicals in current use. Aldicarb usage is restricted to certain states in USA following concern over groundwater contamination. It is one of only four pesticides selectively monitored in food residues by the FDA (Anon 1993b)). The insecticide Carbofuran which has partly replaced Aldicarb in some markets is also under environmental scrutiny in Canada (Anon 1993c). The future of nematicides is uncertain given that their application to soil is likely to remain necessary. This procedure is likely to become an increasingly discouraged basis for crop protection.

Current limitations to control ensure that the case of new technologies for plant nematode control is arguably the strongest within crop protection. Plant biotechnology offers an approach of high and flexible potential. It can be applied to facilitate conventional plant breeding or to develop novel defences. The latter approach is the focus of our current work; our objectives are simply stated but ambitious. We aim to design novel defences against nematodes that will be effective, environmentally acceptable and applicable to all crops in the developing and developed world for which transformation technology is available. The work summarized is based on two research programes, both underpinned by distinct patent applications (Gurr et al., 1992; Hepher and Atkinson, 1992). This account will be limited to an overview of the emerging technology as it applies to cyst and root-knot nematodes.

THE FEEDING OF CYST AND ROOT-KNOT NEMATODES

Nematode Establishment in Host Plants

The mode of feeding has particular relevance to opportunities for disrupting nematode development. The sedentary habit of both cyst and root-knot nematodes is correlated with the induction of specific feeding sites. The two groups also show differences in their biology that may prove significant in designing appropriate novel control strategies. Therefore aspects of their biology will be summarised before approaches to their novel control are considered.

The cyst of *Heterodera* and *Globodera* spp. is the tanned body wall of the former female, enclosing some or all of her eggs and providing protection both from harsh environmental conditions and many predators. Diffusates from host crops, and/or the season help synchronise emergence of the parasite with the availability of host root systems. The infective-stage (J2) hatches from its egg and then it emerges from the cyst. The J2 is about 0.5mm in length and typically it moves only a few centimetres to a plant root. In many species, the specificity of hatching helps to ensure that the invaded root is that of a host plant. The animal typically invades at the zone of root elongation and moves directly towards the vascular cylinder by intracellular migration using a mouth stylet to cut through plant cell walls (Atkinson and Harris, 1989). Once close to the vascular tissue, behaviour changes and an initial feeding cell is selected. The animal releases secretions and subsequently the plant cell is modified over several days to become a syncytial cell with transfer cell-like attributes (Wyss and Grundler, 1992). The feeding cell is formed by progressive cell wall dissolution between the initial cell attacked and those adjacent to it. The animal develops through two more stages before becoming an adult. Females feed at each developmental stage and become much larger than males. They require a stable biotrophic relationship with the fully formed syncytium over several weeks to become gravid and it is their feeding that causes much of the economic effect of the pathogen.

The females of *Meloidogyne* species lay eggs into a gelatinous matrix and do not form cysts. The eggs are less persistent than those of most cyst nematodes and they are not responsive to host root diffusate. The animals are, however, able to invade and develop on a wide range of plants. Since most species occur in warm soils, they can complete several generations per season and even low densities invading a highly susceptible crop can reach yield-damaging densities. The invasion of the plant is very different from the process shown by cyst nematodes. The individual of a *Meloidogyne* species enters a root behind its tip, normally migrates in an intercellular manner towards the meristem and then turns around and moves up the root axis into the developing vascular tissue (Wyss and Grundler, 1992). Here the J2 induces mitosis without cytokinesis in several cells and feeds from these multinucleate giant cells in turn throughout its subsequent development. There is a characteristically prolonged lethargus of several days after initial feeding. In this time the animal moults three times and a diminutive adult female is formed. This stage is then responsible for most of the feeding and growth upon which subsequent fecundity is based. Most of these species are parthenogenetic and the proportion of males in a population is normally low.

The Function of Feeding Cells

An interesting feature of both giant cells and syncytial cells is that there are few plasmodesmata between them and those cells lying adjacent to the feeding cell complex (Jones, 1981). In contrast there are considerable wall ingrowths between the feeding cell and the xylem. Cyst and root-knot nematodes should be added to a recent list of parasitic angiosperms and some homopteran insects that exploit the xylem stream (Press and Whittaker, 1993). The feeding cells of these parasitic nematodes possess proton (H^+) pumps localised particularly in the plasma membrane at wall ingrowths between the cell and the xylem vessels. The presence of such a pump has been demonstrated physiologically using fluorescent dyes between the apoplast and the giant cells of *M. incognita* (Dorhout et al., 1992). Plants possess three or four amino acid porters for different amino acids that act in symport with the proton pump (Bush, 1993). This amino acid transporter system and additional systems for sugar uptake are essential for normal nutrition of plant heterotrophic tissues. The parasitic feeding demand of the nematode far exceeds that required by healthy root cells. The increase in surface area of plasmalemma made possible by the wall ingrowths suggest an increase in the number of pump and porter molecules involved in meeting the uptake requirements of the modified cell. It follows that other proteins involved in uptake of such compounds and their subsequent utilisation are also likely to be expressed at abnormally high levels in the modified plant cell. Nitrogen moves in the xylem principally as amino acids, nitrates, ureides and the amides asparagine and glutamine rather than as just amino acids. In radish this is the predominate form of nitrogen in the stream (Pate, 1973). A second brassica, rape, supports an increase in prevalence of female *H. schachtii* among established parasites (Betka et al., 1991) when provided with glutamine but not the corresponding amino acid (glutamic acid). Once in the symplast, glutamine is readily converted to glutamate (Emes and Fowler, 1979), transaminated and then used to replenish the amino acid pool before protein synthesis. Active transamination would ensure all essential amino acids are available to the parasites. This avoids dependence on import through the limited symplastic route provided by the low density of plasmodesmata between the feeding cells and those surrounding it. It follows that high expression is to be expected for both the structural proteins and the enzymes that are involved in this enhanced protein uptake and metabolism by the modified plant cell.

Up-regulated genes could be identified in feeding cells from an understanding of the physiology and biochemistry of the host-parasite interaction including the nutritional requirements of the nematode. Such information could underpin the design of novel defences. This approach has not been used extensively to-date because current knowledge of the system remains incomplete.

Gene Expression in Nematode Feeding Cells

Considerable morphological changes are caused in cells during their modification into a nematode feeding site. There must also be cellular responses to the subsequent and continual withdrawal of nutrients by the parasite. Therefore radical changes in gene expression in such cells may be anticipated involving both up and down regulation of gene expression relative to their counterparts in healthy plants. A few such changes may also be so considerable that they seem qualitative in nature.

Changes in gene expression in feeding cells of *Globodera* were first investigated using *in vitro* mRNA translation (Hammond-Kosack et al., 1989). They suggested limited changes in gene expression occurred but interpretation was limited by the high proportion of unresponding cells that was necessarily collected for analysis. Therefore we developed an approach able to lessen this technical limitation involving the use of synchronously infected roots (Atkinson and Harris, 1989) and development of a protocol for PCR-directed cDNA library construction. The approach was successful in allowing cDNA libraries to be based on limited amounts of syncytially-enriched material, allowing genes abundantly expressed at feeding sites to be identified (Gurr et al., 1991). cDNA library construction has subsequently been achieved by others interested in studying molecular responses to nematode infection of roots (e.g., Van der Eycken et al., 1992; Niebel et al., 1992). Its principal limitation is that library construction is still based on a low proportion of nematode feeding cells in the total mass of harvested tissue. As a result, we have found differential screening efficient for identifying abundant but not rare transcripts that arise from feeding cells.

We have alleviated problems inherent in differential library screening of infected roots in two ways. We have explored the use of gene tagging in *Arabidopsis* (see later). Secondly, we have prepared cDNA libraries from which transcripts common to healthy roots and nematodes have been subtracted. Magnetic bead technology has allowed facile cDNA synthesis, subtraction cycles and library construction. These protocols were applied to extend our range of libraries to tobacco roots infected by *M. javanica*. We have also used more conventional single-strand subtractive techniques to construct libraries from both potato and sugar beet roots infected with *G. rostochiensis* and *H. schachtii* respectively. The results for a cDNA library produced using tobacco roots infected with *M. javanica* provide an example of the value of the approach. Six rounds of subtraction were completed and 90 clones identified using a reverse northern analysis. Subsequently, 25 clones were selected by northern analysis for further study. One of these clones from a subtractive approach is induced by both *Meloidogyne* spp. and cyst nematodes in a range of crops. Interesting clones from these libraries have been identified and characterised further by *in situ* hybridisation and as promoter:*gus*A fusions.

ISOLATION OF PROMOTERS

cDNA clones up-regulated in nematode-infected tissue were selected for further study. Specific oligonucleotide primers to the 5' region of each cDNA clone of interest were used for rapid amplification of their 5' termini (Frohman et al., 1988). Promoters were isolated from genomic plant DNA by vector-ligated PCR using primers specific to the 5' RACE products. These promoters have been linked to a GUS reporter and effector gene constructs and tested in transgenic plants.

PLANT TRANSFORMATION

Promoter Tagging in *Arabidopsis*

The advantage of *Arabidopsis* as a model host has been strongly advocated and offers advantages for molecular analysis of nematode-infected plants (Sijmons et al., 1991). We were kindly provided with transgenic *Arabidopsis* lines containing a promoterless *gusA* gene, encoding the β-glucuronidase (GUS) reporter by Dr Keith Lindsey. These lines had been transformed with the binary vector pΔgusBin19 (Lindsey et al.,1993; Topping *et al.*,1991). Transcriptional activation of the 'interposon' GUS-reporter sequence can occur when the construct is inserted downstream of a native plant gene promoter. This occurs as a relatively random process. Our interest was to determine its value as an alternative to subtractive cDNA library screen as a strategy for isolating non-abundant transcripts.

Transgenic lines were grown on agar under sterile conditions for 10-14 day old seedlings before challenge by sterile *M. incognita* juveniles. Seedlings were tested for GUS activity at 3-5 days following infection in both agar culture and in nematode infested soil. As an example of our results, one line has been identified that shows up-regulation of GUS activity associated with the giant cells and the gall tissue surrounding *M. incognita*. The temporal expression of this line has been studied in detail. Increased GUS activity is observed within 48 hours following infection on agar. A high level of GUS activity is maintained during the first few weeks of development. At later time points it declines in galled regions and becomes undetectable. This suggests that the promoter is no longer active when the mature female stops feeding. This spatial distribution of staining was confirmed using cryosections of 5-bromo-4-chloro-3-indolyl β-D-glucuronide (X-Gluc)treated galled tissue. The line also shows GUS activity in aerial parts of the plant and much weaker activity in the vascular cylinder of young rather than older roots.

Agrobacterium rhizogenes-mediated transformation

Promoter tagging plays only a limited role in our work on design of novel defences since promoters of interest have already been obtained from screening cDNA libraries. Instead, we have favoured a model system based on *A. rhizogenes*-mediated transformation to form hairy roots for stages of our work beyond promoter identification. The advantages we value are a period of only 4 weeks from transformation to challenge of roots with nematodes and an ability to use the crop and even the cultivar of interest for all our studies. The approach can accommodate study of a range of nematodes including *Globodera* spp.

and *H. glycines* that do not parasitise *Arabidopsis*. Furthermore whole plants can be regenerated from interesting root lines for further analysis. Such plants can serve as prototypes prior to transformation of target crop species with constructs of interest using *Agrobacterium tumefaciens*.

Under the conditions we have optimised, nematodes will develop normally and as quickly on hairy roots as on the corresponding whole plant. We currently produce hairy roots from commercially important cultivars of potato, tomato, tobacco and sugar beet using *Agrobacterium rhizogenes* strain LBA9402 harbouring pRi1855 (Hamill et al., 1987) plus a binary vector with a chimaeric fusion gene of interest. These fusion genes were constructed by linking *gusA* reporter gene to promoters that are either constitutive or are activated following nematode challenge of the transgenic plants.

Agrobacterium tumefaciens -mediated transformation

We have transformed *Arabidopsis* in tandem with hairy root culture to test constructs of particular interest to gain confidence that results from hairy roots are comparable to those achieved by whole plant transformation with *A. tumefaciens*. To date this latter system is principally being used for study of efficacy of proteinase inhibitors (see later).

Transformation of crop species by our industrial collaborators is beyond the scope of this account and involves different protocols depending upon crop plant, the effector gene and the promoter of interest. As an example, we have introduced the proteinase inhibitor from rice, oryzacystatin I (*Oc*-I Kondo et al., 1989) with and without its intron into the plant transformation vector pBIN19 under the control of various promoters and the *A. tumefaciens* nopaline synthetase terminator. Each of the constructs has been transferred to *A. tumefaciens* LBA4404 for transformation of tomato cv Ailsa Craig. It is envisaged that *Oc*-I without its intron will be the more highly expressed of the two constructs since monocot introns are generally spliced with low efficiency in dicots. These constructs are currently being evaluated following nematode challenge.

CHARACTERISATION OF PROMOTERS

In its simplest form, a novel defence requires an element (promoter) to regulate expression of a gene (effector) encoding for a protein with biological activity that lessens success of the parasite. Genes that are up-regulated in feeding cells have been identified from understanding the physiology and biochemistry of the nematode feeding cells. The value of such an approach depends upon the defence envisaged. The promoter of such a gene is also likely to be expressed at variable levels elsewhere in healthy tissues. This may be acceptable for use with an effector gene whose product does not harm plant cells.

Monitoring Promoter Activity in Hairy Roots

GUS analysis demonstrated a degree of variation in the amount and sometimes the pattern of GUS expression among different root transgenic lines for the same fusion gene. Therefore it is necessary to compare several lines transformed with the same construct and to select an interesting line for study. This line is sub-cultured until sufficient material is available for a time course study for infected and healthy roots. As an example, one promoter isolated from potato showed constitutive expression in the root tips was also strongly induced following nematode invasion and development within the re-differentiating root areas around the feeding sites. Externally, galls developing after *Meloidogyne* infection and sites of infection with *Globodera* appeared completely blue. However, GUS activity was often reduced or even absent within the giant cells or the syncytium. This could be due to a number of possible reasons. The gene may be down-regulated in nematode feeding cells either specifically or as part of a more general down-regulation of gene expression by the nematode as recently proposed from the study of intersposons (Goddijn et al., 1993). Other possibilities relate to activity of the expressed β-glucuronidase, for example, the loss of coloration may be caused by the nematode withdrawing cell contents at a rate that exceeds further synthesis. Another possibility is that the parasite secretes a protease (see later) or other products that destroy or inhibit enzyme activity. We currently see value in distinguishing between these possibilities. One approach is the application of mathematical models to feeding based on a recently developed approach (van Haren et al., 1993). This may be necessary to determine likely effector concentrations within feeding cells (see later).

Non-lethal Monitoring of GUS activity

The feeding cell is a dynamic system experiencing a high physical extraction rate of up to 50% of its volume/day (Muller et al 1981; Atkinson and Harris, 1989). This effect is not normally experienced by plant cells and so the flux within such cells merits particular attention. The temporal study of GUS-reporter gene expression (Jefferson *et al.*, 1987) in individual, transgenic plants is not readily studied even in a semiquantitative manner using the most commonly used GUS substrate, X-Gluc. It is relatively cytotoxic and destructive microtome sectioning is necessary to visualise its distribution in inner cell layers. Our work required study of reporter gene expression levels during a prolonged time course associated with nematode development in transgenic roots. We required a substrate that is not appreciably phytotoxic and that could be accurately monitored using *in toto* root mounts. We used novel GUS substrates which met our criteria. These substrates, ImaGene Red and ImaGene Green (Molecular Probes, Inc., Eugene, OR, USA), are lipophilic analogues of resorufin β-D-glucuronide and fluorescein di-β-glucuronic acid respectively, and they diffuse freely across membranes of plant protoplasts and single cells (Naleway et al., 1991). β-glucuronidase produces a fluorescent product with both substrates. As a result, whole roots can be examined using a scanning laser confocal microscope (Leica). This instrument can form images of a selected region that excludes fluorescence emitted from other planes of focus. Therefore it provides optical sections that localise GUS activity to internal cell layers of the specimen. Although images derived from internal tissues can be formed, the technique is neither destructive nor results in toxicity to nematodes or roots. It is possible to repeat GUS assays on individual plants following loss of the fluorophore from the root during further culture.

We use ImaGene Red for analysis of GUS activity in both hairy roots and in those of transgenic *Arabidopsis thaliana*. ImaGene Green suffers from greater non-specific fluorescence in roots. The principal value of the latter to us is for detecting unwanted GUS expression in aerial tissues. In such tissues, chloroplasts autofluoresce when excited by the green light required to visualise ImaGene Red and so mask light output from the probe. GUS expression can be localised to single cells of internal cell layers using the fluorescent probes and a confocal microscope. It is now possible to monitor the expression of *gus*A fusion transgenes in all parts of individual transgenic plants both during normal development and in response to invasion, plant cell modification and feeding by nematodes. We are currently developing a basis for semiquantification of GUS expression during the time course of infection.

Range of promoters of interest

As a result of characterisation, we have isolated promoters that provide constitutive expression in plants and others that respond to nematode challenge. Some of those responding to nematode infections are specific to *Meloidogyne* or cyst nematode infection whereas others are reactive when challenged by either nematode group.

EFFECTORS

A wide range of proteins have the potential to disrupt feeding cell formation or nematode development. The selection of those with particular potential is centred around efficacy and those without any toxicological hazard to non-target organisms. Nematodes could be disorientated or killed within the rhizosphere but it is more feasible and practical to limit attention to proteins with biological activity against the parasites after host invasion. Such effectors might act directly against the nematode or indirectly against the pathogen by modifying the host. Depending upon the mode of action, both types of effector could be expressed during invasion, establishment or subsequent development of the parasite.

DISRUPTION PRIOR TO ESTABLISHMENT

We have identified proteins that are effective against nematode invasion. In addition we have characterised a promoter that is responsive to nematode wounding but we are currently uncertain if the upregulation achieved would limit an active, migratory nematode. If not, a constitutive root promoter could be used to provide a preformed defence against such nematodes. Such strategies may reduce invasion by cyst and root-knot nematodes but they have particular relevance to nematodes such as *Pratylenchus*, *Helicotylenchus* and other plant-invading species that do not establish permanent feeding sites. They are likely to lack efficacy against cyst and root-knot nematodes which show considerable intraspecific competition at population densities above their economic threshold. Consequently, a partial mortality caused by a defence that merely suppresses invasion may have a lower than expected effect on the number of established parasites.

DISRUPTING PLANT GENE EXPRESSION AT THE FEEDING SITE

Given that the majority of cyst and root-knot nematodes fail to reach a feeding site there is a strong argument for concentrating initial research on novel defences that affect the established parasites. The best natural paradigm is offered by a single dominant gene for resistance (H_1) to *G. rostochiensis* which is available in some commercial cultivars of potato such as cv Maris Piper. The resistant response in this cultivar is expressed only after establishment of the nematode. Each animal invades in the usual way and initiates an apparently normal feeding site, loses its locomotory muscles and becomes developmentally committed to that site. Subsequently the resistant plant successfully isolates the feeding cell from other cells including vascular elements. In this condition, the feeding site is unable to support female development and only males mature. Therefore an objective at the start of our work was to develop analogous systems that limit feeding cell development or its effective utilisation by the parasite.

It is possible to develop strategies that attenuate changes in gene expression in the feeding cells without a damaging limitation to lower levels of expression of the same genes under non-pathological conditions. An example of the required effect is shown by the host range of *Meloidogyne incognita*. It can parasitise many hundreds of plant species but there is a range of host suitabilities. There is variation in the size of the giant cell complex formed by the female on different host plants and a clear correlation exists between feeding cell size, body size and fecundity. Interestingly, comparison of a rank order of size of the females formed by two species of root-knot nematode on nine host species is broadly similar except soybean is a less satisfactory host for *M. incognita* than *M. hapla* (Al-Yahya, 1993).

Study of natural infections by *Meloidogyne* led us to consider cell attenuation without cell death is a practical objective and one that may reduce the impact of a promoter that allows a low level of unwanted expression in healthy tissues. An obvious effector for prototype development is the RNase, barnase (Hartley 1988, 1989). It has proven efficacy for selective killing plant cells and its use lacks any patent restrictions. Under control of promoters that are active in male flower parts it has been used to emasculate plants (Mariani et al., 1990).

Barnase is of value for use in prototype resistant plants because of its efficacy but there are appreciable disadvantages for an effector that renders the plant highly intolerant to misdirected expression. This problem can be avoided by using an effector that is directed against the parasite and not its feeding cells (see later). A second approach is to use an effector that attenuates feeding cells with their high metabolic activity without inducing plant cell death as occurs naturally for *Meloidogyne* on marginal hosts. This requires an inherently less toxic effector protein than barnase expressed at levels that critically impair metabolism of feeding cells. Such an effector protein could be tolerated at low levels in healthy plant cells without detriment to them. If required, additional protection could be provided using a constitutively expressed restorer gene with a role similar to that of barstar in countering unwanted expression of barnase. Such plants have two bases for protection from unexpected expression of the effector protein outside of feeding cells.

ANTI-NEMATODE STRATEGIES

Gene products that are effective against nematodes by the oral route may be targeted with advantage to the feeding cells. This should enhance efficacy, limit unnecessary protein synthesis and restrict expression to parts of the plant that lack economic value.

We have made a particular study of plant protease inhibitors. They are expressed naturally in plants and reduce the ability of certain insects to use dietary protein. This delays development and may reduce fecundity. These effects are due in part to direct effects on digestive proteases but other mode of actions also occur (Ryan, 1990). Ten distinct families of protease inhibitors from plants are known showing specificity according to the mechanistic class of proteolytic enzymes across a range of organisms. Given the known action and the range available, they appear to have considerable potential for control of crop pests including nematodes.

We established the working hypothesis that established nematodes use proteases during feeding. Both females of *G. pallida* (Koritsas and Atkinson, 1994) and juveniles possess proteolytic activity. Exudates of the J2 of *G. pallida* show peaks of protease activity at pH 5.7 and pH 7. The activity at pH 5.7 was inhibited *in vitro* by both cowpea trypsin inhibitor (CPTI) and soybean trypsin inhibitor. This provided a rational basis for testing if transgenic potatoes expressing a proteinase inhibitor perturb feeding by the early parasitic stages of *G. pallida*.

Expression of approx. 0.8% soluble protein as CPTI had no significant effect on invasion and establishment of J2 *G. pallida* and it did not delay emergence of males. However expression of the protease inhibitor caused a reduction in size of both sexes during early growth. In addition the female:male ratio was 2:5 and not 2:1 as found on transformed, control plants. Even if the unreasonable assumption is made that all unsexed individuals are female, the adjusted sex ratio of 1:1 is significantly lower than for corresponding controls. It seems that the sexual fate of at least some individuals was determined by nutritional status but we do not know at present if CPTI acts within the animal or against proteases of plant or animal origin within the feeding cell. Expression of only 0.1-0.5% soluble protein as the inhibitor also caused a shift in sex ratio in favour of males.

Our data shows that *G. pallida* is unaffected by CPTI expression during its invasion and that the inhibitor has its effect only after the animal is developmentally committed to a feeding site. As with insects, effects may not be simply due to inhibition of digestive proteases. Growth rate is reduced and the sex of some *G. pallida* individuals is directed to males. They require less food to reach a sexual maturity than females which are about 100x the body size of their mates. Apparently CPTI expression influences mechanisms that normally enable the animal to match its growth rate, sex and development to that supportable by the feeding site. *Caenorhabditis elegans*, requires continual maintenance of sexual commitment during development (Kuwabara and Kimble, 1992). This is likely to be a general principle in nematodes. Suppression of females is also caused for *G. rostochiensis* by a resistant potato (H1; cv Maris Piper). Perhaps this resistance is durable because it evokes an important regulatory route of benefit on susceptible plants in ensuring less favourably placed animals reach maturity by developing as males. This may ensure that phenotypes responsive to this effect are prevalent in field populations. This suggests description of feeding or nutrient utilization provides a powerful basis on which to engineer plant resistance to nematodes.

ENGINEERING IMPROVED BASIS FOR RESISTANCE

We envisage several bases for modifying effector genes to alter the efficacy of their products. One approach is that based on engineering antibodies in plants now that this is feasible (Hiatt *et al.*, 1989; Düring *et al.*, 1990). An engineered antibody fragment could be used to carry a toxin incorporated into the Fc region of its heavy chain (Hiatt and Ma, 1992). Another approach would be to target antigen that has an essential role for the pathogen *in planta*. Clear possibilities are the antibodies raised previously to the dorsal pharyngeal gland secretions of *H. glycines* (Atkinson et al., 1988) and *M. incognita* (Hussey, 1989). A principal limitation of this approach is that bioassays for efficacy are not currently available to demonstrate biological efficacy before antibody engineering is attempted. An antibody disrupting an essential plant protein such as an enzyme could also provide a basis for novel defences against nematodes. However, this approach seems unnecessarily complex given that plants possess genes such as protease inhibitors that can be used as effectors.

Protein engineering of cysteine protease inhibitors has potential for nematode control. Study of proteases in *Globodera* (Koritsas and Atkinson, 1994) established that the adult female possesses principally cysteine protease activity. We envisage disruption of female development is likely to provide a more efficacious defence than one directed principally at the establishing juvenile. Therefore part of our current attention is focused on cysteine protease inhibitors. Published sequence information exists for a number of such genes such the two cystatins from rice, oryzacystatins (*Oc*-I and *Oc*-II). We have isolated *Oc*-I from genomic DNA by the PCR approach of splicing by overlap extension (SOEing) to remove its intron (Horton and Pease 1991). We have shown that *Oc*-I can be used as a heterologous probe for Southern analysis and for library screening of other species. Given that these cystatins are not normally expressed in plant roots there is no reason to assume that they are optimally effective against the proteases of root-nematodes. We have cloned *Oc*-I into an expression vector pQE to provide a virtually unlimited supply of inhibitor for further work. Currently we are modelling the structure of oryzacystatin I protein (Oc-I) against published data for other cystatins to assist selection of sites for directed mutagenesis aimed at enhancing activity. Expressed Oc-I is being used for biochemical assays and as an immunogen to raise a polyclonal antibody for later use in monitoring expression levels. More importantly, sufficient Oc-I have also been obtained for crystallisation trials that should ultimately lead to resolution of its three-dimensional structure. We intend extending our structural studies to a nematode protease. The objective is to reduce the Ki of the inhibitor against a target protease. Our current progress with mutagenesis is encouraging. Removal of the N-terminal 21 amino acids of Oc-I has been reported to cause a marginal improvement in Ki (Abe et al., 1988). Our modelling studies suggest this effect is unlikely and we find no such improvement for the corresponding protein. However our modelling studies predicted that deletion of one particular codon for Oc-I should enhance its Ki. This deletion caused a 10 fold improvement in Ki value against papain. We hope for still further improvements to Ki before correlating the extent of such changes with enhanced efficacy of a corresponding transgene *in planta*. This work establishes one strategy for designing nematode resistant plants by protein engineering. It is likely that a number of other possible approaches will be developed in the future.

CONCLUDING REMARKS

Plant biotechnology for nematode control would be highly beneficial if it overcomes current limitations on the wider use of resistant cultivars. The approach offers clear advantages to the grower and the consumer. The cost is only that of any premium associated with the seed carrying the resistant trait without additional costs associated with application or changes in agronomy that characterise use of pesticides and biocontrol.

Experimentally effective lines are already available from our work. We are generating distinct and successful prototypes using the approaches summarised above. However additional safeguards against unwanted expression may be provided for additional safety in use. There will be a subsequent time-lapse before the systems are commercially available because of the need to evaluate them thoroughly both to ensure commercial acceptability and to meet all regulatory requirements. It is possible to envisage feeding cell attenuation approaches that do not require expression of additional proteins within the plant. Expression of protease inhibitors in non-harvested roots seems an inherently safe approach given these proteins occur in widely eaten seeds. Our objective is to provide novel resistant cultivars that gain widespread favor with growers and so reduce nematicide use and any environmental damage they cause. We are confident that both of our principal approaches can provide bases for nematode control. Feeding cell attenuation provides a highly specific defence against key pests of world crops. A natural or engineered protease inhibitor under control of constitutive promoter may prove able to protect a crop such as rice from many different nematode species even when concurrent infection occurs. It is important to us that the new approaches offer an inexpensive, efficacious and environmentally acceptable alternative to chemical control of plant parasitic nematodes in both the developed and developing world.

Acknowledgements

We thank Dr. M. Bevan, for providing *A. rhizogenes* strain LBA9402, Dr. K. Lindsey, for transgenic *Arabidopsis* lines and Dr. J.J. Naleway, (Molecular probes Inc), for free samples of Imagene GUS substrates. We gratefully acknowledge financial support from SERC Biotechnology Directorate, AFRC, ODA, Advanced Technologies (Cambridge), Maribo Seed Biotechnology and Nickerson BIOCEM.

REFERENCES

Abe, K., Emori, Y., Kondo, H., Arai, S. and Suzuki, K., 1988, The NH_2-terminal 21 amino acids residues are not essential for the papain - inhibitory activity of oryzacystatin, a member of the cystatin super family, *J. Biol. Chem.* 263:7655.

Al-Yahya, F.A., 1993, Aspects of the Host-Parasite Interactions of *Meloidogyne* spp. (Root-Knot Nematodes) and Crop Plants. PhD thesis, The University of Leeds, 109pp.

Anon (1993a) *Agrow* No 179 p 16.

Anon (1993b) *Agrow* No 190 p 17

Anon (1993c) *Agrow* No 190 p 19

Atkinson, H.J. and Harris, P.D., 1989, Changes in nematode antigens recognised by monoclonal antibodies during early infection of soya beans with the cyst nematode *Heterodera glycines*, *Parasitology* 98:479.

Atkinson, H. J., Harris, P. D., Halk, E. J., Novitski, C., Leighton-Sands, J., Nolan, P. and Fox, P.C., 1988, Monoclonal antibodies to the soya bean cyst nematode *Heterodera glycines*, *Ann. Appl. Biol.* 112: 459.

Betka, M., Grundler, F., and Wyss, U., 1991, Influence of changes in the nurse cell system (syncytium) on the development of the cyst nematode *Heterodera schachtii*, Single amino acids, *Annu. Rev. Phytopathol.* 81:75.

Bush, D.R., 1993, Proton-coupled sugar and amino acid transporters in plants, *Annu. Rev. Plant Physiol. Plant Mol. Biol.* 44:513.

Dorhout, R., Kolloffel, C., and Gommers, F.J. 1992. Alteration of distribution of regions with high net proton extrusion in tomato roots infected with *Meloidogyne incognita*, *Physiol. and Mol. Plant Pathol.* 40:153.

Düring, K., Hippe, S., Krezaler, F.and Schell, J., 1990, Synthesis and self-assembly of a functional monoclonal antibody in transgenic *Nicotiana tabacum*, *Plant Mol. Biol.* 15:281.

Emes, M.J., and Fowler, M.W., 1979, The intracellular location of the enzymes of nitrate assimilation in the apices of seedling pea roots, *Planta* 144:249.

Frohman , M.A., Dush, M.K., and Martin, G.R., 1988, Rapid production of full-length cDNAs from rare transcripts: Amplification using a single gene-specific oligonucleotide primer, *Proc. Natl. Acad. Sci.*, USA 85:8998.

Goddijn, O.J.M., Lindsey, K., Van der Lee, F.M., Klap, J.C. and Sijmons P.C., 1993, Differential gene expression in nematode-induced feeding structures of transgenic plants harbouring promoter-*gus*A fusion constructs, *The Plant Journal* 4:863.

Gurr, S. J., McPherson, M. J., Atkinson, H. J., and Bowles, D. J., 1992, Plant Parasitic Nematode Control Patent Corp. Treaty/GB91/01540.

Gurr, S.J., McPherson, M.J., Scollan, C., Atkinson, H.J. and Bowles, D.J., 1991 Gene expression in nematode-infected plant roots. *Mol. Gen. Genet.* 226:361.

Hamill, J.B., Prescott, A., and Martin, C., 1987, Assessment of the efficiency of co-transformation of the T-DNA of disarmed binary vectors derived from *Agrobacterium tumefaciens* and the T-DNA of *Agrobacterium rhizogenes*, *Plant Mol. Biol.* 9:573.

Hammond-Kosack, K., Atkinson, H.J. and Bowles, D.J., 1989, Local and systemic changes in gene expression in potato plants following root infection with the cyst nematode *Globodera rostochiensis*, *Physiol. Mol. Plant Pathol.* 37:339.

Hartley, R.W., 1988, Barnase and barstar. Expression of its cloned inhibitor permits expression of a cloned ribonuclease, *J. Mol. Biol.* 202:913.

Hartley, R.W., 1989, Barnase and barstar: two small proteins to fold and fit together, *Trends Biochem. Sci.* 14:451.

Hepher, A., and Atkinson, H.J., 1992, Nematode control with proteinase inhibitors. European patent application number, 92301890. 7 publication number 0502730A1

Hiatt, A., Cafferkey, R., and Bowdish, K., 1989, Production of antibodies in transgenic plants, *Nature* 342:76.

Hiatt, A., and Ma, J.K.C., 1992, Monoclonal antibody engineering in plants, *FEBS Letters* 307:71.

Horton , R.N. and Pease, L.R., 1991, Recombination and mutagenesis of DNA sequences using PCR. *In*: "Directed Mutagenesis: A Practical Approach", M.J. McPherson, ed., IRL Press, Oxford.

Hussey, R. S., 1989, Monoclonal antibodies to secretory granules in esophageal glands of *Meloidogyne* spp, *J. Nematol.* 21:392.

Jefferson, R.A., Karanagh, T.A., and Bevan, M.W., 1987, GUS fusions - β-glucoronidase as a sensitive and versatile gene fusion marker in higher plants. *EMBO J.* 6:3901.

Jones, M.G.K., 1981, Host cell responses to endoparasitic nematode attack: structure and function of giant cells and syncytia, *Ann. Appl. Biol.* 97:353.

Kondo, H., Emori, Y., Abe, K., Suzuki, K. and Arai, S., 1989, Cloning and sequence analysis of the genomic DNA fragment encoding oryzacystatin, *Gene* 81:259.

Koritsas, V. M., and Atkinson, H.J., 1994, Proteinases of females of the phytoparasite *Globodera pallida*, (Potato cyst nematode) *Parasitology* (*in press*).

Kuwabara, P., and Kimble, J., 1992, Molecular genetics of sex determination in *C. elegans.*, *Trends Genet.* 8:164.

Lindsey, K., Wei, W., Clarke, M.C., McArdle, H.F., Rooke, L.M., and Topping, J.F., 1993, Tagging genomic sequences that direct transgene expression by activation of a promoter trap in plants, *Trans. Res.* 2:33.

Mariani, C., Beuckeleer, M., Truettner, J., Leemans, J. and Goldberg, R. B., 1990, Induction of male sterility in plants by a chimaeric ribonuclease gene, *Nature* 347:737.

Muller, J., Rehbock, K. and Wyss, U., 1981, Growth of *Heterodera schachtii* with remarks on amounts of food consumed, *Revue Nematol.* 4:227.

Naleway, J.J., Zhang, Y.Z., Bonnett, H., Galbraith, D.W., and Haugland, R.P., 1991, Detection of GUS expression in transformed plant cells with new lipophilic, fluorogenic β-glucuronidase substrates. *J. Cell Biol.* 115, 151a (Meeting abstract)

Nickle, W. R., 1991, Editor "Manual of Agricultural Nematology" Marcel Dekker New York.

Niebel, A., Van der Eycken, W., Inzé, D., Van Montagu, M., and Gheysen, G., 1992, Potato genes induced during *Globodera pallida* infection. 21st Int. Nematol. Symp., *Nematologica* 38:426.

Pate, J.S., 1973, Uptake, assimilation and transport of nitrogen compounds by plants, *Soil Biol. Biochem.* 5:109.

Press, M.C., and Whittaker, J.B., 1993, Exploitation of the xylem stream by parasitic organisms, *Phil. Trans. R. Soc. Lond.* 341:101.

Roberts, P. A., 1992, Current status of the availability, development and use of host plant resistance to nematodes, *J. Nematol.* 24:213.

Ryan, C.A., 1990, Protease inhibitors in plants: Genes for improving defenses against insects and pathogens, *Annu. Rev. Phytopathol.* 28:425.

Sasser, J. N., 1979, Economic importance of Meloidogyne in tropical countries. *In*: Root-Knot Nematodes (Meloidogyne species). Systematics, Biology and Control, F. Lamberti and Taylor C.E., eds.,. Academic Press, London.

Sasser, J.N., and Freckman, D.W., 1987, A world perspective on nematology: the role of the Society. *In*: "Vistas on Nematology", J.A.Veech and Dickerson, D.W., eds., Soc. of Nematologists.

Sijmons, P.C., Grundler, F.M.W., von Mende, N., Burrows, P.R. and Wyss, U., 1991 *Arabidopsis thaliana* as a new model host for plant-parasitic nematodes, *The Plant Journal* 1:245.

Topping, J.F., Wei, W.B. and Lindsey, K., 1991, Functional tagging of regulatory elements in the plant genome, *Development* 112:1009.

Van der Eycken W., Niebel, A., Inzé, D., Van Montagu, M. and Gheysen, G., 1992, Molecular study of root knot nematode induction by the nematode *Meloidogyne incognita*, *Med. Fac. Landbouww. Univ. Gent* 57/3a:895.

van Haren, R.J.F., Hendrikx, E.M.L., Atkinson, H.J., 1993, "Growth curve analysis of sedentary plant parasitic nematodes in relation to plant resistance and tolerance", *In* "Predictability and nonlinear modeling in natural sciences and economics", J. Grasman and van Straten, G., eds., Wiley, New York, in press.

Wyss, U., and Grundler, F. M. W., 1992, Feeding behaviour of sedentary plant parasitic nematodes. *Netherlands J. Plant Pathology Supplement* 2 98:165.

MOLECULAR BIOLOGY OF NEMATODE RESISTANCE IN TOMATO

Valerie M. Williamson, Kris N. Lambert, and Isgouhi Kaloshian

Department of Nematology
University of California
Davis, CA 95616
USA

INTRODUCTION

The *Mi* gene

Tomato, *Lycopersicon esculentum,* is a host for several species of root-knot nematodes, and nematode infestation can result in severe yield loss for this crop. The *Mi* gene of tomato confers effective resistance to three root-knot nematode species, *Meloidogyne incognita, M. javanica* and *M. arenaria*, but not to a fourth, *M. hapla* (Gilbert and McGuire, 1956; Braham and Winsted, 1957; Roberts and Thomason, 1986). *Mi* was introduced into tomato from the wild species *Lycopersicon peruvianum* using embryo rescue of an interspecific cross of this wild species with *L. esculentum* (Smith, 1944). Progeny of a single F_1 plant are the sole source of nematode resistance in currently available fresh-market and processing tomato cultivars (Medina-Filho and Tanksley, 1983). Recent restrictions on the use of nematicides have increased reliance on the gene *Mi* for nematode control in tomato.

Mi has been described as dominant, or semi-dominant, and is one of the best characterized nematode resistance genes, thus providing an excellent model for studies on the molecular basis of resistance. The resistance conferred by *Mi* is characterized by the appearance of a localized region of necrotic plant cells, called a hypersensitive response (HR), around the head of the invading nematode (Frazier and Dennett, 1949; Riggs and Winstead, 1959; Dropkin, 1969; Dropkin et al., 1969). Compared to the resistance response in other plant-nematode interactions *Mi*-mediated resistance is among the most rapid (Trudgill, 1991). The earliest visible indications of the HR occur about 12 hours after inoculation of roots with nematode juveniles (Dropkin et al., 1969; Paulson and Webster, 1972).

Early events in the tomato-nematode interaction

It is during the first 12-24 hours after infection by nematodes that resistance or susceptibility is determined by the plant and when the first visible events that discriminate the two responses are visible (Figure 1). In both resistant and susceptible plants, nematodes are attracted to root tips, penetrate behind the root cap and move between cells to the feeding site. It is likely that common changes in gene expression occur in susceptible and resistant plants during the migration of the nematode to its feeding site. In support of this we have found that acid phosphatase levels increase in both responses (Ma and Williamson, unpublished). Induction of genes in response to wounding is an example of a class of genes that may be induced in both susceptible and resistant genotypes. Proteinase inhibitors and other stress or wound-induced genes have been reported to be induced systemically in potato

Advances in Molecular Plant Nematology
Edited by F. Lamberti *et al.*, Plenum Press, New York, 1994

after *H. glycines* infection in both compatible and incompatible interactions (Hammond-Kosack et al., 1990). However, root-knot nematodes cause much less cell damage when they invade so it is not certain that wound induced genes are turned on in this system. It is also possible that, in resistant plants, the early steps in establishing a compatible interaction are initiated before they are "interrupted" by the resistance response.

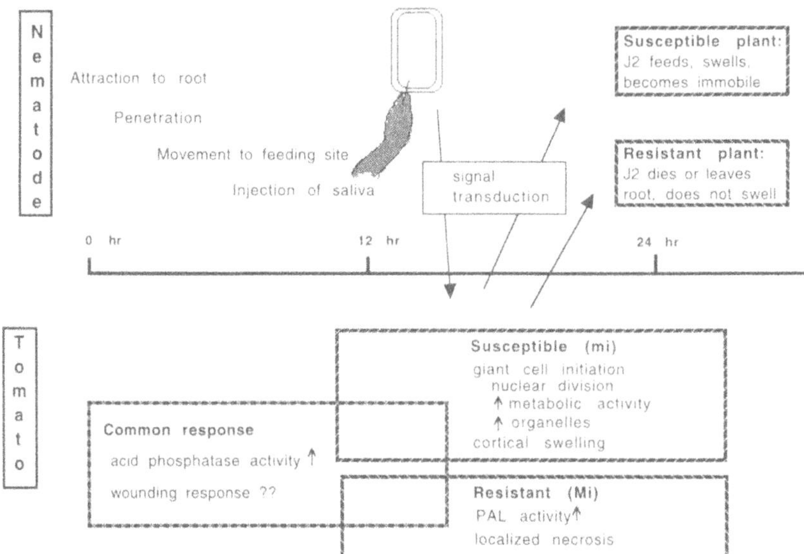

Figure 1. Early steps in the interaction of root-knot nematodes with tomato.

At approximately 12 hours after applying root-knot nematodes to tomato roots (the precise timing varies considerably with experimental conditions), the nematodes reach the site of giant cell initiation. In susceptible plants, signals from the nematode somehow initiate the complex series of events leading to feeding site development (Huang, 1985; Hussey, 1985). In resistant plants, microscopic observation indicates that nematode juveniles do not elicit extensive necrosis while migrating through the root tissue, but do so while attempting to establish a feeding site (Paulson and Webster, 1972). This observation suggests that cell penetration by the nematode stylet or new signals from the nematode may trigger *Mi*-mediated recognition of the nematode and initiation of a signal transduction cascade leading to resistance. The enzyme phenylalanine ammonia lyase (PAL) which is induced early in the resistance response to many other pathogens, is increased in activity level in resistant tomato by 12 hr after nematode inoculation (Brueske, 1980; Bowles, 1990). Within a few days of infection the J2 either die or leave the root to find a new host.

Resistance due to *Mi* is lost at elevated temperatures (Dropkin, 1969). Plants inoculated and maintained at 32 °C for 2 days, and subsequently held at 27 °C for one month, contained abundant galls and eggs (Dropkin, 1969). Dropkin (1969) concluded, based on this result and other temperature shift experiments, that the determination of resistance occurs during the first 24-48 hours after infection, and that once this time period is passed, resistance is not triggered, even at the permissive temperature (i.e., 27 °C). It may be that the *Mi* product is no longer synthesized or is in an inactive state after nematode infection. Alternatively, it could be that the nematode molecules that trigger the response are no longer present after 24 hours. Genes induced in the resistance response may be transcribed concurrently with early genes in the susceptible response and may stop the progression of the susceptible response, perhaps due to the localized cell death in the responding tissues.

We have taken two approaches to investigating the molecular sequence of events that result in resistance to root-knot nematodes in tomato. One is to identify genes that are increased in expression early in the resistance response and the second is to clone the *Mi* gene itself.

IDENTIFICATION OF GENES INDUCED EARLY IN THE RESISTANCE RESPONSE

Seedling assay

A large number of uniformly infected tomato root tips are required to study the changes in gene expression early in the resistance response. For this, a protocol was developed to obtain rapidly hundreds of uniformly infected root tips (see Ho et al., 1992). Basically, surface sterilized seeds are germinated on moist filter paper. After radicles reach a length of 2 cm or longer, seedlings are aligned on moist filter paper on a glass plate, and roots are covered with nylon mesh. The tips are inoculated with nematode juveniles from a hydroponic culture system for continuous production of root-knot nematode juveniles (Lambert et al, 1992). Rows of seedlings are set up in a rack and kept moist by wicking moisture from a reservoir. The localized necrosis, observed as brown spots or regions in the root tips, is easily visible two days after application of approximately 50 juveniles per root tip. At this time susceptible tomato root tips display obvious swelling, early signs of a successful infection. No swelling of resistant root tips is observed. After 2-3 days, the localized necrosis is observed as a brown area around the anterior of the nematodes.

Inoculation of root tips with high numbers of nematodes (150 J2/root tip) results in a large amount of responding tissue in both resistant and susceptible root tips. Microscopic examination of root tips reveals that all tips (resistant and susceptible) inoculated under these conditions contain large numbers of nematode juveniles after 12 hours.

cDNA library construction

Even with the seedling inoculation system, it was not feasible to obtain sufficient mRNA to construct a cDNA library using conventional methods as only about 5 ng of mRNA are produced from 100 root tips (estimated by an assay based on radioactive labeling using reverse transcriptase). For this reason a PCR-based procedure was designed to generate a cDNA library from this tiny amount of mRNA (Lambert and Williamson, 1993). Briefly, mRNA is hybridized to oligo-dT paramagnetic beads (Dynal, Inc., Lake Success, N.Y.). The first strand cDNA is synthesized with the oligo-dT on the beads as primer, creating a solid phase library. Unprimed oligo-dT is removed by exonuclease and the cDNA is A-tailed. Second strand cDNA is synthesized using a primer that has oligo-dT at its 3' end (primer T). Amplification is carried out with primer T and a derivative of primer T lacking oligo-dT stretch at its 3' end. A strategy for library construction modified from Stoker (1990) that prevents self ligation of inserts was used to insert the amplified inserts into the plasmid pBluescript. By this strategy a cDNA library was constructed from resistant tomato root tips 12 hours after inoculation with approximately 150 *M. javanica* J2 per root tip.

Differential screening of cDNA library

Differential screening was carried out to identify cDNAs that differed in transcript level between uninfected root tips and those that had been exposed to root-knot nematodes for 12 hours (Figure 2). A number of clones were observed to change, either increase or decrease, in level after nematode infection. After screening approximately 300 inserts, 7 clones were identified that were increased in expression in the infected root tips. Southern analysis showed that all seven clones were of tomato, rather than nematode, origin and corresponded to either single copy genes or members of small gene families. Each clone was used to probe a Northern blot of RNA extracted from root tips infected with nematodes and RNA from uninfected controls. In all cases induction was confirmed. To determine whether these genes were specifically induced in the resistance response or were also induced in susceptible tomato, the PCR-amplified inserts from each clone were each spotted onto 4 nylon membranes. Membranes were probed with labeled first-strand cDNA from root tips of: 1) infected susceptible tomato; 2) uninfected susceptible tomato; 3) infected resistant tomato; and 4) uninfected resistant tomato.

Five of the seven clones were found to be induced in susceptible root tips upon nematode infection. One clone, called D7, was found to be highly induced in susceptible and resistant plants. The other four were less increased in susceptible than in resistant plants. Clone D7 shows strong sequence similarity (82% identity of amino acid sequence) to tobacco extensin, an abundant cell wall protein (Lambert and Williamson, unpublished; De Loose et al., 1991; Ye and Varner, 1991). Genes that are induced in both susceptible and resistant root tips could represent very early events in the susceptible response, including transiently expressed genes that are involved in initiation of the elaborate feeding site, or could represent genes induced in response to tissue damage upon nematode infection. *In situ* localization of transcripts may help to resolve these possibilities.

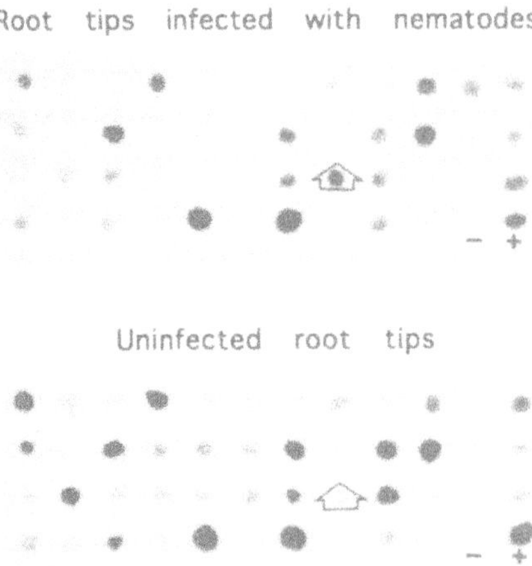

Figure 2. Differential screening to identify cDNA clones induced in the resistance response. Individual cDNA inserts were amplified using PCR and equal amounts of amplified DNA was spotted onto duplicate nylon filters. The filters were probed with ^{32}P-labeled first-strand cDNA from nematode-infected root tips or uninfected root tips from resistant tomato seedlings. Positive control (+) is tomato cyclophilin, a gene believed to be constitutively expressed in plant tissues (Gasser et al., 1990), and the negative control (-) is pBluescript. A clone that appears to be increased in expression after nematode infection is marked by an arrow.

Clones, 23a and 94 are induced only in the resistance response. Clone 23a has been sequenced and contains an open reading frame that is 60% identical in amino acid sequence to miraculin, a protein found in the berries of *Richadella dulchifica*, the miracle bush (Figure 3). Miraculin changes taste perception from sour to sweet and has homology to soybean trypsin inhibitor (Theerasilp et al., 1989). The homology of clone 23a to a protein which changes taste perception and has sequence similarity to a proteinase inhibitor suggests that the 23a product may have a role in resistance, perhaps by binding to amphids of nematodes and altering nematode perception of its host, or perhaps by inhibiting nematode proteinases.

```
           1
      23a  PLTFTPVDPK KGVIRESTDL NIIFSANSIC VQT..TQWK LDDFDETTGQ
miraculin  PLAFFPENPK EDVVRVSTDL NINFSAFMPC RWTSSTVSR LDKYDESTGQ

           51
      23a  YFITLGGDQG NPGVETISNW FKIGKYDRD. .YKLLYCPT VCDFCKVICR
miraculin  YFVTIGGVKG NPGPETISSW FKIEEFCGSG FYKLVFCPT VCGSCKVKCG

           101
      23a  DIGISFRMEL DVWL
miraculin  DVGIYIDQKG RRRLALSDKPFAFEFNKTVYF
```

Figure 3. Comparison of the amino acid sequence of clone 23a to miraculin. The deduced amino acid sequence of the clone 23a insert is aligned with the C-terminal amino acid sequence of miraculin (Theerasip et al., 1989). Identical amino acids are boxed.

CLONING *Mi*

Map-based cloning

Because the presence or absence of the gene *Mi* determines whether or not a tomato plant is resistant to root-knot nematodes, it is likely that the *Mi* gene product is involved in recognition of the nematode by the plant, and in triggering the signal transduction process of the resistance response. For this reason, a clone of *Mi* has been targeted for isolation using a map-based cloning approach. Map-based cloning requires the identification of DNA markers surrounding the target gene followed by the physical mapping and cloning of sequences between the two closest flanking markers. The final step is to identify the fragment carrying the resistance gene by transforming a susceptible host with the putative clone and testing for the acquisition of resistance.

Tomato has several advantages as a plant for a map-based cloning approach. Genetic analysis in this plant is extremely straightforward and a large number of morphological and DNA markers have already been mapped (reviewed in Rick and Yoder, 1988). It has a modest genome size (950 megabases), and a large resource of resistance genes have been identified in tomato and its wild relatives. A high density RFLP map of the tomato genome with over 1000 markers has been generated (Tanksley et al., 1992), and a yeast artificial chromosome (YAC) library that contains large insert clones spanning the tomato genome has been constructed (Martin et al., 1992). Techniques for isolation and analysis of megabase size DNA from tomato have been published (van Daelen et al., 1989; Wing et al., 1993). DNA transformation of tomato by *Agrobacterium*-based vectors is relatively efficient.

Genetic map of the *Mi* region

The alien origin of *Mi* and the surrounding DNA sequences was the key to obtaining DNA markers linked to *Mi*. Few polymorphisms are seen between different cultivars of the domestic tomato, *Lycopersicon esculentum,* but when one compares *Lycopersicon* species, RFLPs are abundant (Bernatzky and Tanksley, 1986; Helentjaris et al., 1986; Miller and Tanksley, 1990). Thus, polymorphisms between inbred nematode-resistant and susceptible cultivars are likely to map to the region of DNA containing *Aps-1* and *Mi* as this region is derived from the wild tomato species *L. peruvianum.* We identified closely linked DNA markers as RFLPs by probing Southern blots of DNA from selected tomato lines with random clones from a cDNA library constructed from a nematode resistant tomato line (Ho et al., 1992). Analysis of DNA from an F_2 population segregating for nematode resistance was carried out to generate a high-resolution genetic map of this region (Messeguer et al., 1991; Ho et al., 1992). Additional information on gene order was obtained by comparing the size of the introgressed *L. peruvianum* segment in a collection of nematode resistant tomato lines (Figure 4). The related nematode-resistant cultivars Motelle, Motaci and Mossol were found to carry the smallest introgressed region which was identified by only one RFLP marker, LC379. This marker has been so far genetically inseparable from *Mi.* The cDNA clone LC379 hybridizes to a gene family of about thirty members, one of which was on a 1.1 kb EcoRI fragment present only in nematode resistant tomato (Ho et al., 1992).

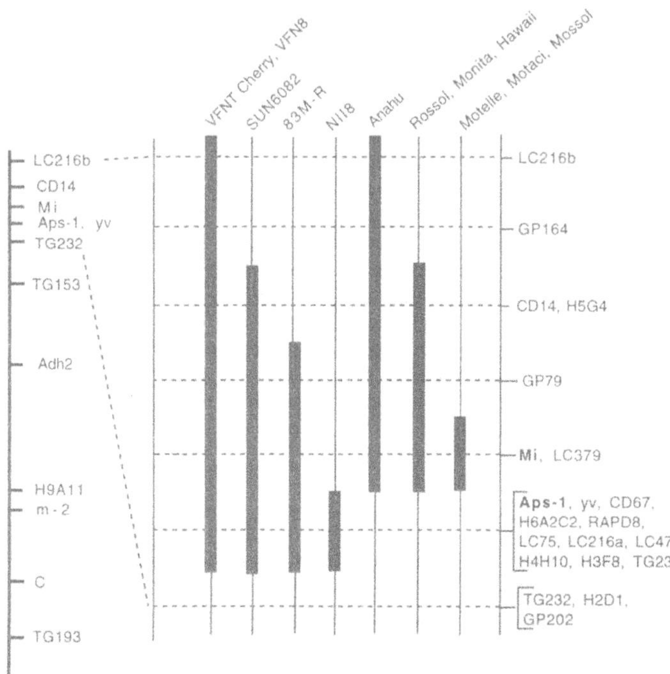

Figure 4. Map of the *Mi* region. A map of chromosome 6 with selected RFLP and morphological markers (Zabel et al., 1993) is shown at the left. The *Mi* region map is shown in more detail on the right. The extent of the introgressed region in selected tomato cultivars is indicated (Ho et al., 1992). Chromosomal regions of *L.esculentum* origin are represented as thin vertical lines and regions of *L.peruvianum* origin are represented by thick black bars. Junctions between regions from different species are depicted as midway between flanking markers though they do not necessarily represent the same crossover in each cultivar.

A PCR-based marker tightly linked to *Mi*

Random amplified polymorphic DNA (RAPD; Williams et al., 1990) markers were sought as an additional approach to obtaining markers linked to *Mi* (Williamson et al., 1994). DNA extracts from nearly isogenic pairs of tomato lines were screened with over 300 decamer primers. Several markers were identified that mapped near *Mi*, but only one amplified product, REX-1, obtained using a pair of decamer primers, was identified in all nematode resistant lines tested including Motelle (Williamson et al, 1994). REX-1 was cloned and the DNA sequences of its ends were determined and used to develop 20-mer primers. PCR amplification with the 20-mer primers produced a single amplified band in both susceptible and resistant tomato lines. The amplified bands from susceptible and resistant lines were distinguishable after cleavage with the restriction enzyme TaqI (Figure 5).

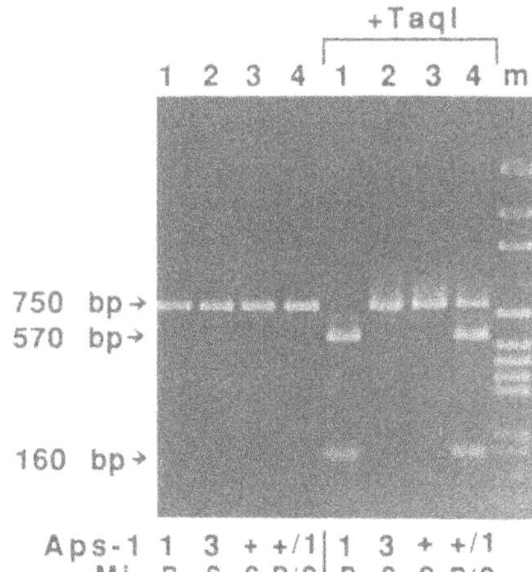

Figure 5. REX-1 amplification products obtained using 20-mer primers, REX-F1 and REX-R2 (Williamson et al., 1994). A band of approximately 750 bp is amplified from both resistant (R) and susceptible (S) tomato lines. After digestion with TaqI two bands are seen in the R line and three are seen in a line heterozygous (R/S) at the *Mi* locus. Lane 'm' contains size markers.

The linkage of REX-1 to *Mi* was verified in an F_2 population. This marker has been so far genetically inseparable from *Mi* in all lines tested. Because PCR assay with these primers is easy and can be done with crude extracts from a tiny amount of plant tissue, REX-1 has been incorporated into screening programs at university and private laboratories, replacing the acid phosphatase isozyme test.

Physical map

A YAC library containing three haploid genomic equivalents of tomato DNA (22,000 clones) was constructed by Martin et al (1992). DNA preparations from pools of 96 YAC colonies were screened by PCR with the REX-1 primers, and a YAC clone, YM1, containing REX-1 was identified. By pulsed field gel electrophoresis (PFGE), the tomato DNA insert in YM1 was determined to be about 160 kb. The ends of the insert were cloned by inverse PCR using primers and protocols in Putterill et al. (1993). By probing Southern blots of tomato genomic DNA that contained recombination events near *Mi* we were able to orient YM1 on the chromosome and determined that the right end was located below *Mi*. PFGE studies have suggested that the introgressed region in the line Motelle is approximately 600 kb in length. A summary of our current view of the introgressed region is presented in Figure 6.

CURRENT STATUS AND PREDICTIONS

Where the *Mi* gene product is localized and how it triggers the resistance response are questions of great interest. We have clones within 450 kb of *Mi* and hope to have the gene cloned within the next year or so. The availability of a clone of *Mi* together with genes that are transcriptionally induced early, and specifically, in the resistance response will assist us in deciphering the signal transduction pathway from *Mi*-mediated recognition to the transcriptional changes that occur specifically in the resistance response. Our strategy is to work forward from the *Mi* gene and backward from the induced transcripts to deduce the pathway for signal transduction. For example, if *Mi* is a protein kinase, as the tomato resistance gene *Pto* appears to be (Martin et al., 1993), then changes in protein phosphorylation may be involved in the signal transduction pathway. From the other end, investigation of promoters of genes, such as the miraculin analog, that are up-regulated specifically in the resistance response should allow identification of promoter elements, and, in turn, transcription factors, involved in the resistance response.

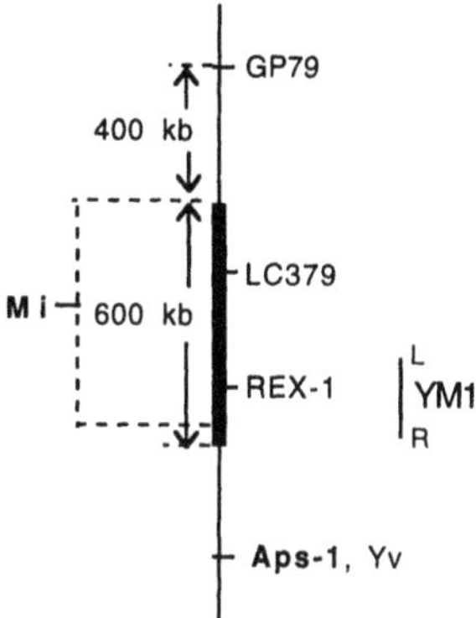

Figure 6. Current physical and genetic map of the *Mi* region. The introgressed region containing *Mi* in cultivar Motelle is indicated by a black bar. The size of this region has been estimated to be approximately 600 kb. The relative positions of markers and the orientation of YAC YM1 were determined by recombinants in an F_2 population segregating for *Mi*. The location of *Mi* in the introgressed region has not been determined because the key recombinant died before it could be tested for resistance.

Another key to understanding resistance may be provided by isolates of *M. incognita* that are able to infect resistant plants (Bost and Triantaphyllou, 1987; Dalmasso et al., 1991; Jarquin-Barberena et al., 1991). It is possible that these virulent nematodes lack, or have alterations in, the molecules that are recognized by resistant plants. Alternatively, they may have evolved a different means of evading the response. Analysis of molecular differences between nematode populations virulent and avirulent on *Mi* may provide additional information on how *Mi* works.

Characterization of genes induced specifically in response to nematode infection of resistant plants may lead to novel approaches to nematode resistance. Genes expressed in response to other pathogens have been shown, when under the control of highly expressed promoters, to confer increased resistance to a number of pests and pathogens (Ryan, 1990; Broglie et al., 1991; Alexander et al, 1993). Thus, genes that are induced early in the nematode resistance response, such as miraculin, may have a direct role in nematode defense and may provide the basis for novel defense strategies when engineered so that they are expressed at the site of nematode infection.

Acknowledgments

The research described in this paper was supported by USDA grants 91-373-6339 and 92-37302-7635. The YAC library screen was partially supported by NSF Cooperative Agreement BIR-8920216 to CEPRAP, a NSF Science and Technology Center, and by CEPRAP corporate associates Calgene, Inc., Ciba Geigy Biotechnology Corporation, Sandoz Seeds, and Zeneca Seeds.

REFERENCES

Alexander, D., Goodman, R.M., Gut-Rella, M., Glascock, C., Weymann, K., Friedrich, L., Maddox, D., Ahl-Goy, P., Luntz, T., Ward, E., and Ryals, J., 1993, Increased tolerance to two oomycete pathogens in transgenic tobacco expressing pathogenesis-related protein 1a, *Proc. Natl. Acad. Sci. USA* 90:7327.

Bernatzky, R., and Tanksley, S.D., 1986, Majority of random cDNA clones correspond to single loci in the tomato genome, *Mol. Gen. Genet.* 203:8.

Bost, S.C., and Triantaphyllou, A.C., 1987, Genetic basis of the epidemiological effects of resistance to *Meloidogyne incognita* in the tomato cultivar Small Fry, *J. Nematol.*44:540.

Bowles, D.J., 1990, Defense-related proteins in higher plants. *Annu. Rev. Biochem.* 59:873.

Braham, W.S., and Winstead, N.N., 1957, Inheritance of resistance to root-knot nematodes in tomatoes, *Proc. Am. Soc. Hort. Sci.* 69:372.

Broglie, K., Chet, I., Holliday, M., Cressman, R., Biddle, P., Knowlton, S., Mauvais, C. J., and Broglie, R., 1991, Transgenic plants with enhanced resistance to the fungal pathogen *Rhizoctonia solani*, *Science* 254:1194.

Brueske, C.H., 1980, Phenylalanine ammonia lyase activity in tomato roots infected and resistant to the root-knot nematode, *Meloidogyne incognita*, *Physiol. Plant Path.* 16:409.

Dalmasso, A., Castagnone-Sereno, P., Bongiovanni, M., and deJong, A., 1991, Acquired virulence in the plant parasitic nematode, *Meloidogyne incognita*. 2. Two-dimensional analysis of isogenic isolates, *Revue Nematol.* 14:277.

De Loose, M., Gheysen, G., Tire, T., Gielen, J., Villarroel, R., Genetello, C., Van Montagu, M., Depicker, A., and Inze, D., 1991, The extensin signal pepetide allows secretion of a heterologous protein from protoplasts, *Gene* 90:95.

Dropkin, V.H., 1969, The necrotic reaction of tomatoes and other hosts resistant to *Meloidogyne*: reversal by temperature, *Phytopathology* 59:1632.

Dropkin, V.H., Helgeson, J.P., and Upper, C.D., 1969, The hypersensitivity reaction of tomatoes resistant to *Meloidogyne incognita:* reversal by cytokinins, *J. Nematol.* 1:55.

Frazier, W.A., and Dennett, R.K., 1949, Isolation of *Lycopersicon esculentum* type tomato lines essentially homozygous resistant to root-knot, *Proc. Am. Soc. Hort. Sci.* 54:225.

Gasser, C.S., Gunning, D.A., Budelier, K.A., and Brown, S.M., 1990, Structure and expression of cytosolic cyclophilin/peptidyl-prolyl cis-trans isomerase of higher plants and production of active tomato cyclophilin in *Escherichia coli*, *Proc. Natl. Acad. Sci. USA* 87:8519.

Gilbert, J.C. and McGuire, D.C., 1956, Inheritance of resistance to severe root-knot from *Meloidogyne incognita* in commercial-type tomatoes, *Proc. Am. Soc. Hort. Sci.* 63:437.

Hammond-Kosack, K.E., Atkinson, H.J., and Bowles, D.J., 1990, Changes in abundance of translatable mRNA species in potato roots and leaves following root invasion by cyst nematode *G. rhostochiensis* pathotypes, *Physiol. Molec. Plant Path.* 3:339.

Helentjaris, T., Slocum, M., Wright, S., Schaefer, A., and Nienhuis, J., 1986, Construction of genetic linkage maps in maize and tomato using restriction fragment length polymorphisms, *Theor. appl. Genet.* 72:761.

Ho, J.Y., Weide, R., Ma, H., van Wordragen, M.F., Lambert, K.N., Koornneef, M., Zabel, P., and Williamson, V.M., 1992, The root-knot nematode resistance gene (*Mi*) in tomato: construction of a molecular linkage map and identification of dominant cDNA markers in resistant genotypes, *The Plant Journal* 2:971.

Huang, C.S., 1985, Formation, anatomy and physiology of giant cells induced by root-knot nematodes, *in*: "An Advanced Treatise on *Meloidogyne*," Vol.1, J.N. Sasser, and Carter, C.C., eds., North Carolina State University Graphics, Raleigh, N. C.

Hussey, R.S., 1985, Host-parasite relationships and associated physiological changes, *in*: "An Advanced Treatise on *Meloidogyne*," Vol.1, J.N. Sasser, and Carter, C.C., eds., North Carolina State University Graphics, Raleigh, N. C.

Jarquin-Barberena, H., Dalmasso, A., deGuiran, G., and Cardin, M.-C., 1991, Acquired virulence in the plant parasitic nematode *Meloidogyne incognita*. 1. Biological analysis of the phenomenon, *Revue Nematol.* 14:261.

Lambert, K.N., Tedford, E.C., Caswell, E.P., and Williamson, V.M., 1992, A system for continuous production of root-knot nematode juveniles in hydroponic culture, *Phytopathology* 82:512.

Lambert, K.N., and Williamson, V.M., 1993, cDNA library construction from small amounts of RNA using paramagnetic beads and PCR, *Nucleic Acids Res.* 21:775.

Martin, G.B., Brommonschenkel, S.H., Chunwongse, J., Frary, A., Ganal, M.W., Spivey, R., Wu, T., Earle, E.D., and Tanksley, S.D., 1993, Map-based cloning of a protein kinase gene conferring disease resistance in tomato, *Science* 262:1432.

Martin, G.B., Ganal, S. D., and Tanksley, S. D., 1992, Construction of a yeast artificial chromosome library of tomato and identification of cloned segments linked to two disease resistance loci, *Mol. Gen. Genet.* 233:25.

Medina-Filho, H., and Tanksley, S.D., 1983, Breeding for nematode resistance, *in* "Handbook of Plant Cell Culture," Vol. 1, D.A. Evans, Sharp, W. R., Ammirato, P.V., and Yamada, Y., eds., Macmillan, New York.

Messeguer, R., Ganal, M., de Vicente, M.C., Young, N.D., Bolkan, H., and Tanksley, S.D., 1991, High-resolution RFLP map around the root knot nematode resistance gene (*Mi*) in tomato, *Theor. appl. Genet.* 82:529.

Miller, J.C., and Tanksley, S.D., 1990, RFLP analysis of phylogenetic relationships and genetic variation in the genus *Lycopersicon*, *Theor. appl. Genet.* 80:437.

Paulson, R.E., and Webster, J.M., 1972, Ultrastructure of the hypersensitive reaction in roots of tomato, *Lycopersicon esculentum* L., to infection by the root-knot nematode, *Meloidogyne incognita*, *Pysiol. Plant Path.* 2:227.

Putterill, J., Robson, F., Lee, K., and Coupland, G., 1993, Chromosome walking with YAC clones in *Arabidopsis*: isolation of 1700 kb of contiguous DNA on chromosome 5, including a 300 kb region containing the flowering-time gene *CO*, *Mol. Gen. Genet.* 239:145.

Rick, C.M., and Yoder, J.I., 1988, Classical and molecular genetics of tomato: highlights and perspectives, *Annu. Rev. Genet.* 22:281.
Riggs, R.D., and Winstead, N.N., 1959, Studies on resistance in tomato to root-knot nematodes and on the occurence of pathogenic biotypes, *Phytopathology* 49:716.
Roberts, P.A., and Thomason, I.J., 1986, Variability in reproduction of isolates of *Meloidogyne incognita* and *M. javanica* on resistant tomato genotypes, *Plant Disease* 70:547.
Ryan, C.A., 1990, Protease inhibitors in plants: genes for improving defenses against insects and pathogens, *Annu. Rev. Phytpathol.* 28:425.
Smith, P.G., 1944, Embryo culture of a tomato species hybrid, *Proc. Am. Soc. Hortic. Sci.* 44:413.
Stoker, A.W., 1990, Cloning of PCR products after defined cohesive termini are created with T4 DNA polymerase, *Nucleic Acids Res.* 18:4290.
Tanksley, S.D., Ganal, M.W., Prince, J.P., de Vicente, M.C., Bonierbale, M.W., Broun, P., Fulton, T.M., Giovannoni, J.J., Grandillo, S., Martin, G.B., Messeguer, R., Miller, J.C., Miller, L., Paterson, A.H., Pineda, O., Roder, M.S., Wing, R.A., Wu, R.A., Wu, W., and Young, N.D., 1992, High density molecular linkage maps of the tomato and potato genomes, *Genetics* 132:1141.
Theerasilp, S., Hitotsuya, H., Nakajo, S., Nakaya, K., Nakamura, Y., and Kurihara, Y., 1989, Complete amino acid sequence and structure characterization of the taste- modifying protein, miraculin, *J. Biol. Chem.* 264:6655.
Trudgill, D.L., 1991, Resistance to and tolerance of plant parasitic nematodes in plants, *Annu. Rev. Phytopathol.* 29:167.
van Daelen, R.A.J., Jonkers, J.J., and Zabel, P., 1989, Preparation of megabase-sized tomato DNA and separation of large restriction fragments by field inversion gel electrophoresis (FIGE), *Plant Molec. Biol.* 12:341.
Williams, G.G.K., Kubelik, A.R., Livak, K.J., Rafalski, J.A. and Tingy, S.V., 1990, DNA polymorphisms amplified by arbitrary primers are useful as genetic markers, *Nucleic Acids Res.* 18:6531.
Williamson, V.M., Ho, J.-Y., Wu, F.F., Miller, N., and Kaloshian, I., 1994, A PCR-based marker tightly linked to the nematode resistance gene, *Mi*, in tomato, *Theor. appl. Genet.*, in press.
Wing, R.A., Rastogi, V.K., Zhang, H.B., Paterson, A.H., and Tanksley, S.D., 1993, An improved method of plant megabase DNA isolation in agarose microbeads suitable for physical mapping and YAC cloning, *Plant Journal* 4:893.
Ye, Z.-H., andVarner, J.E., 1991, Tissue-specific expression of cell wall proteins in developing soybean tissues, *The Plant Cell* 3:23.
Zabel, P., van Wordragen, M., Weide, R., Liharska, T., Stam, P., and Koornneef, M., 1993, Integration of the classical and molecular linkage maps of tomato chromosome 6, *Genetics* 135:1175.

MOLECULAR BIOLOGY OF NEMATODE RESISTANCE IN TOMATO, DISCUSSION

GEORGI asked whether WILLIAMSON had looked to see what happens to the expression of the cDNA clones in *Mi* - bearing plants infected with resistance - breaking nematode isolates. WILLIAMSON replied that they had not, but that it is a very good idea, and they plan to do the experiment in the future. WILLIAMSON noted that it will be interesting to determine whether the genes induced by these nematodes are the same as those induced by avirulent nematodes on susceptible plants.

BIOENGINEERING RESISTANCE TO SEDENTARY ENDOPARASITIC NEMATODES

Charles H. Opperman[1,2], Gregoria N. Acedo[2], David M. Saravitz[2], Andrea M. Skantar[1], Wen Song[2], Christopher G. Taylor[2], and Mark A. Conkling[2,1]

[1]Department of Plant Pathology
[2]Department of Genetics
North Carolina State University
Raleigh, NC 27695

INTRODUCTION

Bioengineering host resistance to plant parasitic nematodes is still a mostly speculative topic. There have been very few examples of plants engineered to have reduced susceptibility to any species of nematode. This, in part, is due to the lack of key information regarding host-parasite relationships. It is difficult to design strategies to engineer resistance when there is not a readily identifiable molecular mechanism to target. There are now many labs actively pursuing molecular aspects of nematode-host interactions, but the practical applications of this work may still be several years off. Other chapters in this book describe in detail studies on nematode feeding site formation, resistance mechanisms, host responses, and nematode stylet secretions. We will use some of this information to set the stage for potential approaches to bioengineered resistance to plant parasitic nematodes.

Traditional management of plant-parasitic nematodes has relied upon three basic tools, namely, crop rotation, chemical nematicides, and host resistance. Of these, host resistance is the most environmentally and economically sound method (Cook and Evans, 1987). Unfortunately, genetic resistance to plant-parasitic nematodes does not occur in all cultivated host species. In those hosts where resistance is available, it has proved to be an extremely valuable commodity. For example, the introduction of the *Heterodera glycines*-resistant cultivar 'Forrest' has reportedly saved soybean growers in the southern United States over $400 million during a 5 year period (Bradley and Duffy, 1982). The increasing reliance upon non-chemical management of plant-parasitic nematodes has resulted, in some cases, in monoculture of resistant varieties. This has lead to the occurrence of so-called "resistance-breaking" populations of nematodes that are, in reality, simply genetic variants selected by the resistant variety. One of the most limiting aspects of the currently available host resistance is the highly specific nature of the interaction. For example, resistance in tobacco is conferred by a single gene (*rk*), resulting in a hypersensitive response when infected by races 1 or 3 of *Meloidogyne incognita* (Sasser, 1980; Slana and Stavely, 1981). The gene is not effective against the other races of *M. incognita*, nor against any other *Meloidogyne* species. Tomato plants carrying the *Mi* gene may be resistant to *M. arenaria, M. incognita*, and *M. javanica*, but not *M. hapla* (Gilbert and McGuire, 1956). The *Mi* gene appears to be quite stable although resistance may breakdown under temperature stress. In both of these cases, resistance is manifested as a hypersensitive response (Kiraly, 1980). In other crop species, resistance to root-knot nematode is generally limited to a single species. Similarly, resistance to cyst nematodes is generally specific for certain pathotypes (Triantaphyllou, 1987).

Breeding for nematode resistance is complicated by several factors. First, few sources of natural resistance have been identified for most crops. The *Mi* gene in tomato is the result of an interspecific hybridization between *Lycopersicon esculentum* and *L. peruvianum* (Watts, 1947). In some crops resistance is oligogenic and inheritance patterns are complex. This is the case with soybean resistance to *H. glycines* (Triantaphyllou, 1987). Finally, the durability of resistance based on such a narrow germplasm declines with repeated cropping. The lack of broad spectrum resistance to the major plant parasitic nematode species has complicated the development of integrated management strategies that do not rely upon chemicals.

A BRIEF HISTORY OF BIOENGINEERING HOST DEFENSE

Rapid advances during the past ten years in our understanding of both plant molecular biology and host-pathogen interactions have made it possible to design genetically engineered crop plants carrying disease resistance traits. This strategy will prove to be particularly useful for crop-pest interactions in which no native resistance genes are known to occur. In addition, it is sometimes difficult or impossible to move resistance genes from a wild species into a cultivated one. In these cases, the best chance for developing resistant cultivars lies in genetic engineering.

The earliest and probably most intensively pursued approach has been genetically engineering plants to resist infection by viruses. Many of these strategies are based upon the usage of pathogen-derived genes as the foreign DNA sequence to be inserted in the plant. The first report of coat protein-mediated cross protection was from tobacco plants transformed with the coat protein gene of Tobacco Mosaic Virus (TMV) (Abel et al., 1986). These experiments arose from observations that infection with a mild strain of virus could prevent or ameliorate later infection from a more pathogenic strain, the so-called cross protection phenomenon (Fulton, 1986). The transgenic plants expressing the TMV coat protein gene were resistant to systemic infection by TMV. Many other plant-virus combinations have been studied since that time. A number of different strategies have been attempted with many different viruses on numerous host species, including non-structural viral proteins, defective interfering viral sequences, and satellite RNAs, all with varying results (see Wilson, 1993, for a complete review of the topic). In no case is the mechanism of protection well understood at the molecular level.

A different approach to engineering virus resistance has been to construct a system of induced hypersensitivity. In this strategy, a highly active cellular toxin (such as diphtheria toxin A fragment, or ricin) is transformed into plants as an antisense construct with a minus sense viral subgenomic RNA promoter at its 3' end (Wilson, 1993). Upon infection by the intact plant virus, the antisense toxin construct is transcribed into mRNA during replication, resulting in production of the toxin and death of the infected cell. Resistance to potato virus X has been achieved, but the utility of this approach for field applications has been questioned (Wilson, 1993).

A number of genes involved in natural host defense systems have been characterized and cloned. These include genes encoding chitinases, β-glucanases, proteinase inhibitors, and pathogenesis related (PR) proteins (Graham et al., 1985; Metraux et al., 1989; Dixon and Lamb, 1990; Keen, 1990; Cramer et al., 1993). Several of these genes have expressed in transgenic plants to provide either disease or insect resistance (see Lamb et al., 1992, for a review). For example, chitinase genes expressed in transgenic tobacco have been shown to confer an increased level of protection from the fungal pathogen, *Rhizoctonia solani* (Broglie et al., 1991). This transgenic resistance was not effective on another fungal pathogen, *Cercospora nicotianae* (Neuhaus et al., 1991). Insect resistance may be produced by expression of proteinase inhibitor genes. The cowpea trypsin inhibitor gene expressed in transgenic tobacco provided resistance to the tobacco budworm, *Heliothis virescens* (Hilder et al., 1987). The gene encoding the tomato proteinase inhibitor II has been engineered into tobacco and provides protection from the tobacco hornworm, *Manduca sexta* (Johnson et al., 1989). The proteinase inhibitors are wound inducible and are thought to activated by the action of systemin, a systemic polypeptide signal molecule (Graham et al., 1986; Pearce et al., 1991). Both of these proteinase inhibitors act with high specificity on trypsin, an enzyme that may be important in insect digestive processes (Plunckett et al., 1982).

Genetic engineering approaches have also been developed that confer resistance to insect pests with transgenic plants that carry and express the Δ-endotoxin (BT) gene from the bacterium, *Bacillus thuringiensis* (Barton et al., 1987). When susceptible insects feed upon leaves of plants expressing the protein, they suffer gut paralysis, leading to the demise of the insect. Proper expression and function of the gene in plants has required significant remodeling of the DNA sequence to compensate for plant gene expression strategies (Perlak et al., 1991). Although insect-resistant crop cultivars have been developed using this approach, there are potential problems associated with BT-expressing plants. The constitutive expression of the BT gene in a host species places strong selective pressures on the pest population to develop resistance to the toxin. Another potential drawback is that non-target pests or predators of target pests may be exposed to the protein toxin. These factors require that proper management practices be observed in order to maintain the usefulness of transgenic BT plants for insect management.

The previously mentioned studies point out how difficult and time consuming it is to design a strategy to engineer host resistance. Two of the major factors to be decided are what gene to insert (toxins, host defense genes, induced HR, etc.) and how to control expression of the foreign gene (i.e., what promoter to choose). There are few plant gene promoters thus far characterized that respond to nematode infection, but there are a number of strong constitutive promoters that are available. Some of these promoters have been used in the systems described above, but recent evidence (discussed below) suggests that this approach may not be successful for nematodes. The relatively scant amount of information regarding plant gene expression during nematode infection points out the difficulties in designing genetically engineered nematode resistance. Our bias, as will become clear, is towards the use of promoters that are integral to the compatible interaction.

TARGET NEMATODES

Although plant parasitic nematodes collectively cause over $77 billion in crop losses on a worldwide scale, the vast majority of the damage is caused by sedentary endoparasitic forms, including *Meloidogyne* spp. and *Globodera* and *Heterodera* spp (Sasser and Freckman, 1987). Sedentary endoparasitic nematodes establish elaborate feeding sites within the host root. Although feeding sites may range from slightly altered cortical cells to vascular cells that are developmentally altered in their fate and function (Jones, 1981), those induced by the root-knot (*Meloidogyne* spp.) and cyst (*Globodera* and *Heterodera* spp.) nematodes are the most elaborate. There are also many species of plant parasitic nematodes, both ectoparasitic and migratory endoparasitic, that do not form specialized feeding sites. These nematodes will require a somewhat different approach for management using genetically engineered crop plants. This review will focus on production of transgenic plants resistant to sedentary endoparasitic nematodes.

Feeding Site Formation

The specialized nature of the sedentary endoparasitic nematode-host relationship makes it vulnerable to identification and exploitation of weak points in the life cycle. In order to identify potential cellular targets for attack, an understanding of the biology of nematode feeding site formation is necessary.

Root-knot nematodes have an very broad host range, encompassing over 2,000 plant species (Sasser, 1980), and most cultivated crops are attacked by at least one species of *Meloidogyne* (Sasser 1980). A detailed review of *Meloidogyne* feeding site formation appears elsewhere in this volume. A brief review is necessary here, however. The feeding site, or giant cells, of the root-knot nematode is initiated after the infective second-stage juvenile has penetrated the host root, generally near the root tip, and migrates to the developing vascular cylinder. It is believed that the induction of giant cell formation is due to glandular secretions injected via the nematode stylet into several root cells surrounding the head (Hussey, 1989). The 5-7 giant cells formed within or near the developing vascular cylinder become the permanent feeding site for the root-knot nematode. The giant cells undergo repeated nuclear divisions without cytokinesis, nuclei are enlarged and lobate, and may contain 14-16 times more DNA than do normal root tip nuclei (Jones, 1981; Wiggers et al., 1990). The giant cells become greatly enlarged, increased numbers of cellular organelles accumulate, and the cytoplasm becomes very dense and granular. The metabolically-active giant cells act as nutrient sinks to provide food to the developing nematode (Huang, 1985). One characteristic feature of root-knot nematode giant cells is the highly invaginated and thickened cell wall, similar to cell walls observed in transfer cells (Jones, 1981). The nematode is dependent upon the giant cells for its survival and reproduction since it becomes immobile soon after giant cell induction.

Cyst nematode species tend to have a narrower host range than do root-knot nematodes. The feeding site of cyst nematodes is referred to as a syncytium. Syncytia are thought to be induced and formed by very different mechanisms than giant cells of root-knot nematode (Jones, 1981). Although syncytia are multinucleate, this does not occur by repeated nuclear divisions without cytokinesis. Instead, the multinucleate state of syncytia results from cell wall dissolution and coalescence of adjacent cells (Jones and Northcote, 1972), and the nuclei are enlarged and lobate (Endo, 1992). As observed in giant cells, syncytia have thickened and invaginated cell walls, but these tend to be near the periphery of the feeding site. Increased numbers of organelles, and dense cytoplasm also occur. In both giant cells and syncytia, the number of plasmadesmata between the feeding site and the surrounding root cells are greatly reduced (Jones, 1981). The enlarged feeding sites are dependent upon constant stimulation from the nematode or they will begin to degrade.

PLANT GENE EXPRESSION PATTERNS

Plant gene expression patterns are altered during nematode feeding site initiation (Bowles et al., 1991; Sijmons, 1993). It is not yet understood how nematodes such as *Meloidogyne* spp. cause these alterations, but it is suspected that glandular secretions injected into plant cells interact directly or indirectly with the plant nuclear genome (Hussey, 1989). Although there are a number of potential alterations in gene activity during nematode infection, it seems likely that the genes encoding certain constitutive enzymes and structural proteins may be up-regulated in order to support the increased cellular metabolic activity related to nematode feeding. This class of genes might be the most straightforward to identify, but few transcripts up-regulated during nematode infection have been reported. There are a number of types of genes that may be altered in their expression patterns in a different manner from up-regulation. The most obvious is the example of genes that may be down-regulated or even turned off entirely. There are no characterized examples of such genes, but it may be speculated that defense genes may be actively suppressed by the nematode during a compatible interaction. It is also possible that genes involved in maturation and senescence of root tissue may be down-regulated during nematode parasitism. Not all genes that are affected during nematode infection and feeding site establishment are necessarily up- or down-regulated. Certain regulatory genes involved in the interaction may have their expression altered in a temporal sense. These genes may be expressed for a longer duration of time in order to accommodate the increased metabolic needs of the developing feeding site. Alternatively, some genes may be expressed at inappropriate times compared to normal root cell differentiation and development. It is also likely that genes not normally expressed in root cells at all may be utilized during giant cell or syncytia formation The substantial alteration of root vascular system cells towards the nematode's needs guarantees that many plant genes have their normal expression patterns either quantitatively or qualitatively altered to meet the demands of the nematode.

There are many different strategies to isolate genes involved in nematode feeding site formation. Approaches based on differential gene expression between healthy and infected roots, such as subtractive hybridization or differential screening of cDNA libraries, could enable one to isolate up-regulated genes (Conkling et al., 1990; Gurr et al., 1991; Niebel et al., 1992; van der Eycken et al., 1992). Other approaches, such as biochemical analysis of giant cell proteins or development of feeding site specific antibodies, may also result in identification of up-regulated or altered expression patterns (Hussey, 1989; Burrows, 1992). There have been several reports of altered gene expression during potato cyst nematode infection of potato. For example, it has been reported that catalase gene expression is increased in potato during infection by *G. rostochiensis* (Niebel et al., 1992). One gene (pMR1) with unknown function has been isolated from potato roots undergoing infection by *G. rostochiensis* (Gurr et al., 1991). There have been a few preliminary reports of structural or enzyme genes that are up-regulated during either root-knot or cyst nematode feeding site establishment. Structural protein genes related to the extensin family have been isolated from root-knot nematode infected tomato roots in a screen designed to identify up-regulated genes (van der Eycken et al., 1992; Niebel et al., 1993). Recently, a promoter-trapping strategy has been described that enables identification of both up-regulated and down-regulated transcripts during nematode feeding site formation (Goddjin et al., 1993). In this method, a promoterless GUS gene is transformed into plants and integrates randomly into the genome. Transcription will only occur when tagged sequences are impacted by the developing feeding site. Using this technique, a number of down regulated genes were tagged, as well as several genes that appeared to be up-regulated (Goddjin et al., 1993). These genes have not yet been characterized, however.

Control of gene expression in the nematode feeding site is of great significance to design genetically engineered resistance. There have been few studies on promoters involved in feeding site gene expression. One comprehensive study has revealed that many genes are down-regulated during feeding site formation (Goddjin et al., 1993). In both transgenic *Arabidopsis* and tobacco plants, down regulation of many constitutive promoters was observed during both cyst and root-knot nematode feeding site formation, including the Cauliflower Mosaic Virus 35S, *rol* (root loci) A-D, nopaline synthase, γ-TIP, and T-*cyt* (Goddjin et al., 1993). These promoters represent an array of different regulatory sequences and all confer strong expression of reporter gene constructs in uninfected root tissue. There have been several reports of promoters that are strongly up-regulated during feeding site formation by root-knot nematode. In one study, the promoter of the *hmg*2 gene (hydroxymethylglutaryl CoA reductase) fused to GUS was observed to be strongly expressed in the developing giant cells and gall tissue shortly after infection by *M. incognita* or *M. hapla* (Cramer et al., 1993). The tomato *hmg*2 gene is a defense related gene that is induced by both fungal and bacterial pathogens, and is part of the sesquiterpenoid biosynthesis pathway (Chappell et al., 1991; Yang et al., 1991).

The promoters of numerous plant organ-specific and pathogen-related genes have been characterized and models emerging from such studies are that *cis*-acting sequences regulating gene expression in higher plants are often composites of several regions acting as transcriptional enhancers or silencers. Regulated transcription is a result of interaction between these sequences with various nuclear DNA-binding proteins and perhaps one another (Benfey et al., 1989). Computer comparisons of available regulatory region nucleotide sequences involved in nematode-induction or suppression do not reveal any obvious relationships (Goddjin et al., 1993).

The *TobRB7*-Root-Knot Nematode Interaction

We have been studying the interaction of root-knot nematodes with the tobacco root-specific gene, *TobRB7*. Root growth and development are critically important to the success of a plant species, but relatively little characterization of root-specific gene expression has been reported. This is due in part to a lack of easily characterized functions that are specific to root tissues and not other plant parts. *TobRB7* may be the most extensively characterized root-specific gene that has been isolated (Conkling et al., 1990; Yamamoto et al., 1990, 1991). The *TobRB7* gene encodes a protein of 250 amino acids and has conserved structural domains with several membrane spanning proteins (Yamamoto et al., 1990). Studies on the function of *TobRB7* suggest that it may be a water channel (Li, Lucas, and Conkling, unpublished). Expression of *TobRB7* is limited to the root meristematic and immature vascular cylinder regions as revealed by *in situ* hybridization studies (Yamamoto et al., 1991). The high level of expression in these tissues coupled with the lack of expression in mature tissue suggests a developmental role for the TobRB7 protein (pRB7). Fusions were made between a deletion series of the 5'-flanking region and the bacterial reporter gene, β-glucuronidase (GUS) in order to identify sequences controlling the root-specific expression patterns (Jefferson, 1987). *In situ* hybridization patterns and the pattern of the GUS gene expression driven by the full-length flanking sequence were indistinguishable, demonstrating that the GUS reporter constructs were appropriate indicators of native gene expression. The deletion series experiments revealed that *cis*-acting elements necessary for root-specific expression of *TobRB7* are located between 636 and 299 nucleotides 5' of the site of transcription initiation (Yamamoto et al., 1991). *TobRB7* expression was lost when the promoter was deleted to the -299 position (Δ0.3).

GUS reporter gene expression patterns in root-knot nematode-infected transgenic tobacco plants carrying the promoter deletion-reporter constructs demonstrated that *TobRB7* is one of the plant genes that is affected by nematode parasitism. Within 4 days after infection, significant levels of GUS activity in and around the developing feeding site were detected (Opperman et al., 1994). The tissue around the root-knot nematode infection site matures as the root continues to grow. Although *TobRB7* would not normally be expressed in this region, GUS is observed to accumulate in the developing feeding site throughout the nematode life cycle. This spatial and temporal shift in gene expression indicates that root-knot nematode infection has resulted in significant alterations in the control of *TobRB7* expression.

This finding is further confirmed by results obtained with the deletion series of *cis*-acting sequences. Although no gene expression is observed with deletions containing only the Δ0.3 region, infection with *Meloidogyne* spp. significantly alters this pattern. In those plants carrying the Δ0.3 construction, GUS accumulation was limited to the developing giant cells and appeared to be regulated by the nematode infection (Opperman et al., 1994). Because the TobRB7 protein is predicted to be structural, it is likely that it is at or near the end of the gene expression hierarchy during giant cell formation. Therefore, it seems likely that expression of the truncated Δ0.3 construct during nematode infection is not a direct interaction, but is downstream from the regulatory events involved in giant cell initiation. All races and species of *Meloidogyne* thus far examined induce expression of the Δ0.3 reporter, but the tobacco cyst nematode, *G. tabaccum* does not (Opperman et al., 1994). The specificity for induction by root-knot nematode is indicative of a somewhat different mechanism for giant cell versus syncytia formation. The most significant aspect of these findings is that the nematode-responsive element (NRE) of the *TobRB7* promoter is not the same as the root-specific element, and can be uncoupled (Opperman et al., 1994).

APPROACHES TO ENGINEERING NEMATODE RESISTANCE

There are several different approaches to design nematode-resistant transgenic crop cultivars. The simplest to implement is the expression of a nematode-specific toxin gene under the control of a constitutive promoter. As mentioned, this type of strategy has been used to design plants carrying the BT gene that are resistant to insect pests (Barton et al., 1987). The down regulation of many currently available constitutive promoters in nematode feeding sites casts doubt on the applicability of this approach at the present time (Goddjin et al., 1993). Additionally, very few protein toxins that act specifically upon plant parasitic nematodes have been identified. Even so, nematicidal plants could be produced if a suitable toxin gene and a constitutive promoter could be identified. Nematode natural enemies may be a source of toxin molecules. The fungal endoparasite *Nematoctonus* spp. secretes a nematoxic substance during spore germination that causes rapid paralysis and death of the nematode (Stirling, 1992). The predatory nematode *Seinura* injects a toxin into its nematode prey, causing immobilization within seconds (Stirling, 1992). Other fungal and bacterial antagonists are known to produce toxic substances to nematodes, but very little information is available on the composition of these compounds. Although the nature of these molecules is unknown, they may be potentially useful in the design of nematicidal transgenic plants if any of the toxins are peptides.

A different approach to designing nematicidal plants is to use genes such as collagenase that may disrupt nematode development if expressed in the feeding cells. Collagen is the major structural polymer of the nematode cuticle, which serves as the structural exoskeleton of the organism, and also forms the stylet (feeding apparatus) and the lining of the esophagus and intestine (Bird and Bird, 1991). Nematode-trapping fungi presumably utilize collagenases to help digest the trapped prey. The large number of collagen genes in nematodes and the probable specificity of collagenase for particular isoforms would necessitate isolation of nematode-specific collagenase genes, either from fungi or nematodes. The expression of collagenase should have no effects on plants, which do not contain collagen.

Interference with either feeding or digestion may also be possible. For example, antibodies to nematode stylet secretions may be utilized to disrupt either feeding site formation or feeding behavior (Hussey, 1989). Expression of antibody genes in plants is feasible (Hiatt et al., 1989), and candidate transgenic plants may be available for nematode resistance trials in the near future. Proteinase inhibitors may also be used in an analogous way to the previously described insect resistance (see review in this volume by Atkinson et al.).

The advantage of these strategies is that they are a preformed defense and do not require nematode induction to be activated. Theoretically, any susceptible nematode which feeds for a suitable period upon cells containing the toxin moieties will not survive. There are several disadvantages to this approach, however. Peptide toxins useful in this approach typically are narrow in their toxic spectra, as is the case with the BT toxins. The use of transgenic plants expressing toxins with activity on only certain nematode species may end up selecting for non-sensitive species, resulting in narrow resistance. The constitutive expression of any "toxin" gene may place upon the sensitive nematode population very strong selective pressure for resistance, placing the durability of this type of defense in doubt. In addition, the global, constitutive expression of toxin genes guarantees that non-target species, including humans, will be exposed to the protein products.

The combinatorial nature of plant gene promoters so far characterized suggests that induction during nematode infection could be uncoupled from normal gene expression control. This is a key factor in designing any pathogen-specific, transgenic, resistant plant. Studies on induction of defense related genes from other plant species may also provide insights into pathogen induction and specificity. As previously suggested in this review, constitutive expression of a protein toxic to nematodes is the simplest approach to implement but is the least desirable. Using sequences conferring tissue-specific, or even nematode-inducible expression of a protein toxic to nematodes could improve this approach, but the selective pressures for toxin-resistant nematodes remain. The most durable method is based upon nematode-induced expression of a protein toxic to the giant cells, i.e., nematode-induced hypersensitivity. This strategy is the most desirable of the three, in that selective pressure for toxin resistance is alleviated. For induced hypersensitivity to be viable, however, the promoter must be very specific and the action of the peptide toxin must remain confined to the host cells expressing the gene. In addition, the cell-specific toxin should not represent a health hazard for human consumption. Molecules such as DNases, RNases, or proteases could be utilized to perturb cellular function necessary for root-knot nematode feeding site initiation. A different approach might be to interfere with the plant gene expression patterns in the developing giant cells, disrupting the proper formation and function of the feeding site. Ribozymes or antisense constructs to nematode-responsive plant genes could be used in this approach. A nematode attempting to feed on cells carrying a nematode-inducible promoter fused to a gene encoding one of these molecules would initiate gene expression resulting in the degradation of the feeding site. This type of engineered resistance is not dissimilar to the natural hypersensitive response to nematode parasites in resistant plants. In cases where a highly specific promoter is not available, it may be possible to make use of a two-component system (Sijmons, 1993). In this approach, the plant is transformed with the induced hypersensitivity cartridge and a neutralizing agent under control of one of the down-regulated constitutive promoters. The expression of the neutralizing agent in non-infected cells acts to counteract any leakiness or nonspecific expression from the inducible construct. Upon nematode infection, the inducible promoter drives expression of the cellular toxin and the constitutive neutralizing factor expression is down-regulated, resulting in death of the developing feeding site cells (Sijmons, 1993). Although this type of approach may counteract leakiness from a feeding site-specific promoter, a potential drawback may be that it requires two separate regulatory events to function properly.

Application

We have taken two different approaches to designing transgenic nematode resistant plants using the *TobRB7* NRE. An induced hypersensitivity approach previously has been utilized to produce transgenic tobacco plants exhibiting male sterility (Mariani et al. 1990). In these experiments, expression of a gene encoding an RNase (*Barnase*) was directed by a tapetal cell-specific promoter. Expression of *Barnase* in the tapetal cells resulted in cell death, abortion of pollen development and consequently, male sterility. Histochemical analysis of these transgenic plants showed that the surrounding cells were unaffected. The apparent specificity of the *TobRB7* NRE suggested that we could use the same approach to produce root-knot nematode induced hypersensitivity. Constructs of *Barnase* driven by either the Δ0.6 or the Δ0.3 deletions were transformed into tobacco plants. As expected, the Δ0.6

plants did not root with high frequency, because this construct was probably expressed during vascular tissue development. In contrast, viable plants were obtained with the Δ0.3 NRE driving *Barnase*. These plants have no observable phenotype when uninfected. A screen of the transformants obtained revealed several lines with altered responses to root-knot nematode infection (Taylor, Opperman, and Conkling, unpublished). In one particular line, no galls and 2 small egg masses were detected, in contrast to the heavy galling and large numbers (>300) of egg masses recovered on the control plants. Histochemical examination of the roots revealed that many second-stage juveniles had penetrated the Δ0.3 NRE-*Barnase* roots, but had arrested in development after 4-6 days. Control plants contained mostly adult female nematodes that had already deposited large numbers of eggs on the root surface. We have now repeated these transformations on two more occasions, and have recovered resistant plants from each independent experiment. These results indicate that the Δ0.3 NRE promoter is not leaky under conditions tested so far and can be used in an induced hypersensitivity approach with a highly active cell toxin.

The second approach has been to inhibit proper feeding site development or function using an antisense *TobRB7* construct. As previously mentioned, the *TobRB7* protein appears to function as a water channel, and we believe that its role in giant cells is to maintain osmotic balance. Full length cDNA antisense constructs of *TobRB7* driven by either the Δ0.3 NRE or the 35S CaMV promoter were transformed into tobacco (Opperman, Acedo, and Conkling, submitted; Acedo and Conkling, unpublished). The plants containing constructs driven by the 35S promoter exhibited typical stress-like phenotypes, including pointed and narrow leaves, long internodal distances, and early flowering. When these plants were infected with root-knot nematode, however, a substantial reduction in galling and egg production was observed. Histochemical examination of roots revealed that numerous male nematodes had formed, and the feeding sites were small compared to the control plants. Some variation in response was observed, possibly due to the previously described down-regulation phenomenon observed with the 35S promoter (Goddjin et al., 1993). Transgenic plants containing the antisense construct driven by the Δ0.3 NRE showed no observable phenotype compared to the wild type. Greenhouse and field trials with the Δ0.3 NRE-*TobRB7* antisense plants provided strong evidence that nematode infection was substantially reduced. In both trials, root galling was reduced by approximately 70% compared to control treatments. The field trial indicated that protection lasted for the entire growing season (approximately four months). Galls that were observed on the roots of antisense plants were small and tended to appear as discrete entities compared to the large and clustered galls on the controls. Although these results provide only circumstantial evidence that *TobRB7* is essential to giant cell formation, they are one of the first indications that interference with feeding site establishment may be a viable approach to engineering transgenic nematode resistance.

The identification of the Nematode-Responsive-Element of *TobRB7* is represents a significant step in the move towards production of transgenic nematode resistant crops. As more genes are isolated that are involved in nematode-feeding-site formation, their control elements will be available for manipulation. The limitation of foreign gene expression to the developing nematode feeding site will allow the greatest flexibility and power in designing control strategies, and will overcome many of the problems associated with less specific constructs. As more information is obtained regarding other genera of plant parasitic nematodes, new methods of multiple nematode resistance will be possible. The identification of genes necessary in a normal susceptible interaction is particularly significant. The control regions of these sequences coupled to proteins detrimental to the nematode feeding site should make it very difficult for the nematode to overcome the engineered resistance. Ultimately, the durability of transgenic nematode resistance may depend upon the specificity of the regions controlling foreign gene expression.

REFERENCES

Able, P.P., Nelson, R.S., De, B., Hoffman, N., Rogers, S.G., Fraley, R.T., and Beachey, R.N., 1986, Delay of disease development in transgenic plants expressing the tobacco mosaic virus coat protein gene, *Science* 232:738.

Barton, K.A., Whitely, H.R., and Yang, N.-S., 1987, *Bacillus thuringiensis* Δ-endotoxin expressed in transgenic *Nicotiana tabacum* provides resistance to lepidopterous insects, *Plant Physiol.* 85:1103.

Benfey, P.N., Ren., L., and Chua, N.-H., 1989, The CaMV 35S enhancer contains at least two domains which can confer different developmental and tissue-specific expression patterns, *EMBO J.* 8:2195.
Bird, A.F., and Bird, J., 1991, "The Structure of Nematodes," 2nd edition, Academic Press, San Diego.
Bowles, D.J., 1990, Defense related proteins in higher plants, *Ann. Rev. Biochem.* 59:873-907.
Bowles, D.J., Gurr, S.J., Scollan, C., and Atkinson, H.J., 1991, Local and systemic changes in plant gene expression following root infection by cyst nematodes, *in*: "Biochemistry and Molecular Biology of Plant-Pathogen Interactions," C.J. Smith, ed., Oxford Science Publications, Oxford.
Burrows, P.R., 1992, Molecular analysis of the interactions between cyst nematodes and their hosts, *J. Nematol.* 24:338.
Bradley, E. B., and Duffy, M., 1982, The value of plant resistance to the soybean cyst nematode: A case study of Forrest soybean, USDA-ARS Staff Report AGES820929, Washington, D. C.
Broglie, K, Chet, I., Holliday, M., Cressman, R., Biddle, P., Knowlton, S., Mauvias, C.J., and Broglie, R., 1991, Transgenic plants with enhanced resistance to the fungal pathogen Rhizoctonia solani, *Science* 254:1194.
Chappell, J., Von Lanken, C., and Vogeli, U., 1991, Elicitor-inducible 3-hydroxy-3-methylglutaryl coenzyme A reductase activity is required for sesquiterpene accumulation in tobacco cell suspension cultures, *Plant Physiol.* 97:693.
Conkling, M. A., Cheng, C.-L., Yamamoto, Y.T., and Goodman H.M., 1990, Isolation of transcriptionally regulated root-specific genes from tobacco, *Plant Physiol.* 93:1203.
Cook, R., and Evans, K., 1987, Resistance and tolerance. *in:* "Principles and Practice of Nematode Control in Crops," R.H. Brown and Kerry, B.R., eds., Academic Press, Sydney.
Cramer, C.L., Weissenborn, D.L., Cottinghan, C.K., Denbow, C.J., Eisenback, J.D., Radin, D.N., and Yu, X., 1993, Regulation of defense-related gene expression during plant-pathogen interactions, *J. Nematol.* 25:507.
Dixon, R.A., and Lamb, C.J., 1990, Molecular communication in interactions between plants and microbial pathogens, *Ann. Rev. Plant Physiol. Plant Mol. Biol.* 11:339.
Endo, B. Y., 1992, Cellular responses to infection, *in:* "Biology and management of the soybean cyst nematode," R. D. Riggs and Wrather, J. A., eds., APS Press, St. Paul.
Fulton, R.W., 1986, Practice and precautions in the use of cross protection for plant virus disease control, *Ann. Rev. Phytopath.* 24:67.
Gilbert, J.C., and McGuire, D.C., 1956, Inheritance of resistance to severe root-knot from *Meloidogyne incognita* in commercial type tomatoes, *Proc. Am. Hort. Soc.* 68:437.
Goddjin, O.J.M., Lindsey, K., van der Lee, F.M., Klap, J.C., and Sijmons, P.C., 1993, Differential gene expression in nematode-induced feeding structures of transgenic plants harbouring promoter-*gus*A fusion constructs, *Plant J.* 4:863.
Graham, J.S., Pearce, G., Merryweathers, J., Titani, K., Ericsson, L.H., and Ryan, C.A., 1985, Wound-induced proteinase inhibitors from tomato leaves I. The cDNA-deduced primary structure of pre-inhibitor I and its post-translational processing, *J. Biol. Chem.* 260:6555.
Graham, J.S., Hall, G., Pearce, G., and Ryan, C.A., 1986, Regulation of synthesis of proteinase inhibitors I and II mRNAs in leaves of wounded tomato plants, *Planta* 169:399.
Gurr, S.J., McPherson, M.J., Scollan, C., Atkinson, H.J., and Bowles, D.J., 1991, Gene expression in nematode-infected plant roots, *Mol. Gen. Genet.* 226:361.
Hiatt, A., Cafferkey, R., and Bowdish, K., 1989, Production of antibodies in transgenic plants, *Nature* 342:76.
Hilder, V.A., Gatehouse, A.M.R., Sheerman, S.E., Barker, R.F., and Boulter, D., 1987, A novel mechanism of insect resistance engineered into tobacco, *Nature* 330:160.
Huang, C.S., 1985, Formation, anatomy, and physiology of giant cells induced by root-knot nematodes. *in:* "An Advanced Treatise on Meloidogyne. Vol. 1. Biology and Control," J.N. Sasser and Carter, C.C., eds., North Carolina State University Graphics, Raleigh.
Hussey, R.S., 1989, Disease-inducing secretions of plant-parasitic nematodes, *Ann. Rev. Phyopath.* 27:123.
Jefferson, R. A., 1987, Assaying chimeric genes in plants: the GUS gene fusion system. *Plant Mol. Biol. Rep.* 5:387.
Johnson, R., Narvaez, J., An, G., and Ryan, C., 1989, Expression of proteinase inhibitors I and II in transgenic tobacco plants: effects on natural defense against *Manduca sexta* larvae, *Proc. Nat. Acad. Sci. USA* 86:9871.
Jones, M.G.K., 1981, Host cell responses to endoparasitic nematode attack: Structure and function of giant cells and syncytia, *Annals Appl. Biol.* 97:353.
Jones, M.G.K., and Northcote, D. H., 1972, Nematode-induced syncytium-a multinucleate transfer cell, *J. Cell Sci.* 10:789.
Keen, N.T., 1992, The molecular biology of disease resistance, *Plant Mol. Biol.* 19:109.
Kiraly, Z., 1980, Defenses triggered by the invader: hypersensitivity. *in:* "Plant Disease-An Advanced Treatise," J.G. Horsfall and Cowling, E.B., eds., Academic Press, New York.
Lamb, C.J., Ryals, J.A., Ward, E.R., and Dixon, R.A., 1992, Emerging strategies for enhancing crop resistance to microbial pathogens, *Bio/Technology* 10:1436.
Mariani, C., De Beuckeleer, M., Truettner, J., Leemans, J., and Goldberg, R. B., 1990, Induction of male sterility in plants by a chimeric ribonuclease gene, *Nature* 347:737-741.

Metraux, J. P., Burkhart, W., Moyer, M., Dincher, S., Middlesteadt, W., Williams, S., Payne, G., Carnes, M., and Ryals, J., 1989, Isolation of a complementary DNA encoding a chitinase with structural homology to a bifunctional lysozyme/chitinase, *Proc. Nat. Acad. Sci. USA* 86:896.
Neuhaus, J.-M., Aho-Goy, P., Hinz, U., Flores, S., and Meins, F., Jr., 1991, High-level expression of a tobacco chitinase gene in *Nicotiana sylvestris*. Susceptibility of transgenic plants to *Cercospera nicotianae* infection, *Plant Mol. Biol.* 16:141.
Niebel, A., van der Eycken, W., Inze, D., van Montagu, M., and Gheysen, G., 1992, Potato genes induced during *Globodera pallida* infection, *Nematologica* 38:426 [Abstr.].
Niebel, A., de Almeida Engler, J., Tiré, C., Engler, G., Van Montagu, M., and Gheysen, G., 1993, Induction patterns of an extensin gene in tobacco upon nematode infection, *Plant Cell* 5:1697.
Opperman, C.H., Taylor, C.G., and Conkling, M.A., 1994, Root-knot nematode directed expression of a plant root-specific gene, *Science* 263:221.
Pearce, G., Strydom, D., Johnson, S., and Ryan, C.A., 1991, A polypeptide from tomato leaves induces wound-inducible proteinase inhibitor proteins, *Science* 253:895.
Perlak, F.J., Fuchs, R.L., Dean, D.A., McPherson, S.L., and Fischhoff, D.A., 1991, Modification of the coding sequence enhances plant expression of insect control protein genes, *Proc. Nat. Acad. Sci. USA* 88:3324.
Plunkett, G., Senear, D.F., Zuroske, G., and Ryan, C.A., 1982, Proteinase inhibitor I and II from leaves of wounded tomato plants: purification and properties, *Arch. Biochem. Biophys.* 213:463.
Sasser, J.N., 1980, Root-knot nematodes: A global menace to crop production, *Plant Dis.* 64: 36-41.
Sasser, J.N., and Freckman, D.W., 1987, A world perspective on nematology. *in:* "Vistas In Nematology," D.W. Dickson and Veech, J.A., eds., Society of Nematologists, Hyattsville.
Sijmons, P.C., 1993, Plant-nematode interactions, *Plant Mol. Biol.* 23:917.
Slana, L.J., and Stavely, J.R., 1981, Identification of the chromosome carrying the factor for resistance to *Meloidogyne incognita* in tobacco, *J.Nematol.* 13:61.
Stirling, G. R., 1991, "Biological Control of Plant Parasitic Nematodes", CAB International, Wollingford.
Triantaphyllou, A.C., 1987, Genetics of nematode parasitism on plants, *in:* "Vistas In Nematology," D.W. Dickson and Veech, J.A., eds., Society of Nematologists, Hyattsville.
van der Eycken, W., Niebel, A., Inze, D., van Montagu, M., and Gheysen, G., 1992, Molecular analysis of the interaction between *Meloidogyne incognita* and tomato, *Nematologica* 38:441 [Abstr.].
Watts, V.M., 1947, The use of *Lycopersicon peruvianum* as a source of nematode resistance in tomatoes, *Proc. Am. Soc. Hort. Sci.* 49:233.
Wiggers, R.J., Starr, J.C., and Price, H.J., 1990, DNA content and variation in chromosome number in plant cells affected by the parasitic nematodes *Meloidogyne incognita* and *M. arenaria, Phytopath.* 80:1391.
Wilson, T.M.A., 1993, Strategies to protect crop plants against viruses: Pathogen-derived resistance blossoms, *Proc. Nat. Acad. Sci. USA* 90:3134.
Yamamoto, Y.T., Cheng, C.-L., and Conkling, M.A., 1990, Root-specific genes from tobacco and *Arabidopsis* homologous to an evolutionarily conserved gene family of membrane channel proteins, *Nuc. Acids Res.* 18:7449.
Yamamoto, Y.T., Taylor, C., Acedo, G.N., Cheng, C.-L., and Conkling, M.A., 1991, Characterization of *cis*-acting sequences regulating root-specific gene expression in tobacco, *Plant Cell* 3:371.
Yang, Z., Park, H., Cramer, C.L., and Lacy, G.H., 1991, Differential regulation of 3-hydroxy-3-methylglutaryl CoA reductase genes by wounding and pathogen challenge, *Plant Cell* 3:397.

PART IV

HOST-PARASITE INTERACTIONS: THE NEMATODE

MELOIDOGYNE STYLET SECRETIONS

Richard S. Hussey, Eric L. Davis, and Celeste Ray

Department of Plant Pathology
University of Georgia
Athens, GA 30602-7274 USA

INTRODUCTION

The feeding relationships parasitic nematodes have evolved with susceptible plants are diverse. Relationships may be relatively simple, such as those of certain migratory ectoparasitic nematodes which feed on epidermal cells of plant roots without killing the cells. Or relationships may be complex, such as those of the sedentary endoparasitic nematodes which modify plant cells into highly specialized feeding sites. Bioactive molecules synthesized in the esophageal glands and secreted through the nematode's protrusible stylet regulate plant-nematode interactions. These stylet secretions may function in penetration and migration of nematodes in plant tissue, modification and maintenance of plant cells as feeding sites, formation of feeding tubes, and/or digestion of host cell contents to facilitate nutrient acquisition by the nematode. Secretory molecules from sedentary endoparasites are particularly intriguing because of the complex changes in plant cell phenotype and function that they modulate.

The sedentary endoparasitic root-knot nematodes (*Meloidogyne* species) are among nature's most successful parasites. They parasitize more than 2,000 plant species and represent a tremendous threat to crop production world-wide (Sasser, 1980). These obligate parasites have evolved a very specialized and complex feeding relationship with their host plants. A successful host-parasite relationship requires these nematodes to elaborately modify several plant root cells to obtain nourishment necessary for their development and reproduction (Hussey, 1985; Jones, 1981). Infective second-stage juveniles migrate in the soil and are attracted to root tips where they penetrate directly behind the root cap. Juveniles migrate intercellularly in the root to the region of cell differentiation and locate cells in the vascular cylinder (Endo and Wergin, 1973; Wyss et al., 1992). The juveniles inject stylet secretions into five to seven cells to transform these plant cells into specialized feeding sites called giant-cells (Bird, 1962; Hussey, 1989a). These multi-nucleate giant-cells become the permanent feeding site for the parasite throughout its life-cycle.

Giant-cell formation is one of the most complex responses elicited in plant tissue by any parasite. Cells parasitized by juveniles undergo repeated karyokinesis uncoupled from cytokinesis (Huang, 1985). Each host cell fed upon increases dramatically in size, the central vacuole diminishes in size, the cytoplasm increases in volume and density, and cell wall ingrowths form, giving the cells a phenotype similar to that of transfer cells (Jones and Payne, 1978). Giant-cells are induced and maintained in susceptible hosts only by the feeding activities of *Meloidogyne* species. Stylet secretions of root-knot nematodes appear to regulate, directly or indirectly, specific host genes affecting protein synthesis, nuclear division, cell growth and differentiation, and cell wall synthesis. Characterization of *Meloidogyne* stylet secretions and the genes encoding them will lead to a better understanding of the molecular events and regulatory mechanisms involved in plant parasitism by root-knot nematodes and should provide new knowledge to develop target-specific tactics to limit crop damage by these pathogens.

NEMATODE ESOPHAGEAL GLANDS

Adaptations of nematodes for plant parasitism involved development of a protrusible stylet, which is used to penetrate cell walls and feed from the cell, as well as marked morphological and physiological modifications of the esophagus (Bird, 1971; Maggenti, 1987). The esophageal glands enlarged considerably as plant parasites evolved from free-living nematodes in the Rhabditida, and presumably the function of the secretions from these glands likewise changed (Hussey, 1987).

Esophageal Gland Morphology

Infective second-stage juveniles of *Meloidogyne* have three large and complex esophageal gland cells, one dorsal and two subventrals, where secretory proteins are synthesized and sequestered in membrane-bound secretory granules (Fig. 1). The complexity and secretory function of these glands are revealed in ultrastructural studies of the esophagi of preparasitic and parasitic second-stage juveniles and adult females (Endo and Wergin, 1988; Hussey and Mims, 1990). Each esophageal gland overlaps the anterior end of the intestine and contains a large lobed nucleus with a prominent nucleolus, abundant Golgi bodies, rough endoplasmic reticulum, secretory granules, and other organelles typical of secretory cells. The dorsal gland cell has a long cytoplasmic extension that extends anteriorly through the metacorpus to terminate in an ampulla, a collecting reservoir for secretory granules, in the esophagus near the stylet knobs. In contrast, the subventral gland cells have short cytoplasmic extensions that terminate in ampullae at the base of the pump chamber in the metacorpus (Fig. 1). A cuticularized duct and a valve with a membrane-delineated end-sac connects each ampulla to the esophageal lumen and regulates, under the control of the nervous system, the release of glandular secretions into the lumen (Endo and Wergin, 1988; Hussey and Mims, 1990). The secretory granules form in the nuclear region of the gland cells and migrate forward through the extensions to accumulate near the valves in the ampullae prior to their contents being secreted (Hussey and Mims, 1990). The manner in which the contents of the secretory granules enter the valves of the esophageal glands, whether directly by exocytosis or by diffusion after they are released into the ampulla, has not been resolved (Hussey, 1992).

The spherical secretory granules, which arise from Golgi bodies, vary in size and morphology between the dorsal and subventral esophageal glands. Secretory granules produced in the subventral glands are usually less electron dense and larger (ca. 800 nm

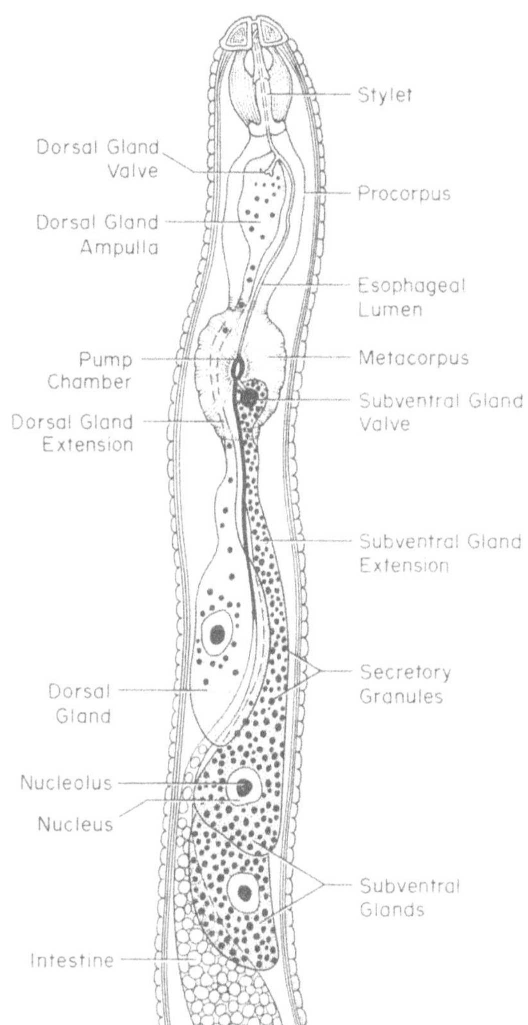

Figure 1. The esophagus of a second-stage juvenile of a plant-parasitic nematode. Reproduced, with permission, from the *Annual Review of Phytopathology* Vol. 27, 1989, by Annual Reviews Inc.

dia.) than those formed in the dorsal gland (ca. 500 nm dia.) (Bird, 1968a; Bird, 1969; Hussey and Mims, 1990) (Fig. 2). In addition, the composition of the matrix of the two types of granules differs. The matrix of subventral gland granules has an electron-transparent core containing spherical vesicles surrounded by a finely granular, electron-dense matrix (Hussey and Mims, 1990; McClure and von Mende, 1987). Dorsal gland granules, on the other hand, contain a homogeneous electron-dense matrix. Secretory granules produced in other organisms are known to contain normally at least two types of molecules: 1) passenger proteins destined for export out of the cell and 2) specific proteins involved in granule transport and exocytosis (Moore et al., 1988). The function and number of different passenger proteins in the secretory granules formed in esophageal glands of *Meloidogyne* species remain to be resolved.

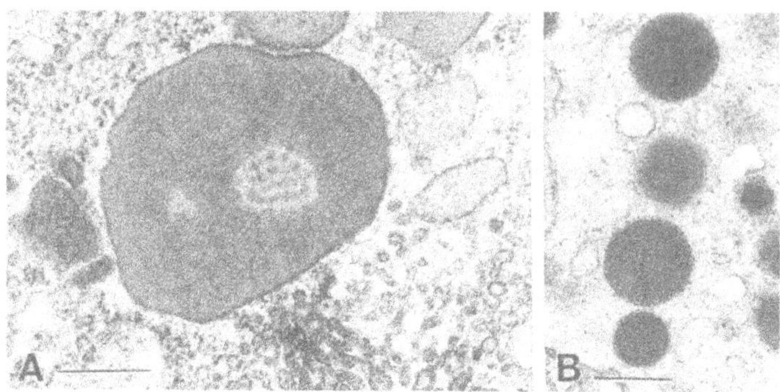

Figure 2. Fine structure of membrane-bound secretory granules formed in the (A) subventral and (B) dorsal esophageal glands of second-stage juveniles of *Meloidogyne incognita*. Bar: 0.25 μm. (Reprinted from Hussey and Mims, 1990.)

Changes In Esophageal Glands During Parasitism

During parasitic development of *Meloidogyne* species, the esophageal glands and secretory granules undergo distinct morphological changes. As second-stage juveniles establish a feeding relationship with host tissue and increase in body width, the dorsal gland increases and the subventral glands decrease in size (Bird, 1969; Bird, 1983). Secretory granules in the subventral glands in parasitic second-stage juveniles decrease in number (Wyss et al., 1992), appear to be in various stages of degeneration and, in adult females, are smaller and greatly reduced in number (Bird, 1968a; Hussey and Mims, 1990). The dorsal gland in preparasitic second-stage juveniles contains few secretory granules (Bird, 1967; Hussey and Mims, 1990). However, soon after juveniles penetrate roots, secretory granules form in the dorsal gland cell and accumulate in its ampulla (Bird and Saurer, 1967; Wyss et al., 1992). In adult females, the dorsal gland predominates whereas the subventral glands are greatly reduced in size (Bird, 1969) (Fig. 3). The changes in esophageal gland morphology, secretory granule morphology, and secretory antigens during parasitism by *Meloidogyne* species (Bird, 1967; Bird, 1983; Davis et al., 1992; Hussey, 1989a; Hussey, 1989b; Hussey and Mims, 1990; Wyss et al., 1992) indicate a changing role for the esophageal glands and their secretions at different stages of the nematode's life cycle.

NATURE AND ORIGIN OF STYLET SECRETIONS

The first evidence that root-knot nematodes inject secretions directly into host cells was published by Linford (1937) who suggested that these secretions were responsible for development of the giant-cells. The most detailed early work on the nature of the contents of esophageal gland secretions was conducted by Bird and co-workers. They detected protein but no nucleic acids in the subventral gland extensions in preparasitic second-stage juveniles (Bird and Saurer, 1967). Chemical changes detected in subventral gland granules in second-stage juveniles following invasion of root tissue led Bird (1968a) to suggest that secretions synthesized in these glands following the onset of parasitism might be responsible for inducing a feeding site in a susceptible host. Ultrastructural cytochemical

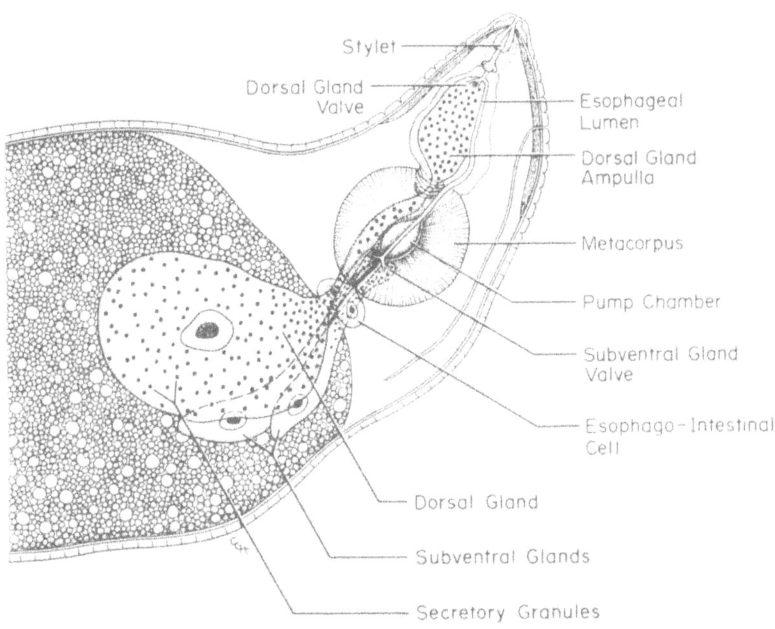

Figure 3. The esophagus of an adult female root-knot nematode.

analyses performed on granules in subventral glands of preparasitic second-stage juveniles of *M. incognita* were positive for acid phosphatase activity and negative for peroxidase, cellulase, DNase, RNase, and nucleic acids (Sundermann and Hussey, 1988). Cellulase (Bird et al., 1975) and a cytokinin-like compound (Bird and Loveys, 1980) have been detected in solutions in which second-stage juveniles of *M. javanica* were incubated.

Adult females of *Meloidogyne* species dissected from galls and placed in an aqueous solution produce stylet secretions that accumulate at the lip region (Bird, 1968b; Hussey and Sasser, 1973; Hussey et al., 1990; Veech et al., 1987). The fine structure of stylet secretions from adult females consists of electron-dense strands that usually surround a granular core (Bird, 1969; Davis et al., 1994). These secretions stain positive for basic protein but stain negative for several enzymes and nucleic acids (Bird and Saurer, 1967). Peroxidase activity was detected in stylet secretions from adult females of *M. incognita* (Hussey and Sasser, 1973), but other results suggest that plant peroxidase may be ingested by adult females and regurgitated during incubation (Starr, 1979; Jones, 1980), which may account for inconsistency in detection of peroxidase in stylet secretions (Veech et al., 1987). Dorsal gland secretory granules stained positive for nucleic acid and peroxidase but not for acid phosphatase, glucuronidase, DNase, catalase, polyphenoloxidase, or cellulase activity (Sundermann and Hussey, 1988). Although the secretory granules were positive for nucleic acids using three reagents, the presence of nucleic acids in stylet secretions of adult females has not been confirmed with other techniques. Granules observed in the lumen of the esophagus of parasitic second-stage juveniles and adult females of *Meloidogyne* species stained positive for DNA, but the origin of these granules was not established (Cardin and Dalmasso, 1984). Stylet secretions collected from adult females of *M. arenaria* and two races of *M. incognita*, separated into nine proteins on SDS-PAGE, three of which were identified as glycoproteins (Veech et al., 1987). No differences in the protein patterns were detected among the *Meloidogyne* species or races.

Monoclonal antibody technology is now being used to identify secretory components produced by plant-parasitic nematodes (Atkinson et al., 1988; Davis et al., 1992; Davis et al., 1994; Hussey, 1989b; Hussey et al., 1990). Development of monoclonal antibodies to secretory granules formed in the esophageal glands represents significant progress toward the isolation and characterization of biologically important stylet secretions (Fig. 4). The specificities of these monoclonal antibodies reveal similarities and differences in the antigens sequestered in the secretory granules formed in the subventral and dorsal glands of *Meloidogyne* species (Table 1) (Davis et al., 1992). Furthermore, several of these antibodies bind different antigens in stylet secretions of second-stage juveniles and adult females of *M. incognita* (Davis et al., 1994).

One secretory protein that is present in both types of glands in preparasitic juveniles but in only the dorsal gland in adult females has been immunopurified (Hussey et al., 1990). This protein also is present in stylet secretions of both life stages of *M. incognita* (Davis et al., 1994). This secretory component is a glycoprotein that has an apparent molecular mass, estimated by electrophoresis, of >210,000 daltons. The glycoprotein composition of secretory proteins is supported by ultrastructural studies which show that esophageal glands are cells in which large quantities of proteins are synthesized and sequestered in Golgi-derived secretory granules for export out of the gland cell (Hussey and Mims, 1990). Ultrastructural immunogold localization of this secretory protein in ultrathin sections of secretory granules in esophageal glands revealed the antigen to be confined to a specific domain of the granule matrix (Hussey et al., 1990).

The proximity of the dorsal gland valve to the base of the stylet enables fluid secretions released from secretory granules of this gland to be secreted through the stylet. Labeling of stylet secretions produced by adult female *M. incognita* with a monoclonal antibody specific for dorsal gland granules (Hussey et al., 1990) and video-enhanced observations of nematode secretory activity *in vivo* (Wyss and Zunke, 1986) provide evidence that dorsal gland secretions can be secreted through stylets of plant-parasitic nematodes. In contrast, location of the subventral gland valves at the base of the metacorpal pump chamber and the rigid circular lumen of the esophagus anterior to the pump chamber should both restrict anterior flow of subventral gland secretions during

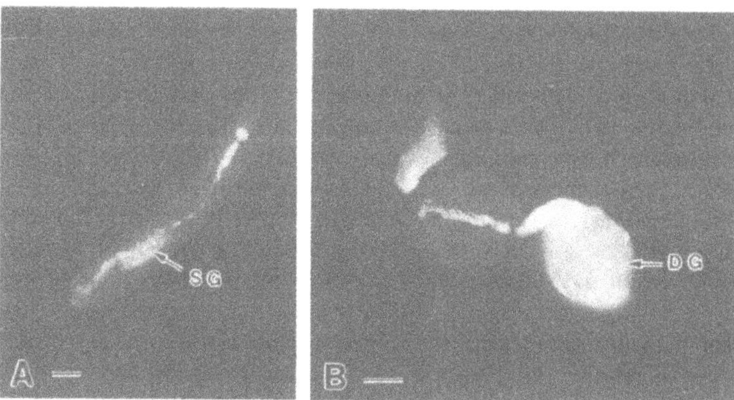

Figure 4. Micrographs of indirect immunofluorescent staining of (A) subventral glands (SG) of second-stage juvenile and (B) dorsal gland (DG) of adult female of *Meloidogyne incognita* with selected monoclonal antibodies and fluorescein isothiocyanate-conjugated second antibody. Bar: 10 μm. (Reprinted from Davis et al., 1992.)

Table 1. Binding specificities[a] of monoclonal antibodies (MAb) to stylet secretions and secretory granules within subventral (SvG) or dorsal (DG) glands of *Meloidogyne* second-stage juveniles (J2) and adult females (AF), and to other nematode species[b] (after Davis et al., 1992; Davis et al., 1994).

MAb code	J2		AF		Mi stylet secretions	Mi Ma Mj	Mh	Hg	Ce
	SvG	DG	SvG	DG					
$3F_4$	+[c]	-	+	-	J2, AF	+	-	-	-
$3H_{11}$	+	-	-	-	J2	+	-	-	-
$7A_9$	-	-	+	-	-	+	-	-	-
$6D_4$	+	+	-	+	J2, AF	+	-	-	-
$12H_7$	-	-	-	+	AF	+	+	-	-
5 MAbs	-	-	-	+	-	+	-	-	-

[a]Determined by indirect immunofluorescence microscopy.
[b]Mi - *M. incognita*, Ma - *M. arenaria*, Mj - *M. javanica*, Mh - *M. hapla*, Hg - *Heterodera glycines*, Ce - *Caenorhabditis elegans*.
[c]Antibody binding localized (+) to dorsal or subventral esophageal glands or no binding observed (-).

pumping of the metacorpus (Doncaster, 1971). It has been suggested that secretions from the subventral glands only pass posteriorly in the esophagus to the intestine and, thus, only function in intracorporeal digestion (Hussey, 1989a; Wyss et al., 1992). However, the subventral glands in *Meloidogyne* second-stage juveniles begin to shrink with the onset of parasitism and appear less active in adult females during the most active feeding period, a time when demand for digestive secretions should be maximal. Atkinson and Harris (1989) have suggested that subventral gland secretions of second-stage juveniles of cyst nematodes may be important in mobilization of lipid reserves while the intestine changes from a storage to an absorptive organ. Wyss et al., (1992) observed the release of contents of subventral gland granules just prior to and during intermittent pumping of the metacorpus in *M. incognita* second-stage juveniles infecting *Arabidopsis thaliana*. They suggested that secretions from subventral glands may assist in juvenile invasion and migration within roots. Incubation of *M. incognita* preparasitic juveniles in resorcinol stimulates vigorous stylet thrusting and accumulation of stylet secretions at the lip region (McClure and Von Mende, 1987). Associated with this secretory activity is forward movement and accumulation of secretory granules in the cytoplasmic extensions and ampullae of the subventral glands, but not in the dorsal gland. This observation led McClure and Von Mende (1987) to suggest that the secretions may have originated in the subventral glands.

Monoclonal antibodies recently have been used to conclusively establish that secretions from subventral esophageal glands can be secreted through a nematode's stylet (Davis et al., 1994). Stylet secretions collected from *M. incognita* adult females incubated in an antibiotic saline solution or from preparasitic juveniles incubated in resorcinol were placed on dialysis membrane and used for immunofluorescence assays. Monoclonal antibodies that only bound secretory granules in the subventral glands in sections of juveniles and adult females specimens also labeled stylet secretions produced by these life stages (Table 1). The specificities of these antibodies clearly establish that secretions synthesized in the subventral esophageal glands can be secreted through the stylet of *M. incognita*. In addition, antibodies with specificity to secretory granules in the subventral glands have been generated by immunizing mice with stylet secretions collected from *M. incognita* adult females (Davis et al., 1992) and preparasitic juveniles (Davis and Hussey, unpublished). Furthermore, the predicted morphological resistance to anterior flow of subventral gland secretions in the esophageal lumen during maximal pumping of the metacorpus (Doncaster, 1971) may be minimal during the secretion phase of a feeding cycle. Production of stylet secretions *in vitro* by preparasitic juveniles and adult females involves very little movement of the metacorpal pump chamber. Also, when the pump chamber is closed there is a slight gap between the sclerotized walls of the triradiate pump chamber that might allow for anterior flow of subventral gland secretions during the secretion phase (Endo and Wergin, 1988). Stylet thrusting of preparasitic juveniles producing stylet secretions when incubated in resorcinol is accompanied by movement of granules in the subventral gland ampullae. Whether this activity represents true secretory activity *in planta* has not been confirmed. However, exposure to resorcinol has limited, if any, detrimental effects on the juveniles, since nematodes treated with resorcinol for several hours are able to infect tomato roots.

Important information on the temporal synthesis of specific secretory components by the two types of esophageal glands in *M. incognita* during parasitism of plants has resulted from studies using monoclonal antibodies (Davis et al., 1994). One antibody that bound to the subventral glands and stylet secretions of preparasitic juveniles did not bind to esophageal glands in early parasitic juveniles (approximately 6 days after inoculation) or in any later parasitic stage. The expression of this antigen correlates with the reduction in subventral gland contents observed after penetration and migration of juveniles of *M. incognita* in *A. thaliana* (Wyss et al., 1992). An antigen recognized in stylet secretions by another antibody is present in both the dorsal and subventral glands in preparasitic and early parasitic juveniles but is absent in the subventral glands of late parasitic second-stage juveniles (approximately 12 days after inoculation) and increases in the dorsal gland in later stages of the nematode's life cycle. This antigen could possibly function in juvenile migration in roots and subsequently in food utilization or possibly have a role in the initiation and maintenance of giant-cells. Another antibody binds an antigen present in the subventral glands of only later parasitic stages of *M. incognita*. This esophageal gland antigen may be important in the secretory process or internal digestion of food since this antibody does not bind to stylet secretions even though it was derived from immunizations with stylet secretions collected from adult females.

Several monoclonal antibodies with specificity to the dorsal gland are female-specific and bind to secretory granules in *M. incognita*, *M. javanica*, *M. arenaria*, but not to those in *M. hapla* (one exception), the soybean cyst nematode (*Heterodera glycines),* or *Caenorhabditis elegans* (Table 1) (Davis et al., 1992). The specificity of these antibodies for esophageal gland antigens only in adult females suggests that these antigens could be involved directly in food utilization or giant-cell maintenance. Interestingly, one dorsal gland specific antibody also bound granules in this gland in *M. hapla*. One possible function of this antigen that might be common to the four *Meloidogyne* species is formation of a feeding tube (Hussey and Mims, 1991). The lack of binding of the rest of the antibodies to esophageal glands in *M. hapla* further supports the notion of more distant phylogenetic and parasitic relationships of this species to the three other common *Meloidogyne* species (Hussey, 1985; Hyman and Powers, 1991).

Feeding Tubes

In addition to injecting into the cytoplasm of host cells stylet secretions that modify the recipient cells, some plant-parasitic nematode species inject secretions that form unique tube-like structures that are called feeding tubes. These structures were first observed in 1911 by Nemec (1911) who later, without any biochemical evidence, described them as proteinaceous threads (Nemec, 1932). Feeding tubes formed in giant-cells by *M. incognita* are straight to slightly curved structures just less than 1 μm wide and up to 110 μm long (Hussey and Mims, 1991). The electron-dense wall of a feeding tube is crystalline and forms a lumen which is 350 nm in diameter, a dimension that is relatively constant

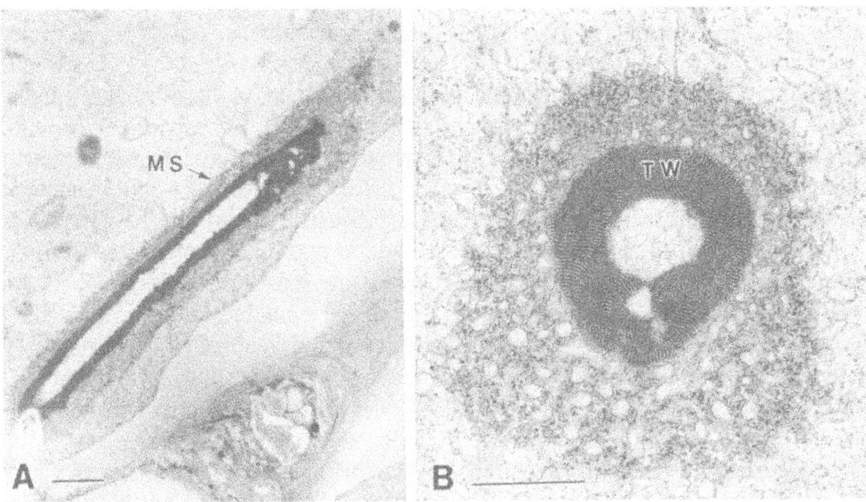

Figure 5. Fine structure of feeding tubes formed by *Meloidogyne incognita*. (A) Longitudinal section through a feeding tube with its proximal end attached to a wall ingrowth of a giant-cell in a tomato root. The feeding tube is enveloped by a compact membrane system (MS). Bar: 1 μm. (B) Cross-section of electron-dense crystalline feeding tube wall (TW) surrounded by a dense membrane system. Bar: .5 μm. (Reprinted from Hussey and Mims, 1991.)

throughout its length (Fig. 5). The distal end of a feeding tube is sealed with wall material (Fig. 5A). In addition, a compact membrane system envelops newly formed and functional feeding tubes. Selective staining of the endoplasmic reticulum and nuclear envelope of giant-cells with osmium tetroxide-potassium ferricyanide revealed that the membrane system enveloping a feeding tube is interconnected with the endomembrane system of the giant-cell (Hussey, unpublished). The stimulus for reorganization of the giant-cell endoplasmic reticulum to form the compact membrane system appears to be associated with stylet secretions that form the feeding tube (Hussey and Mims, 1991). The periphery of the membrane system is composed of rough endoplasmic reticulum while smooth endoplasmic reticulum is adjacent to the tube wall. The intimate association of the complex membrane system with the feeding tube indicates that it might function in synthesizing and/or transporting soluble assimilates in the metabolically active giant-cells to the feeding tube for withdrawal by the parasite. Root-knot nematodes feed in cycles from giant-cells and a new tube is formed each time the nematode removes its stylet and reinserts it into a giant-cell to initiate a new feeding cycle. Consequently numerous feeding tubes are present in each giant-cell (Rumpenhorst, 1984; Hussey and Mims, 1991). The crystalline walls of feeding tubes no longer used by feeding nematodes persist in the cytoplasm of giant-cells but the membrane systems that enveloped these tubes degenerate (Hussey and Mims, 1991).

The manner by which a nematode's stylet removes nutrients via the 'feeding tube' or, indeed, if it does so at all, remains unclear. For plant-parasitic nematodes that establish prolonged feeding relationships with a host cell, the plasma membrane becomes invaginated around the stylet tip in the parasitized cell (Hussey et al., 1992a). Callose deposition commonly forms between the plasma membrane of the parasitized cell and the stylet of the feeding nematode, but is absent around the stylet orifice which is tightly appressed to the plasma membrane (Hussey et al., 1992b). Nevertheless, a direct association of the stylet orifice on the outside of the plasma membrane with the lumen of a feeding tube located on the inside of the plasma membrane would be necessary to facilitate withdrawal of nutrients from the cytoplasm of the parasitized cell via the feeding tube. Feeding tubes frequently project from short wall ingrowths through which the nematode's stylet is inserted, possibly facilitating alignment of the stylet orifice with the feeding tube lumen. Noteworthy, walls of giant-cells, which are generally thickened by secondary wall formation, remain thin opposite the lip region of root-knot nematodes so that their small stylets can penetrate the cell wall (Hussey and Mims, 1991). However, the manner in which *Meloidogyne* species withdraw nutrients from cells has not been observed directly. Feeding root-knot nematodes are sensitive to tissue manipulation during fixation for electron microscopy. As a result not a single example is available that shows a root-knot nematode with its stylet inserted into a giant-cell. Wyss et al. (1984) considered feeding tubes to serve as a means for facilitating the flow of soluble nutrients in the cytoplasm of the food cells toward the stylet orifice. They also reported that the cytoplasm around feeding tubes appeared modified and suggested that enzymes from the tubes predigested the cytoplasm to promote entry of nutrients into the tubes. Since the distal ends of feeding tubes apparently are sealed and their walls do not have noticeable pores, the walls must be permeable to cell fluids. Razak and Evans (1976) proposed that these structures might function as cytoplasmic filters that prevent cell organelles from obstructing the stylet orifice (ca. 150 nm diam.). Utilization of feeding tubes to acquire nutrients from food cells by certain parasitic nematodes may diminish the need for secretion of hydrolytic enzymes to modify the cytoplasm of the parasitized cell prior to ingestion of nutrients. Also if the stylet orifice needs to be appressed to the lumen of the tube for proper function, digestive enzymes would have to enter the host cell cytoplasm through the tube wall. The dense membrane system which envelops functional feeding tubes formed by *Meloidogyne* species might limit access of secreted enzymes passing through the tube to the cytoplasm of the cell. Regardless of the manner in which feeding tubes function, more than likely they are essential for efficient withdrawal of soluble assimilates from the feeding site by the parasitic nematode. Conclusive evidence of an essential role for feeding tubes in nutrient acquisition by nematodes awaits experiments demonstrating that perturbation of feeding tube formation has a detrimental effect on nematode development.

While there have been some studies on the morphology of feeding tubes formed by different nematodes (Rebois, 1980; Wyss et al., 1984; Hussey and Mims, 1991), no data are available on the nature of the nematode secretions that contribute to feeding tube formation. Furthermore, it has not been definitively determined where these secretions are synthesized in the nematode, even though the dorsal esophageal gland has been implicated. An ultrastructural study showed that a crystalline secretory component, whose fine structure was similar to that of the feeding tube wall, accumulated in an open dorsal gland valve in an adult female of *M. incognita* (Hussey and Mims, 1990). This observation, coupled with observations that 1) the fine structure of feeding tubes formed by *Meloidogyne* species is identical in giant-cells formed in different plant species (Hussey and Mims, 1991), 2) subcellular structures similar to feeding tubes have not been reported in healthy cells of any plant species or plant cells parasitized by other organisms, 3) feeding tubes form rapidly in parasitized cells following insertion of a nematode's stylet and initiation of the secretion phase of a feeding cycle (Wyss and Zunke, 1986), and 4) the fine structure of feeding tubes formed by different nematode genera parasitizing the same plant species is dissimilar (e.g. *M. incognita* and *Heterodera glycines* on soybean, Hussey unpublished). Together these observations provide compelling evidence that feeding tubes form directly from nematode stylet secretions injected into the cytoplasm of parasitized cells. Interestingly, stylet secretions produced *in vitro* by adult females of *Meloidogyne* species form a very viscous, insoluble precipitate that adheres to the lip region (Davis et al., 1994), but when these same secretions are injected into the cytoplasm of giant-cells all or some secretory components polymerize into the unique feeding tube. Apparently interaction of the secretion with some component of the giant-cell cytoplasm, or perhaps the pH of the cytoplasm, induces the polymerization of the secretion into a tube.

Secretions Of Animal Parasitic Nematodes

Recent studies of gland secretions of the animal parasitic nematode, *Trichinella spiralis*, suggest that these secretions might have a regulatory function similar to that of *M. incognita* secretions (Bird, 1992). Secretions from the *T. spiralis* stichosome, a large multicellular esophageal gland, transform muscle cells into nurse cells (Despommier, 1990). Some of the changes induced in the parasitized muscle cells resemble the changes induced in plant cells by *M. incognita*. The nurse cells are characterized by several hypertrophied nuclei with enlarged nucleoli and proliferation of subcellular organelles. Like the esophageal gland proteins of *M. incognita*, secretions of *T. spiralis* are thought to be involved in the transformation of muscle cells since they have been immunolocalized in the nucleoplasm of the infected cells (Despommier et al., 1990). This notion is supported by experiments showing that injection of secretions collected in vitro into rat muscles mimics cellular changes that occur *in vivo* (Ko et al., 1992). This suggests the possibility of cell modifications when isolated secretions of *Meloidogyne* are injected into healthy plant root cells.

NEMATODE GENES ENCODING SECRETIONS

Cloning genes that encode esophageal gland secretory proteins of root-knot nematodes is an integral part of characterization of these secretions and will enhance our understanding of the molecular genetics of plant-nematode interactions. Nucleotide sequences of cloned secretion genes can be searched in data bases for homologies with known sequences and to deduce protein structures. Furthermore, cloned secretion genes can be overexpressed *in vitro* to yield sufficient quantities of secretory proteins for functional and structural assays. The availability of nematode secretion genes also will enable studies to be conducted with transgenic plants to determine the effect of the products of expressed secretion genes on host cell phenotype. Cloned genes also may permit the genetic basis for host range specificity to be established by comparing the structure of secretory protein genes from different *Meloidogyne* species.

Isolation of a Secretion Gene

Monoclonal antibodies can be used to immunoscreen cDNA expression libraries to isolate nematode genes encoding secretory proteins. Screening of *M. incognita* cDNA expression libraries with monoclonal antibodies with specificity for esophageal gland secretory proteins has been initiated in our laboratory (Ray and Hussey, unpublished). cDNA libraries have been constructed from preparasitic second-stage juveniles and adult females of *M. incognita*. One lambda clone has been isolated from a cDNA library from adult female nematodes using a monoclonal antibody specific for secretory granules in the subventral esophageal glands of this life stage. The monoclonal antibody was generated by twice implanting nitrocellulose carrying stylet secretions collected from *M. incognita* females in the spleen of a BALB/c mouse (Davis et al., 1992). The cDNA insert has been shown by Southern blot analysis to be derived from *M. incognita* genomic DNA, and also has been used as a probe to isolate the homologous gene from a *M. incognita* genomic library (Ray and Hussey, unpublished). Hybridization to other Southern blots containing *M. incognita*, *M. arenaria*, *M. javanica*, *M. hapla*, *H. glycines*, and *C. elegans* genomic DNA indicated homologous sequences were present in the four *Meloidogyne* species but not in *H. glycines* or *C. elegans*.

The availability of this gene provides a unique opportunity to study the structure and function of a gene encoding an esophageal gland protein. Approximately 60% of the genomic DNA sequence of the secretion gene has been determined to date. Comparison of this partial sequence with the nucleotide sequence of the cDNA clone revealed the presence of five introns. These introns share several similarities with introns commonly found in *C. elegans* including their small size, high AT content (60-80%), and nucleotide sequences at the intron borders (Emmons, 1988). Such similarities imply that mRNA splicing mechanisms have been conserved between *C. elegans* and root-knot nematodes.

Expression of this gene is developmentally regulated and tissue specific in *M. incognita*. The antibody used to isolate the cDNA clone binds an antigen associated with the secretory granules in the SvG of adult females but not in those of preparasitic second-stage juveniles. These results suggest that gene expression and antigen synthesis is regulated in the SvG and occurs following the onset of parasitism. Interestingly, the same antibody binds an antigen in the somatic longitudinal muscles of the juveniles. Furthermore, in preliminary northern blot analysis, more pCRH10-specific poly(A)$^+$ mRNA was detected in adult females than in second-stage juveniles. This difference in transcript accumulation could be due to stage-specific regulation of transcription levels (Ray and Hussey, unpublished). While further characterization of this gene is in progress, other clones are being isolated by immunoscreening the cDNA libraries.

CONCLUSIONS AND FUTURE RESEARCH

Little is known about the molecular basis of parasitism by any plant-parasitic nematode. A key to understanding parasitism of plants by these nematodes is to identify and characterize stylet secretions and the genes encoding them. Particularly intriguing are the stylet secretions of *Meloidogyne* species that are responsible for the remarkable transformation of parasitized root cells. The complex cellular modifications suggest that root-knot nematode secretions interact, directly or indirectly, with the host's genome, modifying the pattern of gene expression in root cells to give rise to the unique giant-cells.

Although important progress has been made in the study of nematode secretions, our knowledge is still rather fragmentary. Monoclonal antibodies have been critical tools for identifying the presence of different secretory antigens. The binding specificities of the antibodies also make them invaluable as biochemical reagents for isolating secretory components either via immunoaffinity purification or identifying specific secretions separated by column chromatography. The nature and number of different components in stylet secretions must still be resolved. Most studies have focused on large macromolecules, but small molecules, e.g. peptides, also could be important components of stylet secretions. In addition, some secretory proteins critical to pathogenesis may be expressed only in the initial stages of parasitism. To identify such proteins, early parasitic stages can be used to generate monoclonal antibodies and as a source of RNA for constructing a stage-specific cDNA library. This library can then be screened with the antibodies specific for the early parasitic stage or by differential hybridization with radiolabeled probes derived from early and late parasitic mRNA's. Even though new evidence from *in vitro* production of stylet secretions supports a possible role for subventral gland secretions in the host-parasite interaction, immunolocalization of these secretions *in planta* would firmly establish their importance in pathogenesis. In addition, immunolocalization studies could provide insights into the nature of the molecular and cellular mechanisms by which nematode stylet secretions affect host cells by delineating the subcellular sites where putative secretions accumulate and regulate cell activities. Attempts to localize *Meloidogyne* secretions in giant-cells using monoclonal antibodies have not been successful (Davis and Hussey, unpublished). The concentration of the secretions may be below a detectable level or epitope accessibility *in planta* may be compromised when nematode secretions are injected into the cytoplasm of a plant cell or polymerize to form a feeding tube. Overexpression of secretory protein genes would allow the development of polyclonal antibodies that might be more suited for immunolocalization studies *in planta*. Some nematode secretions might bind to receptors on the plasma membrane to modulate host cell changes via signal transduction pathways. However, since nematodes possess a hollow, protrusible stylet, secretions are probably delivered directly into the cytoplasm via the stylet through minute perforations in the plasma membrane at the stylet orifice (Hussey et al., 1992a). In addition to stylet secretions that regulate host-nematode interactions, secretions that form the unique feeding tubes are of interest. Clearly, secretions that form these large tubes must be a principal secretory component of stylet secretions of the life-stages that produce feeding tubes.

A model genetic system has not been developed for any plant-parasitic nematode. Future studies of the function and transcriptional regulation of putative secretion genes involved in plant parasitism at the molecular level clearly will require the development of a DNA transformation system for plant-parasitic nematodes. Several aspects of *M. incognita* biology make classical genetic studies difficult with this organism. Since *M. incognita* reproduces by obligatory mitotic parthenogenesis, the opportunity to perform genetic crosses is not available. Consequently, new genetic material cannot be easily introduced, allelic relationships such as complementation and epistasis cannot be addressed, and the genetic basis of different phenotypes cannot be established. One way to circumvent the lack of genetic tools for studying parasitic nematode biology is to microinject cloned parasite DNA into *C. elegans*, where it can be subjected to classical genetic methods in well-defined genetic backgrounds (Wood, 1988). The feasibility of this approach is supported by several lines of evidence indicating that many basic molecular and developmental mechanisms are conserved across diverse nematode genera (Riddle and Georgi, 1990; Grant, 1992). The introduction of several animal parasitic nematode genes into *C. elegans* also indicates that this approach is valid (Grant, 1992). Although the esophageal gland cells are less developed in *C. elegans* than in plant parasites (Albertson and Thomson, 1976; Bird and Bird, 1991), transformation of *C. elegans* with putative secretion genes from plant parasites using *lacZ* transcriptional fusions will provide information on regulatory sequences required for expression of genes in the esophageal glands. In addition, if homologues of secretion genes are present in *C. elegans*, this model system also may provide some insights into the function of the proteins encoded by the parasite genes.

The biological activity of nematode secretory proteins needs to be determined before the role of any secretion in plant parasitism can be considered. Several experimental approaches may be utilized to elucidate the biological activity of *Meloidogyne* stylet secretions. When secretion genes have been cloned, overexpression of these genes will simplify purification of secretory proteins for use in assays of biological activity. Once sufficient quantities of a secretory protein are available, the protein can be microinjected into root protoplasts from susceptible and resistant tomatoes. Injected protoplasts can be monitored with time-lapse video for cellular changes that mimic early stages of giant-cell formation (susceptible response) or a hypersensitive reaction (resistant response). An alternative approach is to determine the effect of expression of cloned secretion genes on the phenotype of transgenic plants. The availability of plant transformation techniques will enable nematode secretion genes, under the control of various promoters, e.g. a root specific promoter (Yamamoto et al., 1991), to be easily transferred into plants. The effects of expression of cloned root-knot nematode secretory protein genes on cell phenotype in transgenic plants could provide important information on the ability of the gene product to modify plant tissue. However, these approaches would have limitations if transformation of plant cells requires more than a single nematode secretory protein. Gene transfer techniques also have been used to achieve expression and assembly of mouse immunoglobulins (Hiatt et al., 1989) and a synthetic gene encoding an antigen-binding single-chain F_v protein (Owen et al., 1992) in transgenic plants. This technology could be exploited to determine whether specific nematode secretory proteins are critical for infection of plants by expressing genes encoding for monoclonal antibodies specific for nematode stylet secretions in transgenic plants. Neutralization of a nematode secretion following injection into the cytoplasm of the antibody-producing root cell could inhibit, depending on the function of the antigen, giant-cell formation and/or nematode feeding and subsequently suppress nematode development on the transgenic plant if the secretory protein is essential for a susceptible interaction.

Concerns for sustainable and environmentally safe nematode management strategies, combined with the tools of modern biology, have stimulated a research emphasis on the basic biology of nematodes and their interactions with plants. This is well-illustrated by the recent development of a novel system for incorporating specific resistance to root-knot nematodes in transgenic plants (Opperman et al., 1994). Understanding the molecular signals from the nematodes and the molecular responses elicited in plant tissues will facilitate their exploitation for target-specific nematode management and provide unprecedented opportunity for limiting nematode damage to crops.

Acknowledgements

The research of the authors has been supported by grants from the USDA CSRS National Research Initiative Competitive Grants Program (Grants 90-37153-5676 and 91-37303-6387) and the University of Georgia Biotechnology Grants Program and by state and Hatch Funds allocated to the Georgia Agricultural Experiment Stations.

REFERENCES

Albertson, D.G., and Thomson, J.N., 1976, The pharynx of *Caenorhabditis elegans*, *Philos. Trans. R. Soc. Lond. B Biol. Sci.* 275:299.

Atkinson, H.J., and Harris, P.D., 1989, Changes in nematode antigens recognized by monoclonal antibodies during early infections of soya beans with the cyst nematode *Heterodera glycines*, *Parasitology* 98:479.

Atkinson, H.J., Harris, P.D., Halk, E.U., Novitski, C., Leighton-Sands, J., Nolan, P., and Fox, P.C., 1988, Monoclonal antibodies to the soya bean cyst nematode, *Heterodera glycines*, *Ann. Appl. Biol.*, 112:459.

Bird, A.F., 1962, The inducement of giant-cells by *Meloidogyne javanica*, *Nematologica* 8:1.

Bird, A.F., 1967, Changes associated with parasitism in nematodes. I. Morphology and physiology of preparasitic and parasitic larvae of *Meloidogyne javanica*, *J. Parasitol.* 53:768.

Bird, A.F., 1968a, Changes associated with parasitism in nematodes. III. Ultrastructure of the egg shell, larval cuticle, and contents of the subventral esophageal glands in *Meloidogyne javanica*, with some observations on hatching, *J. Parasitol.* 54:475.

Bird, A.F., 1968b, Changes associated with parasitism in nematodes. IV. Cytochemical studies on the ampulla of the dorsal esophageal gland of *Meloidogyne javanica* and on exudations from the buccal stylet, *J. Parasitol.* 54:879.

Bird, A.F., 1969, Changes associated with parasitism in nematodes. V. Ultrastructure of the stylet exudation and dorsal esophageal gland contents of female *Meloidogyne javanica*, *J. Parasitol.* 55:337.

Bird, A.F., 1971, Specialized adaptation of nematodes to parasitism, *in:* "Plant parasitic nematodes, Vol. II," B.M. Zuckerman, Mai, W.F., and Rohde, R.A. eds., Academic Press, New York.

Bird, A.F., 1983, Changes in the dimensions of the esophageal glands in root-knot nematodes during the onset of parasitism, *Int. J. Parasitol.* 13:343.

Bird, A.F., and Saurer, W., 1967, Changes associated with parasitism in nematodes. II. Histochemical and microspectrophotometric analysis of preparasitic and parasitic larvae of *Meloidogyne javanica*, *J. Parasitol.* 53:1262.

Bird, A.F., Downton, W.J.S., and Hawker, J.S., 1975, Cellulase secretion by second stage larvae of the root-knot nematode (*Meloidogyne javanica*), *Marcellia* 38:165.

Bird, A.F., and Loveys, B.R., 1980, The involvement of cytokinins in a host-parasite relationship between the tomato (*Lycopersicon esculentum*) and a nematode (*Meloidogyne javanica*), *Parasitology* 80:497.

Bird, A.F., and Bird, J., 1991, The structure of nematodes. Academic Press, San Diego.

Bird, D.Mc.K., 1992, Mechanisms of the *Meloidogyne*-host interaction, *in*: "Nematology from molecule to ecosystem," F.J. Gommers, and Maas, P.W.Th., eds., Eur. Society of Nematologists, Inc., Dundee, Scotland.

Cardin, M.C., and Dalmasso, A., 1985, Reactivity of esophageal granules of *Meloidogyne* to stain for DNA, *Nematologica* 32:179.

Davis, E.L., Aron, L.M., Pratt, L.H., and Hussey, R.S., 1992, Novel immunization procedures used to develop monoclonal antibodies that bind to specific structures in *Meloidogyne* spp., *Phytopathology* 82:1244.

Davis, E.L., Allen, R., and Hussey, R.S., 1994, Developmental expression of esophageal gland antigens and their detection in stylet secretions of *Meloidogyne incognita*, *Fun. Appl. Nematol.* (In press).

Despommier, D.D., 1990, *Trichinella spiralis*: the worm that would be a virus, *Parasitol. Today* 6:193.

Despommier, D.D., Gold, A.M., Buck, S.W., Capo, V., and Silberstein, D., 1990, *Trichinella spiralis*: secreted antigen of the infective L_1 larva localizes to the cytoplasm and nucleoplasm of infected host cells, *Expt. Parasitol.* 71:27.

Doncaster, C.C., 1971, Feeding in plant parasitic nematodes: mechanisms and behavior, *in:* "Plant parasitic nematodes," B.M. Zuckerman, Mai, W.F., and Rohde, R.A., eds., Academic Press, New York.

Emmons, S.W., 1988, The genome, *in:* "The nematode *Caenorhabditis elegans*," W.B. Wood, ed., Cold Spring Harbor Laboratory Press, Cold Spring Harbor.

Endo, B.Y., and Wergin, W.P., 1973, Ultrastructural investigation of clover roots during early stages of infection by the root-knot nematode, *Meloidogyne incognita*, *Protoplasma* 78:365.

Endo, B.Y., and Wergin, W.P., 1988, Ultrastructure of the second-stage juvenile of the root-knot nematode, *Meloidogyne incognita*, *Proc. Helminth. Soc. Wash.* 55:286.

Grant, W.N., 1992, Transformation of *Caenorhabditis elegans* with genes from parasitic nematodes, *Parasitol. Today* 8:344.

Hiatt, A., Cafferkey, R., and Bowdish, K., 1989, Production of antibodies in transgenic plants. *Nature* 342:76.

Huang, C.S., 1985, Formation, anatomy, and physiology of giant-cells induced by root-knot nematodes, *in:* "An advanced treatise on *Meloidogyne*. Volume I. Biology and control," J.N. Sasser and Carter, C.C., eds., N. C. State University Graphics, Raleigh.

Hussey, R.S. 1985, Host-parasite relationships and associated physiological changes, *in:* "An advanced treatise on *Meloidogyne*. Volume I. Biology and control," J.N. Sasser and Carter, C.C., eds., N. C. State University Graphics, Raleigh.

Hussey, R.S., 1987, Secretions of esophageal glands of Tylenchida nematodes, *in*: "Vistas on nematology," J.A. Veech and Dickson, D.W., eds., Society of Nematologists, Hyattsville.

Hussey, R.S., 1989a, Disease-inducing secretions of plant parasitic nematodes, *Annu. Rev. Phytopath.* 27:123.

Hussey, R.S., 1989b, Monoclonal antibodies to secretory granules in esophageal glands of *Meloidogyne* species, *J. Nematol.* 21:392.

Hussey, R.S., 1992, Secretions of esophageal glands in root-knot nematodes, *in*: "Nematology from molecule to ecosystem," F.J. Gommers, and Maas, P.W.Th., eds., Eur. Society of Nematologists, Inc., Dundee, Scotland.

Hussey, R.S., and Sasser, J.N., 1973, Peroxidase from *Meloidogyne incognita*, *Physiol. Pl. Path.* 3:223.

Hussey, R.S., and Mims, C.W., 1990, Ultrastructure of esophageal glands and their secretory granules in the root-knot nematode *Meloidogyne incognita*, *Protoplasma* 165:9.

Hussey, R.S., Paguio, O.R., and Seabury, F., 1990, Localization and purification of a secretory protein from the esophageal glands of *Meloidogyne incognita* with monoclonal antibodies, *Phytopathology* 80:709.

Hussey, R.S., and Mims, C.W., 1991, Ultrastructure of feeding tubes formed in giant-cells induced in plants by the root-knot nematode *Meloidogyne incognita*. *Protoplasma* 162:55.

Hussey, R.S., Mims, C.W., and Westcott, S.W., 1992a, Ultrastructure of root cortical cells parasitized by the ring nematode *Criconemella xenoplax*, *Protoplasma* 167:55.

Hussey, R.S., Mims, C.W., and Westcott, S.W., 1992b, Immunocytochemical localization of callose in root cortical cells parasitized by the ring nematode *Criconemella xenoplax*, *Protoplasma* 171:1.

Hyman, B.C., and Powers, T.O., 1991, Integration of molecular data with systematics of plant parasitic nematodes. *Annu. Rev. Phytopathol.*, 29:89.

Jones, M.G.K., and Payne, H.L., 1978, Early stages of nematode-induced giant-cell formation in roots of *Impatiens balsamina*, *J. Nematol.* 10:70.

Jones, M.G.K., 1980, Micro-gel electrophoretic examination of soluble proteins in giant transfer cells and associated root-knot nematodes (*Meloidogyne javanica*) in balsam roots, *Phys. Pl. Path.* 16:359.

Jones, M.G.K., 1981, Host cell responses to endoparasitic nematode attack: structure and function of giant-cells and syncytia, *Ann. Appl. Biol.* 97:353.

Ko, R.C., Fan, L., and Lee, D.L., 1992, Experimental reorganization of host muscle cells by excretory/secretory products of infective *Trichinella spiralis* larvae, *Trans. Royal Soc. Trop. Med. and Hyg.* 86:77.

Linford, M.B., 1937, The feeding of the root-knot nematode in root tissue and nutrient solution, *Phytopathology* 27:824.

McClure, M.A., and von Mende, N., 1987, Induced salivation in plant-parasitic nematodes, *Phytopathology* 77:1463.

Maggenti, A.R., 1987, Adaptive biology of nematode parasites, *in:* "Vistas on Nematology," J.A. Veech and Dickson, D.W., eds., Society of Nematologists, Hyattsville

Moore, H.H., Orci, L., and Oster, G.F., 1988, Biogenesis of secretory vesicles. *in*: "Protein transfer and organelle biogenesis," R.C. Das and Robbins, P.M., eds., Academic Press, New York.

Nemec, B., 1911, Über die Nematodenkrankheiten der Zuckerrübe, *Zeit. Pflanzen.* 21:1.

Nemec, B., 1932, Über die Gallen von *Heterodera schachtii* auf der Zuckerrübe, *Studies from the Plant Physiological Laboratory of Charles University, Prague. Vol. IV* 2:1.

Opperman, C.H., Taylor, C.G., and Conkling, M.A., 1994, Root-knot nematode-directed expression of a plant root-specific gene, *Science* 263:221-223.

Owen, M., Grandecha, A., Cockburn, B., and Whitelam, G., 1992, Synthesis of a functional anti-phytochrome single-chain F_v protein in transgenic tobacco, *Bio/Technol.* 10:790.

Razak, A.R., and Evans, A.A.F., 1976, An intracellular tube associated with feeding by *Rotylenchulus reniformis* on cowpea root, *Nematologica* 22:182.

Rebois, R.V., 1980, Ultrastructure of a feeding peg and tube associated with *Rotylenchulus reniformis* in cotton, *Nematologica* 26:396.

Riddle, D.L., and Georgi, L.L., 1990, Advances in research on *Caenorhabditis elegans*: application to plant parasitic nematodes, *Ann. Rev. Phytopathol.* 28:247.

Rumpenhorst, H.J., 1984, Intracellular feeding tubes associated with sedentary plant parasitic nematodes, *Nematologica* 30:77.

Sasser, J.N., 1980, Root-knot nematodes: a global menace to crop production, *Plant Dis.* 64:36.

Starr, J.L., 1979, Peroxidase isozymes from *Meloidogyne* spp. and their origin, *J. Nematol.* 11:1.

Sundermann, C.A., and Hussey, R.S., 1988, Ultrastructural cytochemistry of secretory granules of esophageal glands of *Meloidogyne incognita*, *J. Nematol.* 20:141.

Veech, J.A., Starr, J.L., and Nordgren, R.M., 1987, Production and partial characterization of stylet exudate from adult females of *Meloidogyne incognita*, *J. Nematol.* 19:463.

Wood, W.B., and the community of *C. elegans* researchers, eds., 1988, The nematode *Caenorhabditis elegans*. Cold Spring Harbor Laboratory Press, Cold Spring Harbor.

Wyss, U., Stender, C., and Lehmann, H., 1984, Ultrastructure of feeding sites of the cyst nematode *Heterodera schachtii* Schmidt in roots of susceptible and resistant *Raphanus sativus* L. var. *oleiformis* cultivars, *Phys. Pl. Path.* 25:21.

Wyss, U., and Zunke, U., 1986, Observations on the behavior of second stage juveniles of *Heterodera schachtii* inside host roots, *Rev. Nematol.* 9:153.

Wyss, U., Grundler, F.M.W., and Munch, A., 1992, The parasitic behavior of second-stage juveniles of *Meloidogyne incognita* in roots of *Arabidopsis thaliana*, *Nematologica* 38:98.

Yamamoto, Y.T., Taylor, C.G., Acedo, G.N., Cheng, C.-L., and Conkling, M.A., 1991. Characterization of *cis*-acting sequences regulating root-specific gene expression in tobacco, *Plant Cell* 3:371.

CHEMORECEPTION IN NEMATODES

Paolo Bazzicalupo[1], Laura De Riso[1], Francesco Maimone[2],
Filomena Ristoratore[1] and Marisa Sebastiano[1]

[1]Istituto Internazionale di Genetica e Biofisica,
Consiglio Nazionale delle Ricerche,
via G. Marconi, 10, 80125 Napoli, Italy
[2]Isituto di Genetica, Università di Bari,
CIRPS, Università di Roma

INTRODUCTION

Chemical signals from the environment are the most important sensory inputs for nematodes. The ability to receive and interpret chemical signals from the environment is essential for nematodes, parasitic or free living, to complete their life cycle. Chemoreceptors are needed: (i) to find food and good environmental conditions, for instance to find the host in parasitic species; (ii) to reach the host's organs in which development can take place and to interpret signals from the host that trigger important developmental switches; (iii) to avoid predators and dangerous or toxic chemicals; (iv) to interpret signals from other individuals of the species that signal crowding; (v) to find the opposite sex for reproduction. If better understood at the cellular and molecular level, chemoreception would have enormous potential as a target for the control of nematodes.

C.aenorhabditis elegans represents the ideal system in which to study some of the basic features of chemoreception. First, a variety of experimental approaches, including extensive genetic analysis, can be applied to the study of *C. elegans*. In addition the study of chemoreception can directly benefit from the work done on the nervous system of this organism. A complete anatomical and developmental description of the nervous system at the light and electron microscopic levels has been obtained (White et al., 1986). Each neuron and most of its connections have been described. On the basis of this work it has been possible to define 118 neuronal classes and most of the circuits. These results are a fundamental resource in the interpretation of pharmacological, genetic and behavioral data resulting from the study of chemoreception. For a review of this work see also "The nematode *Caenorhabditis elegans*" Wood et al., eds. 1988.

Nematodes differ from each other in size, life cycle, and other important aspects, but the general plan on which they are constructed is highly conserved. The anatomy of the chemoreception organs is also highly conserved in various nematode species. It is likely that the conservation at the ultrastructural and anatomic levels derives, at least partially, from conservation at a molecular level of the protein domains forming these structures. Thus it should be possible to use the knowledge, and the molecular reagents and probes derived from the study of chemoreception in C. *elegans*, to identify the homologous components in other nematode species.

Although in a sense chemoreception is a property of all cells, which modify their metabolism, activity and development in response to extracellular cues, it is clear that metazoan organisms have set aside a specialized group of cells to sense the environment. These sensory cells are connected and send signals to the rest of the nervous system which in turn influences behavior, metabolism and development of the organism. One can describe chemoreception based behavior as responses of effector cells to an external chemical signal, present in the environment and received by specialized sensory cells. The effector cells are often body muscle, (scored in chemotaxis, avoidance, repulsion or attraction assays) but can be other cell types, some of which are unknown, as in the case of dauer formation. The chemosensory neurons bear the receptors for the chemical signals. As far as we know the signals from the sensory neurons do not, in general, go to the effector cells directly but to interneurons in the nerve ring where some level of integration with other signals occurs and from where a signal is sent to the effector cells.

ANATOMY

As in other nematodes, sensory organs in C. *elegans* are each made up of two non-neuronal support cells, the sheath cell and the socket cell (Ward et al., 1975; Perkins et al., 1986; Chalfie and White, 1988). These form a channel around a bundle of ciliated neuronal endings which, in the case of chemosensory neurons, reach the external medium through a gap in the cuticle. The amphidia, a pair of laterally symmetrical sensilla in the head, are the main chemosensory organs in all nematodes. Of the 12 neurons sending processes to the amphidia, 8 (ADF, ADL, ASE, ASG, ASH, ASI, ASJ and ASK) have ciliated endings that reach the external environment. Three cells, AWA, AWB and AWC form wing-shaped invaginations in the sheath cell cytoplasm; they are called wing cells. One neuron AFD forms finger like structures in the sheath cell cytoplasm and is called the finger cell. (Ward et al., 1975; White et al., 1986; Perkins et al., 1986). The other sensilla that have a chemosensory function are the two phasmids in the tail, the six inner labial sensilla around the buccal opening and, in the male, the cephalic sensilla. The finding of ultrastructural alterations of various elements of the amphids and of these organelles in many chemoreception mutants supports the notion that they in fact have a chemosensory function (Lewis and Hodgkin, 1977; Perkins et al., 1986). The precise ancestry of the cells contributing to all the sensilla mentioned has been described and experiments using precursor laser ablation during development have shown no compensatory regulation on the part of the surviving cells (Sulston et al., 1983). The connections of all these neurons have also been described in detail (White et al., 1986)

BEHAVIOR

C. elegans behaviors which depend on the reception of chemical signals and which have been studied to some extent include:

1) Chemotaxis. Worms can orient their movement sensing the concentration of a number of chemicals present in the medium acting as attractants or repellents (Ward, 1973; Lewis and Hodgkin, 1977).

2) *C. elegans* is also attracted or repelled by some volatile chemicals (Bargmann et al., 1993).

3) Avoidance. On encountering a repellent chemical, *C. elegans* backs up; avoidance responses to high osmotic strength, and other noxious stimuli have been studied (Culotti and Russell, 1978; Bargman et al., 1990; personal communications from several authors, R. Herman, R. Horwitz, J. Thomas and our own unpublished work).

4) Entry and exit from the dauer larva developmental pathway is determined by the ability of worms to sense a pheromone, whose concentration is a measure of crowding, and chemical(s) signalling the presence of food (Riddle, 1988; Bargmann and Horvitz, 1991a).

5) Egg laying is affected, among other things, by environmental stimuli of which the presence of food is one. Upon removal of bacteria, hermaphrodites stop laying eggs very rapidly but resume laying when food is present again (Trent et al., 1983).

6) Males but not hermaphrodites are attracted to hermaphrodites, (Sulston and Hodgkin, personal communication).

The study of the behavior of worms after ablation of single or multiple sensory cells has been used to assign a specific function to many of the chemosensory neurons present in *C. elegans*, especially those of the amphids (Bargman and Horvitz, 1991 a, b; Bargmann et al., 1993; Kaplan and Horvitz, 1993, and personal communication from these authors and from J. Thomas). Table 1 summarises most of the results of the laser ablation experiments.

Some conclusions regarding the basic organization of chemical sensitivity in nematodes can be drawn from these experiments.

1 - Chemosensory neurons of the amphid are multi-functional, i.e. each neuron bears more than a single type of receptor. An extreme case is shown by the two ASH neurons which seem to mediate avoidance of various repellent stimuli including mechanical stimuli (Kaplan and Horvitz, 1993).

2 - There is a certain degree of redundancy. For instance, the receptors for some chemoattractants and those for dauer formation are present on more than one chemosensory neuron (Bargmann and Horvitz, 1991 a, b).

3 - The integrity of the sheath cell which secretes the matrix surrounding the amphid's cilia is necessary for chemosensory function.

4 - While water soluble chemicals seem to have receptors on the neurons whose cilia are exposed to the outside, attraction and repulsion to volatile

chemicals is mostly mediated by receptors present on the amphidial wing cells AWA and AWC, (Bargmann et al., 1993).

5 - Receptors for avoidance responses to water soluble chemicals seem to be present only on the amphidial neurons ASH and ADL (Bargmann et al., 1990; Thomas and Horvitz, personal communication).

Table 1. Function of amphid neurons: summary of the results of laser ablation experiments*.

Amphid Neuron					Behavior
					CHEMOTAXIS
ASE	ADF	ASG	ASI		cAMP, biotin, Na^+, Cl^-
ASE		ASG	ASI ASK		lysine
					DAUER FORMATION
	ADF	ASG	ASI	ASJ	any one of them can prevent constitutive dauer formation
	ADF			ASJ	recovery from dauer
					AVOIDANCE
ASH					high osmotic strength
	ADL				garlic and avoidance factor avoidance
					VOLATILE ODORANTS DETECTION
AWC					benzaldehyde, butanone, isoamyl alcohol
		AWA			diacetyl, pyrazine

*The data were collected from various authors and the relevant references can be found in the text.

AVOIDANCE

In our laboratory we have concentrated our studies on the avoidance response. When it encounters a repellent chemical, *C. elegans* stops and reverses. This avoidance reflex seems, at present, to be the simplest response to a chemical signal that can be studied in *C. elegans*. It is probably sustained by a very simple circuit viz. the chemosensory neurons ASH and ADL whose ciliated endings terminate in the amphidial channel, the interneurons AVA and AVE which drive backward movement and the VA and DA motorneurons. All other behaviors require some level of temporal, spatial and discriminatory integration. Chemotaxis requires the worm to monitor the concentration of the attractant that the two amphids sense at the two sides of the mouth. Moreover, in order to move along the gradient in the appropriate direction, the animal has to integrate the change in the attractant concentration over time. Dauer formation also requires the integration of at least two opposite signals, the pheromone and the food signal. Avoidance, on the other hand, may be thought of as entailing the binding of enough ligand molecules to specific receptors on the sensory cells to produce an effective signal capable of activating the interneuron AVA which in turn activates DA and VA motoneurons so that the animal moves backward. ASH and ADL, which are the two amphidial neurons implicated in documented avoidance responses, are the only two amphidial neurons that synapse directly onto AVA and AVE (White et al., 1986). The avoidance response to water soluble repellents is, in this sense, probably different from repulsion, which has been studied for volatile chemicals and probably could better be described as negative chemotaxis (Bargmann et al., 1993).

The stimuli so far identified, capable of triggering the avoidance reflex are: high osmotic strength (Culotti and Russell, 1978), a factor present in aqueous extracts of garlic, a factor present in *C. elegans* extracts, low pH, SDS, Cu^{++}(reviewed in Bargmann et al., 1990), quinine, quinidine and other antimalarial drugs, (our own unpublished results). Avoidance has been tested according to a scheme devised originally by Culotti and Russell (1978) and consists of placing animals in a small ring of repellent on a plate. Animals sensing the repellent will stay within the ring until either they adapt or the concentration of the repellent becomes lower than the threshold to trigger the avoidance response. The assay has been used to identify repellents, to describe the behavior of mutant strains, to select for mutants which, unable either to sense a repellent or respond to it, cross the ring. However, when used to determine the genotype of single worms in genetic crosses, the assay has proved, with few exceptions, either unreliable or cumbersome. The study of the cellular elements involved is being pursued in laser ablation experiments (Bargmann, Thomas and Horvitz, personal communication). The genetic and molecular analysis of avoidance promises to yield interesting information on chemical sensitivity.

GENETIC STUDIES

Our study of avoidance has entailed primarily a mutational analysis in an attempt to identify the components necessary for this response. Genetic mutational analysis is, in *C. elegans*, an extremely powerful tool and has therefore already been used to identify genes necessary for chemoreception. The molecular genetics tools developed for the study of *C. elegans* make it possible to start cloning projects once a gene has been properly defined by formal genetic analysis. The identification of mutants is the first necessary step. The chemoreception mutants that have been isolated so far can be divided into two main groups:

i) mutants isolated on the basis of behavioral phenotypes. These include mutants abnormal for chemotaxis, *che* (Dusenbery et al., 1975; Lewis and Hodgkin, 1977), osmotic avoidance, *osm* (Culotti and Russell, 1978), dauer formation, *daf* (Riddle, 1988; Vowels and Thomas, 1992), volatile chemical sensitivity, *odr* (Bargman et al., 1993).

ii) mutants isolated on the basis of a cellular phenotype. Perkins et al. (1986) have described the property of chemosensory neurons of live wild type *C. elegans* to take up fluoresceine isothiocyanate (FITC) from the medium. They found conditions in which only six neurons of each amphid and the two neurons of each phasmid take up the dye, transport it retrogradely and become fluorescent.

Mutants with alterations in this staining pattern have been isolated, *dyf* , for dye filling abnormality (Perkins et al., 1986; R. Herman, D. Riddle, J. Thomas, personal communication; our own results, unpublished). The Dyf phenotype is largely dependent on specific cellular properties of some chemosensory neurons and thus, in general, has generated chemoreception-specific mutants. In addition, since the Dyf phenotype can be detected on single live worms, it is easier to use this assay to identify amphid mutants than to use a behavioral screen.

Dyf Mutants

Many *dyf* mutants have been isolated through screenings that involved an enrichment step such as collecting worms unable to avoid high osmotic strength or unable to form dauers and then testing their dye filling phenotype. We screened directly for staining because we were interested in determining which behavioral phenotypes would accompany the lack of staining. In our screen we concentrated on the amphids and chose only mutants in which none of the cells, that normally do, would stain. We isolated 19 independent recessive mutants. We have tested them for chemotaxis to NaCl, for dauer formation, and for avoidance of various repellent stimuli. All the mutants are normal for sensitivity to light touch. Behavioral tests for some of them and for some mutants isolated by others are reported in Table 2.

Table 2. Behavior of *dyf* mutants

Strain	Dyf	Osm	Gar	Afa	Che	Daf	Mat	Mec
N2 w.t.	+	+	+	+	+	+	+	+
gb 294	+/-	+	+	+	-	+	+	+
gb 287	-	-	+	+	-	+	+	+
mn 335	-	-	+	+	-/+	+	+	+
gb 283	-	-	+	+	-/+	+	+	-/+
gb 284	-	-	+	+	-	+	+	+
gb 285	-	-	+	+	-	+	+	+
gb 293	-	-	+	+	-/+	+	+	+
gb 297	-	-	+	+	-/+	-/+	+	+
gb 305	-	-	+	+	-	+	+	+
gb 308	-	-	+	-/+	-	+	+	+
gb 281	-	-	-	-	-	+	+	+
gb 286	-	-	-	-	-	-/+	+	+
gb 302	-	-	-	-	-	-/+	+	+
gb 288	-	-	-	-	-	+	+	+

Dyf = FITC staining; Osm, Gar and Afa = avoidance of high osmotic strength, garlic extract and worm extract, respectively; Che = chemotaxis to NaCl; Daf = dauer formation; Mat = male mating; Mec = sensitivity to light touch. *gb* strains were isolated in our laboratory in Naples; *mn 335* is an allele of *dyf-1*

The most common defect in our *dyf* mutants is lack of osmotic avoidance, 16 of them being Osm. Six of the Osm mutants also lack or have reduced sensitivity to the factor present in garlic extracts, Gar phenotype, and to the 'avoidance factor', present in worm extracts, Afa phenotype. None of the mutants is Gar and not Afa or vice versa.

We have also studied in more detail *dyf-1*, a new gene identified by a mutant *mn335*, isolated by Todd Starich at the University of Minnesota. In addition to the lack of staining of chemosensory cells, *mn335* is partially unable to sense high osmotic strength, SDS, Cu^{++} and quinine. However, its ability to avoid garlic and avoidance factor are unaltered as is the ability to respond to light touch. E.M. reconstruction of the anterior sensory organs shows that the cilia of the amphids are shorter than in the wild type with an amphidial channel almost completely collapsed in its terminal part. There is also an accumulation of a dense matrix between the cilia and in the sheath cell cytoplasm. It is possible that the primary defect lies in the capacity of sheath cells to secrete the matrix. A possible explanation for the phenotypes of *mn335* and of most Dyf mutants is that the general architecture and proper functioning of the amphidial channel are necessary to transform the high osmotic strength of the medium into a secondary signal capable of stimulating the ASH neuron, which has been shown to mediate this response (Bargmann et al., 1990). In Dyf mutants this architecture is frequently altered, hence the Osm phenotype. The sensitivity to other repellents such as the factor present in garlic extracts or that in worm extracts may be more directly dependent on specific receptors, and it is at least partially conserved in mutants with altered cilia. It is possible that mutations that are Gar and not Afa or vice versa can only be found in a few genes, such as those coding for the specific receptors, and it is unlikely that mutations in such genes will confer a Dyf phenotype.

The Dyf phenotype is related to abnormalities in the structure, growth and physiology of the cilia and of the support cells of the sensilla, and mutants selected for this phenotype are interesting because they can help identify the relative genes. Cloning projects in several laboratories are leading to the molecular isolation of a few *dyf* genes (unpublished). In our laboratory we have identified and cloned the genomic region containing *dyf-1*. However, the limits of the gene, its structure and its sequence have still to be determined and thus at present we know nothing about the nature of its product.

Garlic Avoidance Mutants

Dyf mutants identify genes required in chemoreceptor neurons for their development, for their general physiology and for the assembly and growth of the cilia. As a consequence their phenotype is pleiotropic and their sensitivity to many chemicals is lost at the same time. In order to identify the genes that may code for specific receptors and for components of the mechanism of transduction of the signal to the post-synaptic neuron, we have started to isolate specific avoidance mutants, hoping to find among them those that have lost the ability to avoid a certain chemical but that are still able to sense and avoid others. We have isolated garlic avoidance mutants by screening for worms that would cross a ring of the repellent and head towards a spot of bacteria functioning as an attractant.

Table 3 shows the result of behavioral assays on these mutants. Garlic avoidance was tested measuring the percent of worms crossing a barrier of the repellent at 15 and 30 minutes. The mutants are, with few exceptions, normal with respect to dauer formation, male mating and sensitivity to light touch. However, as shown in Table 3, more than half of the mutants also fail to avoid high osmotic strength. This is somewhat unexpected since only one mutant (*gb371*) has a weak Dyf phenotype, sometimes staining poorly. It is possible that with this screen we have identified components necessary for the avoidance response and not essential for other behaviors. It is also worth noting that, for several mutants, the percent of worms crossing the garlic barrier early in the assay is not significantly different from wild type and that only at 30 minutes is the mutant phenotype apparent. This indicates that the mutants still sense the repellent, but either the threshold is higher than in the wild type, or the worms adapt more easily. At least in the case of NA371 and NA380, we were able to show that their threshold for garlic as well as for other repellents is higher.

Table 3. Phenotypes of garlic avoidance mutants

Strain	Gar	Osm	Mec	Daf	Dyf
N2 w.t.	+ (1-5)	+ (5)	+	+	+
NA371	- (34-61)	- (44)	+	+	-/+
NA372	- (4-44)	+ (8)	+	+	+
NA373	- (8-50)	- (19)	+	+	+
NA374	- (10-30)	+ (5)	+	+	+
NA375	- (32-55)	+ (7)	+	+	+
NA376	- (22-38)	- (42)	+	+	+
NA377	- (28-45)	- (51)	+	+	+
NA378	- (41-73)	- (47)	-/+	-/+	+
NA379	- (19-34)	+ (6)	+	+	+
NA380	- (8-30)	- (38)	+	+	+
NA381	- (3-41)	+ (8)	+	+	+
NA382	- (21-48)	- (32)	+	+	+

Numbers in parentheses are percent of worms crossing the ring of repellent. In the case of garlic the two numbers represent the percent of worms crossing the barrier after 15 and 30 minutes. Gar and Osm = avoidance of garlic and high osmotic strength, measured after 30 minutes; Mec = response to light touch; Daf = dauer formation which was estimated qualitatively; Dyf = staining of amphid neurons with FITC. NA strains were isolated in our laboratory after EMS mutagenesis.

CONCLUSIONS

In vertebrates various chemicals are sensed by various cells located in various structures. In nematodes the major sensory structure, the amphid, has been shown to be multifunctional. Chemotaxis, chemical avoidance, odorant reception, dauer formation, thermotaxis and, recently, some aspects of mechano-reception have all been shown to involve sensory cells of the amphids.

Repellent substances identified so far are all sensed, in *C. elegans*, by neurons present in the amphids. Laser ablation experiments have shown that the amphid cells involved in sensing volatile repellents (AWA and AWC) are largely different from those sensing water soluble repellents, mainly ASH and ADL (Bargmann et al., 1993; Thomas and Horvitz, personal communication).

Genetic analysis has shown that it is possible to identify chemoreception-specific genes. *dyf* genes code for components necessary for the growth, assembly and general physiology of chemosensory neurons, of their cilia and of their support cells. Some of the mutants selected on the basis of behavioral alterations may identify genes coding for specific receptors and for components of signal transduction mechanisms.

The repellent stimuli capable of triggering the avoidance reflex can be divided into those that are toxic to worms and to a variety of other organisms, e.g. high osmotic strength, SDS, Cu^{++}, low pH, quinine; and those that are not toxic but are presumably signals of dangerous environmental conditions, such as garlic extract and avoidance factor. These two different groups of substances may be sensed by quite different mechanisms operating within the same cells (ASH and ADL). The ability to sense the first group of substances is certainly very primitive and may involve mechanisms in which the stimulus directly affects the polarity of the membrane. The avoidance of the second group of substances may involve more sophisticated mechanisms such as specific receptors and signal transduction pathways to produce the activation of the post-synaptic neuron. The avoidance phenotypes of *dyf* mutants also correlate with this distinction. *dyf* mutants almost always lose the ability to sense high osmotic strength, and usually the ability to sense SDS, Cu^{++} and low pH, whereas they can usually still respond to garlic extract or avoidance factor (for instance alleles of *che-3, che-12, osm-3, osm-6* and *dyf-1*).

It has been known for a long time that parasitic nematodes are repelled by a variety of substances, often produced by plants and naturally present in the environment. The effect of garlic extract on *C. elegans* is just an example of a more common phenomenon. The elucidation at the behavioral and at the molecular level of the mechanisms for chemical avoidance in *C. elegans* may suggest new ways to interfere with the ability of parasitic nematodes to find the host, infect it and produce damage.

Acknowledgements

This work was in part supported by the EEC program STD, contract CT92-0096-I, and by CNR Project, Biotecnologie e Biostrumentazione, to P. B. Support was also received from CIRPS, Università di Roma and by COMIPA, Roma to P.B. and to F.M. We thank the *C. elegans* community for sharing many unpublished results.

REFERENCES

Bargmann, C.I., Hartweig, E. and Horvitz, R.H., 1993, Odorant-selective genes and neurons mediate olfaction in *C. elegans*, *Cell* **74**: 515.

Bargmann, C.I. and Horvitz, H.R., 1991a, Control of larval development by chemosensory neurons in *Caenorhabditis elegans*, *Science* **251**: 1243.

Bargmann, C.I. and Horvitz, H.R.,1991b, Chemosensory neurons with overlapping functions direct chemotaxis to multiple chemicals in *C. elegans*, *Neuron* **7**:729

Bargmann, C.I., Thomas J.H. and Horvitz, H.R., 1990, Chemosensory cell function in the behavior and development of *Caenorhabditis elegans*, *Cold Spring Harbor Symp. Quant. Biol.* **LV**: 529.

Chalfie, M. and White, J.G., 1988, The nervous system In "The Nematode Caenorhabditis elegans", W.B.Wood ed. Cold Spring Harbor Lab. C.S.H. New York.

Culotti, J. and Russell, D., 1978, Osmotic avoidance defective mutants of the nematode *Caenorhabditis elegans*, *Genetics* **90**: 243.

Dusenbery, D.B., Sheridan, R.A. and Russell, R.L, 1975, Chemotaxis defective mutants of the nematode *Caenorhabditis elegans, Genetics* **80**: 297.
Kaplan, J.M. and Horvitz, H.R., 1993, A dual mechanosensory and chemosensory neuron in *Caenorhabditis elegans., Proc. Natl. Acad. Sci. USA* **90**: 2227.
Lewis, J. and Hodgkin, J., 1977, Specific neuroanatomical changes in chemosensory mutants of the nematode *Caenorhabditis elegans, J. Comp. Neurol.* **172**: 489.
Perkins, L.A., Hedgecock, E.M., Thomson, N. and Culotti, J.G., 1986, Mutant sensory cilia in the nematode *Caenorhabditis elegans, Dev. Biol.* **117**: 456.
Riddle, D.L., 1988, The dauer larva In "The Nematode Caenorhabditis elegans", W.B. Wood ed. Cold Spring Harbor Lab. C.S.H. New York.
Sulston, J.E., Schieremberg, E., White, J.G. and Thomson, N., 1983, The embryonic cell lineage of the nematode *Caenorhabditis elegans, Dev. Biol* **100**: 64.
Trent, C.N., Tsung, N. and Horvitz, H.R., 1983, Egg-laying defective mutants of the nematode *Caenorhabditis elegans, Genetics* **104,** 619.
Vowels, J.J. andThomas, J.H., 1992, Genetic analysis of chemosensory control of dauer formation in *Caenorhabditis elegans, Genetics* **130**: 105.
Ward, S., 1973, Chemotaxis by the nematode *Caenorhabditis elegans*: Identification of attractants and analysis of the response by the use of mutants, *Proc. Natl. Acad. Sci. USA* **70**: 817.
Ward, S., Thomson, N., White, J.G. and Brenner, S., 1975, Electron microscopical reconstruction of the anterior sensory anatomy of the nematode *Caenorhabditis elegans, J. Comp. Neurol.* **160**: 313.
White, J.G., Southgate, E., Thomson, J.N. and Brenner, S., 1986, The structure of the nervous system of the nematode *Caenorhabditis elegans, Philos. Trans. R. Soc. London, Ser. B* **314**: 1.
Wood, W.B. and the Community of C. elegans researchers, eds.,1988, "The Nematode Caenorhabditis elegans" Cold Spring Harbor Lab. C.S.H. New York.

GENETICS OF *MELOIDOGYNE* VIRULENCE AGAINST RESISTANCE GENES FROM SOLANACEOUS CROPS

Philippe Castagnone-Sereno

INRA
Laboratoire de Biologie des Invertébrés
BP 2078
06606 Antibes Cédex, France

INTRODUCTION

The genetic basis of plant-pathogen interaction is built on the gene-for-gene concept (Flor, 1942), which states that for each gene conferring resistance in the host plant, there is a matching or complementary gene in the pathogen, called an avirulence gene. In this hypothesis, both plant resistance gene and pathogen avirulence gene are dominant (Gabriel and Rolfe, 1990), and only the confrontation between both dominant alleles results in the hypersensitive reaction in the plant, often associated with a cascade of other defence responses (Figure 1). The gene-for-gene complementarity occurs most frequently in plant-pathogen interactions involving obligate and biotrophic parasites which are highly specialized and have a narrow host range (Heath, 1981; Keen, 1982). Since Flor's work in the 1940s, many avirulence genes have been identified by classical genetic studies in plant-pathogen interactions including viruses, bacteria, fungi, insects and nematodes (Sidhu, 1987), but only recently have they been cloned and characterized in the cases of viruses, bacteria, and more recently fungi. The first bacterial avirulence gene cloned was *avrA* from *Pseudomonas syringae* (Staskawicz et al., 1984), the first viral avirulence gene cloned was the coat protein gene of the tobacco mosaic virus (Culver and Dawson, 1991), and the first fungal avirulence gene cloned was *avr9* from *Cladosporium fulvum* (Van Kan et al., 1991). All of the currently known avirulence genes encode single protein products of various sizes, but except for the coat protein of the tobacco mosaic virus, sequencing of avirulence genes has not yet resulted in determining their functions in the pathogen nor the mechanisms by which they elicit the plant hypersensitive reaction.

In order to overcome the plant resistance, the parasite must evade recognition by the plant, using an as yet unknown mechanism. Within plant pathogenic bacteria, the evasion of recognition can be achieved at least in three different ways: (i) the avirulence gene may be mutated (Kobayashi et al., 1990); (ii) the avirulence gene may be disrupted by a transposable element (Kearney et al., 1988); (iii) the avirulence gene may be deleted (Staskawicz et al., 1984). All these situations lead to the virulence of the pathogen on the resistant plant, without any specific virulence gene being involved. Even if virulence factors that allow pathogenesis on certain cultivars or plant species have been identified, it must be noticed that no virulence gene *sensus stricto* has been cloned so far (Keen, 1992).

Advances in Molecular Plant Nematology
Edited by F. Lamberti *et al.*, Plenum Press, New York, 1994

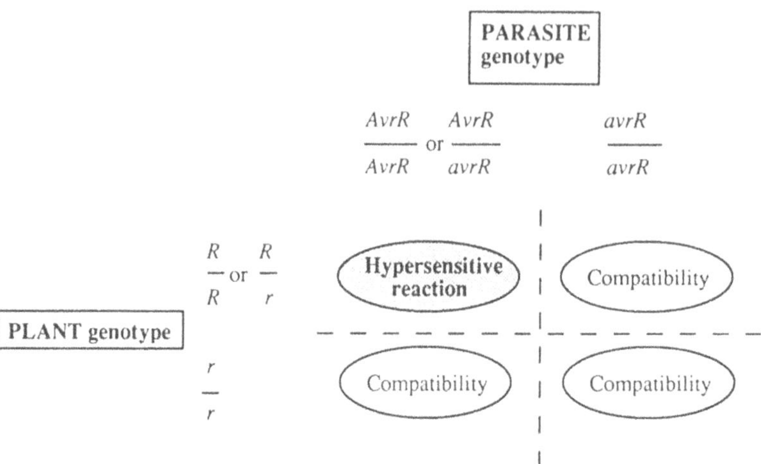

Figure 1. Interaction between the plant resistance gene (*R*) and the complementary avirulence gene (*Avr*) of the parasite. Only the occurrence of both dominant alleles results in the expression of plant resistance (which typical phenotypic expression is the hypersensitive reaction). In any other case, the interaction between the plant and the parasite is compatible, and the parasite can develop and reproduce on the plant.

The concept of virulence, and its associated terminology, has been often discussed among nematologists. In some instances, it has been suggested that 'virulence' should be replaced by 'parasitism' to describe the ability of nematodes to overcome the effects of genes for resistance (Triantaphyllou, 1987), this term measuring rates of development and reproduction. But this definition appears confusing, as parasitism should apply to interactions with both susceptible and resistant plants on which nematodes develop and reproduce themselves. In this paper, the definition of virulence in case of plant-parasitic nematodes will be the one generally accepted among phytopathologists (Shaner et al., 1992), which presents it as the ability of a pathogen to circumvent a given resistance gene in the plant. Moreover, this definition is in agreement with recent reviews dealing with plant resistance to nematodes (Trudgill, 1991; Dalmasso et al., 1992). In any case, the term 'virulence' should not be used on a relative basis, and should refer to a well-identified resistance gene in the plant.

In recent years, many studies on the ability of nematodes to reproduce on plants carrying resistance gene(s) have been developed, but some interactions have been analysed more extensively, mainly due to a better knowledge of the genetic support of the plant resistance. Moreover, research has focused largely on sedentary endoparasites, i.e. root-knot species (*Meloidogyne* spp.) and cyst-forming nematodes (*Heterodera* and *Globodera* spp.), because plants essentially resist them by the hypersensitive reaction, as they do for viruses, bacteria and fungi. In that way, these nematodes can be considered as true endocellular parasites. A gene-for-gene interaction between soybean and the soybean cyst nematode, *Heterodera glycines*, has been documented (Luedders, 1983; Triantaphyllou, 1987), in which both plant resistance genes and nematode avirulence genes surprisingly appear to be recessive (Luedders, 1987; Luedders, 1990). But definitive genetic studies are complicated by the use of soybean lines with many genes for resistance to heterogeneous amphimictic nematode populations (Luedders, 1989). Virulence in the beet cyst nematode (*H. schachtii*) versus resistance genes in *Beta* genotypes has been recently identified (Müller, 1992; Lange et al., 1993). Most work deals with the interaction between *Solanum* spp. and potato cyst nematodes (*Globodera* spp.), which resistance-virulence pattern is of a quantitative nature fitting with a polygenic system (Nijboer and Parlevliet, 1990). Selection for virulence was demonstrated for *G. pallida* on resistant *S. vernei* hybrids (Turner et al., 1983; Whitehead, 1991), and for *G. rostochiensis* against the *H1* resistance gene in *S. tuberosum andigena* (Janssen et al., 1990). The first Mendelian proof for a gene-for-gene relationship involving a plant-parasitic nematode was provided on that model, with virulence being controlled by a single recessive gene (Parrott, 1981; Janssen et al., 1991). The construction of a linkage map with markers flanking the locus of interest has been chosen as a strategy to clone the gene for avirulence to *H1* in *G. rostochiensis* (Gommers et al., 1992; Bakker et al., 1993). On a genetic point of view, it is to be outlined that all these nematodes share amphimixis as their mode of reproduction.

On the other hand, extensive studies have also been conducted on virulence in root-knot nematodes of the genus *Meloidogyne*, which main species reproduce exclusively by mitotic parthenogenesis. Therefore, this paper will focus on plant-nematode interactions involving root-knot nematodes and their solanaceous hosts, with emphasis on the genetic aspects of the relationships between *M. incognita* and the *Mi* resistance gene of tomato. Moreover, emerging molecular strategies for cloning of (a)virulence gene(s) will be discussed, in relation to the characteristic features of this model system.

GENETIC ANALYSIS OF *MELOIDOGYNE INCOGNITA* VIRULENCE

Plant resistance is actually the most efficient and environmentally safe way to control root-knot nematodes of the genus *Meloidogyne*, especially on vegetables, for which many resistance genes are available (Fassuliotis, 1979; Fassuliotis, 1987). Among the most important solanaceous crops, resistance has been identified and characterized in tomato (Bailey, 1941; Gilbert and MacGuire, 1955), pepper (Hare, 1956; Hare, 1957; Hendy et al., 1985), potato (Gomez et al., 1983; Mendoza and Jatala, 1985), egg-plant (Daunay and Dalmasso, 1985) and tobacco (Clayton et al., 1958; Slana and Stavely, 1981). In most, if not all cases, the histological manifestation of the plant resistance is the hypersensitive reaction localised around the infection site, a response common to that elicited byother plant-parasitic agents like viruses, bacteria and fungi (Bell, 1981).

The increasing use of cultivars resistant to *Meloidogyne* spp. has nevertheless been associated with many reports on the occurrence of field populations able to reproduce on these plants. On tobacco, Graham (1969) showed that an *M. incognita* biotype was able to attack the resistant cultivar NC 95. More recently, an *M. arenaria* population was also found able to overcome the resistance of two other tobacco cultivars (Noe, 1992). Because tomato is associated with many agronomical systems worldwide, and because *Meloidogyne* is a major pest for this culture, extensive information on the occurrence of nematode virulence is available for this interaction. In 1973, Sikora et al. found a virulent *M. javanica* population in India. The same observations were reported for African isolates of *M. incognita* and *M. javanica* on the resistant cultivar Rossol (Taylor, 1975; Netscher, 1976). Comparison of ten North American *M. incognita* isolates showed that one was equally attacking two resistant cultivars and a susceptible one (Viglierchio, 1978). Even if monoculture of resistant tomatoes may generate a selection pressure on the nematodes, infestations of resistant tomatoes in fields not usually grown with this crop are noticed. This is especially true for African populations of *M. javanica* (Netscher, 1976) and *M. arenaria* (Prot, 1984). Recently, a large survey in Senegal showed the occurrence of virulent isolates within the three main root-knot nematode species, but also in *Meloidogyne* populations of an undescribed species (Berthou et al., 1989). In France, virulent *M. arenaria* and *M. incognita* populations have been reported (Hendy et al., 1983; Jarquin-Barberena et al., 1991). These data provide evidence that populations collected in the same field may behave dramatically differently on different plant genotypes (such as resistant one), although they are undistinguishable otherwise. In our opinion, the general idea that *Meloidogyne* species can be divided into races according to their pathogenic behaviour in a differential host test (Sasser, 1979) therefore needs to be challenged, as already suggested by the results of earlier observations (Southards and Priest, 1973) and further analysis (Netscher, 1983; Triantaphyllou, 1987).

In addition to observations on naturally occurring resistance-breaking biotypes, many studies have been initiated, under controlled conditions, on the interaction between virulent *M. incognita* populations and the tomato *Mi* resistance gene. The main objectives of such works were i) to collect data on the establishment and stability of the virulence phenotype; and ii) to provide information on the inheritance and the genetic determinism of this character.

Artificial selection for virulence and establishment of near-isogenic lines

Artificial selection experiments have been conducted for many decades, with concordant results on the response of root-knot nematodes to selection on resistant cultivars. After transfer for successive generations on resistant tomato plants, new virulent *M. incognita* or *M. arenaria* populations were developed, designated as 'B populations' while the original populations maintained on susceptible tomatoes were designated as 'A populations' (Riggs and Winstead, 1959). Very few galls were found on the plant roots after initial inoculation, but galls became more numerous after each successive transfer on the resistant cultivar. The same result was obtained with successive transfers of several *M. incognita* isolates on resistant tomato and tobacco plants, which showed increased reproduction rates on resistant plants (Triantaphyllou and Sasser, 1960). However, it has been reported that not all nematode populations do have the ability to adapt on resistant tomatoes (Netscher and Taylor, 1979).

Comparison of juvenile (J2) root penetration, egg-mass production and female fecundity over successive generations on resistant tomatoes showed that virulence in *M. incognita* against the *Mi* resistance gene occurs through a step-by-step increase of the nematode reproduction (Bost, 1982; Jarquin-Barberena et al., 1991). Considering the ameiotic parthenogenetic mode of reproduction of this species, the mechanism(s) underlying the selection of virulent clones remain(s) unclear. On the other hand, the theoretical absence of recombination events during *M. incognita* reproduction should therefore account for the establishment of near-isogenic avirulent and virulent lines when the starting population (before selection) originates from a single J2 or female (Jarquin-Barberena et al., 1991). The availability of such a biological material should be considered an advantage, and its use should provide useful information on the mechanisms involved in nematode virulence, as will be discussed below.

Genetic analysis of the variability and inheritance of *Meloidogyne incognita* virulence against the tomato *Mi* resistance gene

It is clear that, because of obligatory parthenogenesis, no controlled mating experiments can be conducted with *Meloidogyne*. Therefore, the variability and mode of inheritance of virulence in this nematode has never been investigated by means of classical Mendelian genetics, unlike the crossing approach used for the establishement of a gene-for-gene relationship between *G. rostochiensis* virulence and the *H1* resistance gene in potato (Janssen et al., 1991).

Based on studies on inheritance of virulence in several single-egg-mass isolates, and on observations showing that increase of the rate of reproduction on resistant tomatoes was a step-wise process, Triantaphyllou (1987) suggested that each step resulted to small-effect mutation of one or more genes that control virulence, increasing the probability of a given J2 parasitizing the resistant plant over its female parent. As an unusually high frequency (10^{-3}) of such mutations was observed, this author also assumed that virulence should be governed by a polygenic system, but without any strong argument to demonstrate this hypothesis.

To test the hypothesis that a genetic determinism may be involved in *M. incognita* virulence against the *Mi* gene, we used experimental procedures derived from the 'isofemale line methods' of Parsons (1980) (Castagnone-Sereno et al., 1994). Each isofemale line (i.e. family) is founded by a single female. Then the character to be studied (i.e. virulence) is measured on several individuals in each line. Finally the mean values are compared with a standard ANOVA procedure. If the null hypothesis is rejected, the character can thus be considered as a family feature, which strongly suggests the occurrence of a genetic basis for the observed variability. The production of egg-masses on 'Piersol' resistant tomatoes was assessed for two successive generations of J2s that had never been submitted to the selection pressure of the *Mi* gene before (natural avirulent nematodes for the mother generation, nematodes reared on the susceptible cultivar used as control for the daughter generation) (Figure 2). Substantial variations among the 63 isofemale lines analyzed were observed for both generations, and results of the ANOVA performed on these data indicated that, for both generations, the variations observed in nematode virulence appeared to be a family feature. This strongly suggests that this variability is under genetic control. Moreover, there was a highly significant correlation between mothers and their daughters for the egg-mass numbers produced on 'Piersol', which indicated genetic inheritance of this character from one generation to the other. Finally, the occurrence of a continuous variation in the reproduction rates observed on the resistant tomatoes may suggest that a polygenic system is involved.

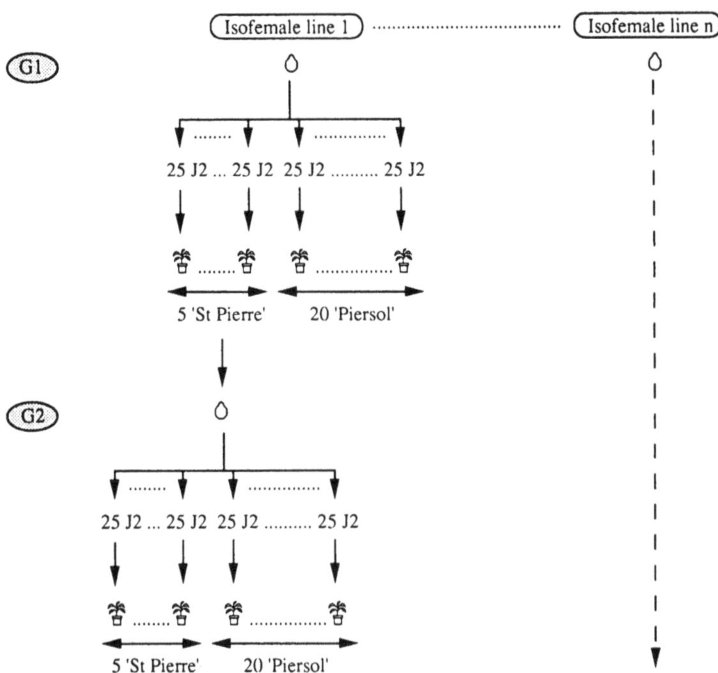

Figure 2. Schematic representation of the experimental procedures designed to test the hypothesis that *Meloidogyne incognita* virulence against the *Mi* gene is under genetic control (Castagnone-Sereno et al., 1994). Each isofemale line was founded by a single female, which had never been selected on the resistant tomato before. In both the first and second generations, virulence was measured on many individuals in each line. 'St Pierre' is a susceptible tomato cultivar and 'Piersol' is a resistant tomato cultivar carrying the *Mi* gene. For more details see text.

Response of *M. incognita* isofemale lines to laboratory selection over four successive generations was also tested (Castagnone-Sereno et al., 1994). Two distinct responses could be observed in the kinetics of change for 31 families independently selected for highest egg-mass production on 'Piersol'. Two of them significantly showed a rapid increase in their ability to reproduce on resistant tomatoes, and both of these lineages produced a significantly higher mean egg-mass number after the first generation on the resistant tomato compared to the other lines. On the other hand, the remaining tested families expressed little or no response to selection, with 41, 52 and 62% of them even being lost after one, two or three generations on the resistant tomato respectively. An ANOVA, used on these data to determine the extent to which selection pressure versus genetic influences affect *M. incognita* virulence, showed that a significant interaction occurred between the isofemale line and generation terms, which indicates that response to the selective pressure of the *Mi* gene over successive generations is not the same for each genotype. This interaction was not significant when the ANOVA was run without the two families that strongly responded to selection, which is in agreement with the occurrence of a set of two families, one responding and one another not responding to selection. This result is in agreement with previous observations that not all nematode populations have the capacity to adapt to resistant tomatoes (Netscher and Taylor, 1979).

All these arguments brought together demonstrate both the occurrence of genetic variability in virulence against the tomato *Mi* resistance gene between isofemale lines and the genetic inheritance of this character from one generation to the other, thus confirming that it is genetically determined in the nematode population studied.

Stability of the virulence phenotype

Few experiments have been conducted to test whether virulent root-knot nematode populations revert back to a predominantly avirulent phenotype (i.e. lose their ability to reproduce on resistant plants) when selection pressure of the resistance gene is removed. Early work showed that virulence to Hawaii 5229 resistant tomatoes was not affected when several laboratory-selected *M. incognita* populations were maintened for three, six, or nine months on a susceptible cultivar (Riggs and Winstead, 1959). Recently, extensive studies, respectively conducted over 9 and 18 successive generations, demonstrated that virulence against the tomato *Mi* resistance gene was a stable character in *M. incognita*, even in the absence of the plant selection pressure (Jarquin-Barberena et al., 1991; Castagnone-Sereno et al., 1993). Over the whole experiment no significant differences in the egg-mass numbers produced on susceptible and resistant tomatoes were observed. These data are also in agreement with those obtained on the interaction between virulent *G. pallida* lines and resistance in *S. vernei* hybrids (Turner, 1990). Moreover, the ability of *M. incognita* to overcome the tomato resistance appeared not to be altered by maintenance on various other susceptible plants such as okra or cucumber (Riggs and Winstead, 1959).

Quite a different behaviour was nevertheless observed between artificially-selected and natural virulent populations propagated for 18 generations on a susceptible tomato cultivar (Castagnone-Sereno et al., 1993). In this experiment, the laboratory-selected *M. incognita* line showed the same ability to overcome the tomato resistance, at both the qualitative and quantitative levels, regardless of the cultivar on which it was reared. On the contrary, a naturally virulent population showed a higher reproduction rate when maintened on the resistant rather than on the susceptible tomato cultivar. This relative loss of reproductive ability might be related to an adverse genetic background associated withvirulence, which decreases the nematode fitness on non-resistant hosts. This process, commonly known as stabilizing selection, has already been proposed in gene-for-gene interactions between plants and pathogenic fungi (Van der Plank, 1982). Whatever the nature of the difference between the wild and selected virulent lines may be, it should be correlated with the way these two populations acquired their ability to overcome the plant resistance. The fact that one lineage was selected for virulence under managed pressure in greenhouse and climate-controlled room conditions may induce genetic changes different from those occurring in a natural *Mi*-resistance breaking *M. incognita* biotype.

Relationships of *Mi*-virulent *Meloidogyne incognita* populations with other plant resistance genes

According to the gene-for-gene theory the virulence of a parasite against a defined plant resistance gene is effective only in the case of this particular interaction. Some authors have tested the ability of *M. incognita* populations selected for virulence against one resistant plant species to reproduce on other resistant plant species. Riggs and Winstead (1959) inoculated resistant cultivars of a wide range of plant species, including many solanaceous crops (e.g. tobacco, pepper), with an *M. incognita* population selected for virulence on resistant tomatoes. None of these interactions appeared to be compatible, suggesting that the virulence was specific to the tomato resistance gene. Quite surprisingly, an opposite result was obtained when *M. incognita* clones built up on resistant tobacco were inoculated on resistant tomatoes (Triantaphyllou and Sasser, 1960). Reproduction of selected *M. incognita* lineages virulent to the *Mi* tomato resistance gene was also assessed on two resistant pepper lines respectively carrying the resistance genes *Me1* and *Me3* (Castagnone-Sereno et al., 1992). As the phenotypic expression of both *Mi* and *Me(s)* genes is characterised by the hypersensitive reaction, and as the gene repertoire of tomato and pepper is highly conserved (Tanksley et al., 1988), one should wonder whether *Mi* and *Me(s)* are homologous genes, which should imply that virulence against one gene would also confer virulence against the other. But as shown by the inability of *Mi*-virulent populations to reproduce on both resistant pepper genotypes, the homology between the two systems had to be rejected. This result nevertheless confirmed Riggs and Winstead's one (1959), both being in favour of the gene-for-gene specificity.

MOLECULAR STRATEGIES FOR THE CLONING OF *MELOIDOGYNE INCOGNITA* (A)VIRULENCE GENE(S)

As stated above, because of the inhability to conduct crosses in parthenogenetic organisms, no classical Mendelian genetic experiments can be developed to identify markers linked to the (a)virulence genes. Therefore, the only way to isolate such gene(s) is to develop a strategy based on differential analysis. For this purpose, the near-isogenicity between avirulent and virulent selected lineages is essential. Assuming that both lines are isogenic apart from their ability to reproduce on resistant tomatoes, one should therefore admit that any biochemical or molecular difference between them should be related to that character.

Molecular evidence for the near-isogenicity of avirulent and selected virulent *M. incognita* lines

A large fraction of repetitive DNA in eukaryotic cells consists of non-coding regions, which are not submitted to any selection pressure, and therefore subject to extensive evolutionary changes. For that reason, search for polymorphisms in the repetitive fraction of *M. incognita* genome appears as an accurate way to assess the isogenicity of avirulent and selected virulent lines at the molecular level.

To compare avirulent and virulent lineages of *M. incognita* (selected according to the procedure of Jarquin-Barberena et al. (1991)), we hybridized Southern blots of genomic DNA from both lines, digested with a set of restriction endonucleases and electrophoresed through an agarose gel, with two kinds of probe (unpublished data). The first one consisted of the whole genomic DNA of each of the two lineages which, by hybridization with itself, allows only the repetitive sequences to appear on the autoradiography. The second kind of probes consisted of a set of middle-repetitive DNA fragments cloned at random from *M. incognita* , which exhibited polymorphism between *M. incognita* geographic populations (Castagnone-Sereno et al., 1991; Piotte et al., 1992; Castagnone-Sereno et al., 1993). In all cases hybridization patterns showed little or no polymorphism between avirulent and virulent lines (Figure 3), thus confirming at least the strong homology between both genomes. This result, in complement to the ameiotic parthenogenetic mode of reproduction of *M. incognita*, strongly suggests that, starting from a single J2, the selection of a virulent line from an avirulent one indeed leads to the establishment of a couple of near-isogenic genotypes. This biological material seems therefore to be suitable for a differential analysis.

Testing the hypothesis of gene amplification related to virulence

Recent studies demonstrated that amplification of specific DNA sequences is a common mechanism for adaptation of eukaryotic cells to a variety of selective conditions. In particular, gene amplification was discovered in drug-resistant mouse cultured cells (Roninson, 1983), and also in whole, normally developed insects as a response to selection by pesticides in field conditions (Mouchès et al., 1986). Variation in chromosome numbers and ploidy levels is not uncommon and well documented in the genus *Meloidogyne* (see Triantaphyllou, 1985, for a review), and could be associated withsome kind of selective amplifications of virulence alleles in some lineages, in relation with the stepwise acquisition of nematode virulence over successive generations. As gene amplification was also found to be correlated with insecticide-resistance level in the genome of a parthenogenetic aphid (Field et al., 1988), the involvement of amplified DNA sequences associated with the virulence phenotype in *Meloidogyne* was to be investigated.

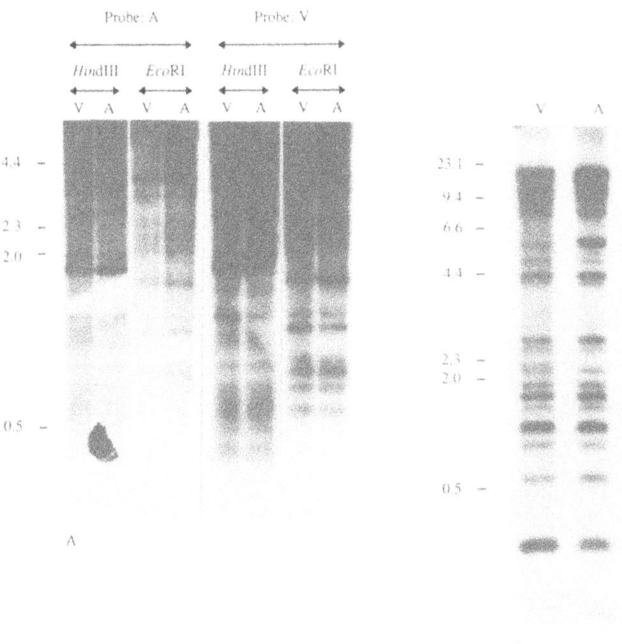

Figure 3. A: Hybridization pattern of genomic DNA from *Meloidogyne incognita* avirulent (A) and virulent (V) near-isogenic lines, digested with *Hin*dIII or *Eco*RI and hybridized with both total genomic DNAs. Molecular weights (on the left) are given in kilobases. B: Hybridization pattern of genomic DNA from *Meloidogyne incognita* avirulent (A) and virulent (V) near-isogenic lines, digested with *Bam*HI and hybridized with the repetitive probe 13.15 (Piotte et al., 1992; Castagnone-Sereno et al., 1993). Molecular weights (on the left) are given in kilobases. In each experiment, the lack of strong polymorphism between the two lines confirms the high homology between their respective genomes.

In that connection, we compared DNAs of near-isogenic avirulent and virulent *M. incognita* lines using an in-gel renaturation technique (Roninson, 1983), to test whether some specific amplified sequences could be correlated with the nematode phenotype (unpublished data). The principle of the technique is the following: DNAs are first cleaved with an appropriate restriction endonuclease. An aliquot of the preparation is then labeled with radioactive deoxynucleotides using phage T4 DNA polymerase and restriction fragments are separated by agarose gel electrophoresis. While in the gel, the DNA fragments are denatured by alkaline treatment and renatured *in situ* under conditions allowing reannealing of only amplified restriction fragments. Non-renatured single-stranded DNA is further degraded with S1 nuclease and eluted by diffusion. The DNA fragments remaining in the gel correspond to repeated sequences, and are visualized after exposure to X-ray film. Whole genomic DNAs from near-isogenic avirulent and virulent *M. incognita* J2s were treated according to that procedure using a wide range of restriction endonucleases. It was thus possible to identify some repeated DNA sequences in the nematode genome, but none of them was preferentially found in one genotype *versus* the other (Figure 4). Even if unconclusive, these preliminary results should not be considered as definitive, at least for experimental reasons (other restriction enzymes to be tested, detection threshold of the technique limited to about one hundred gene copies,etc). Molecular evidence of amplification associated with *Meloidogyne* virulence is not yet available.

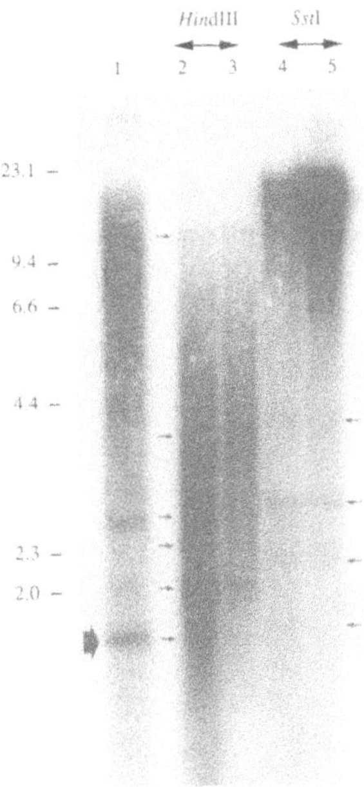

Figure 4. Detection of repeated fragments in *Meloidogyne incognita* genome by the in-gel renaturation technique. Lane 1: DNA of *Caenorhabditis elegans*, strain 'Bergerac', digested with *Eco*RV and used as positive control. The 1.6 kb Tc1 transposable element (large arrow), released by *Eco*RV, is present at 300 copies in the nematode genome. Lanes 2, 4: DNA of *M. incognita* avirulent line digested with *Hin*dIII and *Sst*I respectively. Lanes 3, 5: DNA of the near-isogenic *M. incognita* virulent line digested with *Hin*dIII and *Sst*I respectively. Molecular weights (on the left) are given in kilobases. Arrows indicate repeated sequences common to both *M. incognita* lineages.

Bidimensional electrophoretic comparison of near-isogenic lineages

One possible way to characterise genes involved in the (a) virulence phenotype could start from the isolation of proteins specific to that character. For that purpose, a high-resolving separation power is essential. Two-dimensional gel electrophoresis (2-DGE) is currently the most efficient method, with the report of up to seven thousand proteins separated in one single operation (Klose, 1989). Furthermore, 2-DGE gels can be run at high reproducibility over long periods of time, permitting direct computerized pattern analysis. It has been recently shown that the direct sequence analysis of proteins from 2-DGE gels is now technically feasible (Bauw et al., 1987; Eckerskorn et al., 1988), with the advantage that many low abundance proteins can only be isolated for sequence analysis by the use of polyacrylamide gel electrophoretic methods. The development of improved protein isolation/protein sequencing methods has opened up new approaches for the analysis of biological problems. In particular, the possibility to support gels with digital pattern storage and analysis systems can allow the identification of significant differences in related 2-DGE patterns at both the qualitative and quantitative levels. Subtractive, quantitative 2-DGE combined with protein sequence analysis allows the simultaneous analysis of the thousands of proteins resolved in a single operation, the identification of significant differences between the patterns and, in principle, the determination of partial sequence information on the resolved protein species. These sequence data can then be compared to those from sequence databases, and/or be used to design degenerate synthetic oligonucleotides in order to isolate, clone and sequence the gene coding for the differential protein previously identified.

Soluble proteins from two pairs of avirulent and virulent near-isogenic *M. incognita* lines, selected according to the procedure of Jarquin-Barberena et al. (1991), have been compared using 2-DGE combined with a sensitive silver stain (Dalmasso et al., 1991). Both natural avirulent populations originated from very different geographic locations (Europe and Africa), and each couple of avirulent/virulent lines was derived from a single J2. Of the four hundred spots resolved on the gels, only one half was taken into account for comparative analysis. One protein was found repetitively differential according to the avirulence/virulence of the two populations, independently of their geographic origin (Figure 5). The correlation between the absence of the protein and the nematode virulence in two independent lines suggests that this compound may be involved in the cascade of biochemical events leading to the compatible or incompatible reaction with the plant. From that point of view, this protein appears as a good candidate for the kind of analysis presented above.

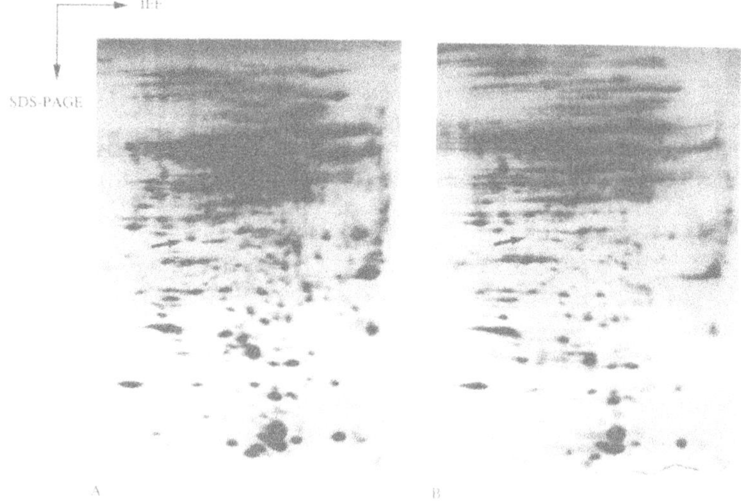

Figure 5. Silver stained gels of soluble proteins from *Meloidogyne incognita* females separated by two-dimensional electrophoresis. A: proteins from the avirulent original line from Africa, and B: proteins from the near-isogenic virulent line. The arrow indicates the protein differential between both lines (adapted from Dalmasso et al., 1991).

Molecular approaches

Some cloning strategies have been recently developed that require no assumptions on the biochemical or molecular nature of the character of interest. Those strategies are mainly devoted to the identification of differences between two closely related genotypes, among which one possesses the phenotype studied and the other not. If the two genomes are close enough, one could therefore hypothesize that most of the differences identified between them are related to that phenotype. From that point of view, couples of near-isogenic avirulent and virulent *M. incognita* lineages do constitute a biological support that seems well adapted to such kind of analysis. Differences may be observed either in the structure of the genomic DNA or in the expression of one or some specific genes.

One hypothesis concerning the acquisition of virulence by a parasite is that the gene governing avirulence may be deleted. In particular, this was demonstrated for the interaction between the bacteria *P. syringae* and soybean (Staskawicz et al., 1984). Methods for identifying and isolating target sequences present in one DNA population ('tester') that are absent or reduced in another ('driver') are called 'difference cloning'. Several variations have been described (Straus and Ausubel, 1990; Wieland et al., 1990), but the general procedure is the following. By many rounds of subtractive hybridization, a large excess of cleaved driver DNA is used to remove sequences common to cleaved tester DNA, enriching the target sequences that are unique to the tester DNA. After avidin/biotin affinity chromatography, the unbound DNA, corresponding in principle to target sequences, is amplified by the polymerase chain reaction (PCR) and subsequently cloned. Ideally, the two lineages compared should be isogenic, so that sequences covered by deletions occurring at irrelevant loci are not recovered. Use of isogenic lines also minimizes the chance of isolating repetitive sequences that are more abundant in one of the two lineages. The potential use of this method to clone virulence genes from a pathogen if a related non pathogenic strain exits has been suggested (Straus and Ausubel, 1990).

The hypothesis that a differential gene expression occurs between avirulent and virulent nematodes should also be considered. In that connection, we isolated poly(A)$^+$RNAs from both avirulent and near-isogenic virulent *M. incognita* J2s, synthesized first-strand cDNAs and constructed the two corresponding cDNA libraries in the phage vector lambda gt11 (unpublished data). From that material, we intend to apply several differential strategies to isolate cDNA clones presenting variations in their respective abundance level (Figure 6). In the first approach, we plan to perform a differential screening procedure on each of the two libraries. In this strategy, the cDNA library to be screened is plated onto a suitable medium in petri dishes, and the DNA of bacteriophages transferred by capillarity to nitrocellulose or nylon membranes. Duplicates for each plate are needed. Then one set of membranes is hybridized with the radioactively-labeled first-strand cDNA population of the avirulent line, and the other with those from the virulent line. Comparison of the hybridization patterns theoretically should allow the selection of clones differentially expressed between the two genotypes. Differential screening proved to be efficient in a number of studies (Cochran et al., 1983; Hara et al., 1988; Weigel and Nevins, 1990). In a second alternative, the objectives are to isolate *Meloidogyne* cDNA clones enriched in target sequences (i.e. sequences that are differential between both genotypes) by use of subtractive screening procedures (Hedrick et al., 1984; Travis et al., 1989), and to further amplify them by PCR after ligation with a choosen primer. Working on cDNA populations instead of genomic DNAs, this technique is very close to the one presented above for the identification of a deletion linked to virulence. The clones amplified may be 1) directly sequenced and characterized, 2) used as probes against both cDNA libraries, 3) cloned to construct a subtractive cDNA library specific to the avirulent or the virulent line.

Whatever the level at which the difference(s) between near-isogenic avirulent and virulent lines may be located, all these approaches, using J2 as biological material for DNA and poly(A)$^+$ RNA extraction, assume that this (or these) difference(s) is (are) expressed at the J2 stage with no induction from the plant needed. As J2 constitute the invading stage in the life cycle of *Meloidogyne*, and as the plant defense hypersensitive reaction is mainly expressed in the early stages of nematode invasion, one should therefore expect that at least some of the virulence factors are active before the J2 penetrate the root tissues. Moreover, from a technical point of view, working on *Meloidogyne* J2 after they have invaded the plant cells would be of considerable difficulty due to the minute amount of material available with the current techniques.

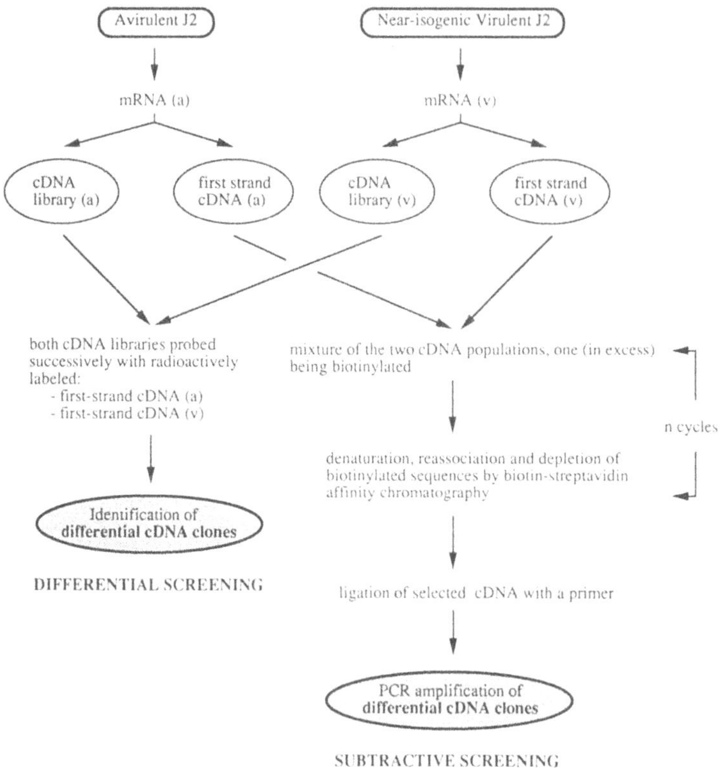

Figure 6. Molecular approaches to identify differential cDNA clones, based on the near-isogenicity between avirulent and selected virulent *Meloidogyne incognita* lines. For more details see text.

CONCLUSIONS AND PERSPECTIVES

Despite their mitotic parthenogenetic mode of reproduction, some *Meloidogyne* populations are able to overcome plant resistance genes. The existence of such virulence phenotypes suggests that avirulent apomictic nematodes may be able to modify their genotype in response to the selection pressure of resistance genes. As no controlled mating experiments can be conducted with this organism, development of new experimental procedures, derived from the 'isofemale line methods', demonstrated that a genetic determinism was indeed involved in the ability of some avirulent *M. incognita* populations to develop virulence against the tomato *Mi* gene, and indicated genetic inheritance of this character from one generation to the other. From that point, and due to the ability to select near-isogenic avirulent and virulent lines, cloning root-knot (a)virulence genes appears as a realistic challenge using difference-based molecular strategies.

Elucidating the molecular nature of (a)virulence genes may have important consequences for the future of management of root-knot nematode populations in the field, in particular for the improvement of the durability of plant resistance. The tomato cultivars commercially available worldwide are all derived from a common source (Medina-Filho and Tanksley, 1983), and the generalization of the use of resistant varieties should be balanced by engineering plants with new forms of resistance, taking into account the way *Meloidogyne* spp. are able to overcome the *Mi* gene. Moreover, the recent advances towards the cloning of the *Mi* gene (Messeguer et al., 1991; Ho et al., 1992) should also provide useful complementary information for the development of strategies to breed durable resistance against root-knot nematodes.

From a more fundamental point of view, understanding how *M. incognita* is able to modify its genotype in response to the selection pressure of the *Mi* resistance gene should provide further data on the possible role of nonmeiotic events responsible for genetic variation, which may facilitate adaptative evolution of such apomictic organisms.

Acknowledgments

The author thanks Drs. A. Dalmasso and G. De Guiran for helpful discussions and comments on the manuscript and C. Slagmulder for the photographic artwork.

REFERENCES

Bailey, D.M., 1941, The seedling test method for root-knot nematode resistance, *Proc. Am. Soc. Hortic. Sci.* 38:573.

Bakker, J., Folkertsma, R.T., Rouppe Van Der Voort, J.N.A.M., De Boer, J.M., and Gommers, F.J., 1993, Changing concepts and molecular approaches in the management of virulences genes in potato cyst nematodes, *Annu. Rev. Phytopathol.* 31:169.

Bauw., G., DeLoose, M., Inze, D., Van Montagu, M., and Vandekerckhove, J., 1987, Alterations in the phenotype of plant cells studied by NH_2-terminal amino acid-sequence analysis of proteins electroblotted from two-dimensional gel-separated total extracts, *Proc. Natl. Acad. Sci. USA* 84:4806.

Bell, A.A., 1981, Biochemical mechanisms of disease resistance, *Annu. Rev. Plant Physiol.* 32:21.

Berthou, F., Ba-Diallo, A., De Maeyer, L., and De Guiran, G., 1989, Caractérisation chez les nématodes *Meloidogyne* Goeldi (Tylenchida) de types virulents vis-à-vis du gène *Mi* de la tomate dans deux zones maraîchères au Sénégal, *Agronomie* 9:877.

Bost, S.C., 1982, "Genetic studies of the *Lycopersicon esculentum-Meloidogyne incognita* interaction," Ph. D. Dissertation, North Carolina State University.

Castagnone-Sereno, P., Bongiovanni, M., and Dalmasso, A., 1992, Differential expression of root-knot nematode resistance genes in tomato and pepper: evidence with *Meloidogyne incognita* virulent and avirulent near-isogenic lineages, *Ann. Appl. Biol.* 120:487.

Castagnone-Sereno, P., Bongiovanni, M., and Dalmasso, A., 1993, Stable virulence against the tomato resistance *Mi* gene in the parthenogenetic root-knot nematode *Meloidogyne incognita*, *Phytopathology* 83:803.

Castagnone-Sereno, P., Piotte, C., Abad, P., Bongiovanni, M., and Dalmasso, A., 1991, Isolation of a repeated DNA probe showing polymorphism among *Meloidogyne incognita* populations, *J. Nematol.* 23:316.

Castagnone-Sereno, P., Piotte, C., Uijthof, J., Abad, P., Wajnberg, E., Vanlerberghe-Masutti, F., Bongiovanni, M., and Dalmasso, A., 1993, Phylogenetic relationships between amphimictic and parthenogenetic nematodes of the genus *Meloidogyne* as inferred from repetitive DNA analysis, *Heredity* 70:195.

Castagnone-Sereno, P., Wajnberg, E., Bongiovanni, M., Leroy, F., and Dalmasso, A., 1994, Genetic variation in *Meloidogyne incognita* virulence against the tomato *Mi* resistance gene: evidence from isofemale line selection studies, *Theor. Appl. Genet.* in press.

Clayton, E.E., Graham, T.W., Tood, F.A., Gaines, J.G., and Clark, F.A., 1958, Resistance to the root-knot disease of tobacco, *Tobacco Sci.* 2:53.

Cochran, B.H., Reffel, A.C., and Stiles, C.D., 1983, Molecular cloning of gene sequences regulated by platelet-derived growth factor, *Cell* 33:939.

Culver, J.N., and Dawson, W.O., 1991, Tobacco mosaic virus coat protein genes produce a hypersensitive phenotype in transgenic *Nicotiana sylvestris* plants, *Mol. Plant-Microbe Inter.* 4:458.

Dalmasso, A., Castagnone-Sereno, P., and Abad, P., 1992, Tolerance and resistance of plants to nematodes. Knowledge needs and prospects, *Nematologica* 38:466.

Dalmasso, A., Castagnone-Sereno, P., Bongiovanni, M., and De Jong, A., 1991, Acquired virulence in the plant parasitic nematode *Meloidogyne incognita*.II. Two-dimensional analysis of isogenic isolates, *Rev. Nématol.* 14:305.

Daunay, M.C., and Dalmasso, A., 1985, Multiplication de *Meloidogyne javanica*, *M. incognita*, et *M. arenaria* sur divers *Solanum*, *Rev. Nematol.* 8:31.

Eckerskorn, C., Jungblut, P., Mewes, W., Klose, J., and Lottspeich, F., 1988, Identification of mouse brain proteins after two-dimensional electrophoresis and electroblotting by microsequence analysis and amino acid composition analysis, *Electrophoresis* 9:830.

Fassuliotis, G., 1979, Plant breeding for root-nematode resistance, *in*:"Root-knot Nematodes (*Meloidogyne* species). Systematics, Biology and Control," F. Lamberti and Taylor, C.E., eds., Academic Press, London.

Fassuliotis, G., 1987, Genetic basis of plant resistance to nematodes, *in*: "Vistas on Nematology," J.A. Veech and Dickson, D.W., eds., Society of Nematologists, Hyattsville.

Field, L.M., Devonshire, A.L., and Forbe, B.G., 1988, Molecular evidence that insecticide resistance in peach-potato aphids (*Myzus persicae* Sulz.) results from amplification of an esterase gene, *Biochem. J.* 251:309.

Flor, H.H., 1942, Inheritance of pathogenicity in *Melampsora lini*, *Phytopathology* 32:653.

Gabriel, D.W., and Rolfe, B.G., 1990, Working models of specific recognition in plant-microbe interactions, *Annu. Rev. Phytopathol.* 28:365.
Gilbert, J.C., and Mac Guire, D.C., 1955, One major gene for resistance to severe galling from *Meloidogyne incognita*, *Tomato Genet. Coop. Rep.* 5:15.
Gomez, P.L., Plaisted, R.L., and Brodie, B.B., 1983, Inheritance of the resistance to *Meloidogyne incognita*, *M. javanica* and *M. arenaria* in potatoes, *Am. Potato J.* 60:339.
Gommers, F.J., Roosien, J., Schouten, A., De Boer, J.M., Overmars, H.A., Bouwman, L., Folkertsma, R., Van Zandvoort, P., Van Gentpelzer, M., Schots, A., Janssen, R., and Bakker, J., 1992, Identification and management of virulence genes in potato cyst nematodes, *Neth. J. Plant Pathol.* Supp. 2:157.
Graham, T.W., 1969, A new pathogenic race of *Meloidogyne incognita* on flue-cured tobacco, *Tobacco Sci.* 13:43.
Hara, E., Nakada, S., Takehana, K., Nakajima, T., Iino, T., and Oda, K., 1988, Molecular cloning and characterization of cellular genes whose expression is repressed by the adenovirus *E1a* gene products and growth factors in quiescent rat cells, *Gene* 70:97.
Hare, W.W., 1956, Resistance in pepper to *Meloidogyne incognita acrita*, *Phytopathology* 46:98.
Hare, W.W., 1957, Inheritance of resistance to root-knot nematodes in pepper, *Phytopathology* 47:669.
Heath, M.C., 1981, A generalized concept of host-parasite specificity, *Phytopathology* 71:1121.
Hedrick, S.M., Cohen, D.I., Nielsen, E.A., and Davis, M.M., 1984, Isolation of cDNA clones encoding T cell-specific membrabe-associated proteins, *Nature* 308:149.
Hendy, H., Pochard, E., and Dalmasso, A., 1983, Identification de deux nouvelles sources de résistance aux nématodes du genre *Meloidogyne* chez le piment, *Capsicum annuum* L., *C.R. Acad. Agr. Fr.* 817.
Hendy, H., Pochard, E., and Dalmasso, A., 1985, Transmission héréditaire de la résistance aux nématodes *Meloidogyne* Chitwood (Tylenchida) portée par deux lignées de *Capsicum annuum* L. Etude de descendances homozygotes issues d'androgenèse, *Agronomie* 5:93.
Ho, J.Y., Weide, R., Ma, H.M., Van Mordragen, M.F., Lambert, K.N., Koornneef, M., Zabel, P., and Williamson, V.M., 1992, The root-knot resistance gene (*Mi*) in tomato: construction of a molecular linkage map and identification of dominant cDNA markers in resistant genotypes, *Plant J.* 2:971.
Janssen, R., Bakker, J., and Gommers, F.J., 1990, Selection of virulent and avirulent lines of *Globodera rostochiensis* for the H1 resistance gene in *Solanum tuberosum* ssp. *andigena* CPC 1673, *Rev. Nématol.* 13:265.
Janssen, R., Bakker, J., and Gommers, F.J., 1991, Mendelian proof for a gene-for-gene relationship between virulence of *Globodera rostochiensis* and the H1 resistance gene in *Solanum tuberosum* ssp. *andigena* CPC 1673, *Rev. Nématol.* 14:207.
Jarquin-Barberena, H., Dalmasso, A., De Guiran, G., and Cardin, M.C., 1991, Acquired virulence in the plant parasitic nematode *Meloidogyne incognita*.I. Biological anlysis of the phenomenon, *Rev. Nématol.* 14:299.
Kearney, B., Ronald, P.C., Dahlbeck, D., and Staskawicz, B.J., 1988, Molecular basis for evasion of plant host defence in bacterial spot disease of pepper, *Nature* 332:541.
Keen, N.T., 1982, Specific recognition in gene-for-gene host-parasite systems, *Adv. Plant Pathol.* 1:35.
Keen, N.T., 1992, The molecular biology of disease resistance, *Plant Mol. Biol.* 19:109.
Klose, J., 1989, Systematic analysis of the total proteins of a mammalian organism: principles, problems and implications for sequencing the human genome, *Electrophoresis* 10:140.
Kobayashi, D.Y., Tamaki, S.J., Trollinger, D.J., Gold, S., and Keen, N.T., 1990, A gene from *Pseudomonas syringae* pv. *glycinea* with homology to avirulence gene D from *P.s.* pv. *tomato* but devoid of the avirulence phenotype, *Mol. Plant-Microbe Inter.* 3:103.
Lange, W., Müller, J., and De Bock, T.S.M., 1993, Virulence in the beet cyst nematode (*Heterodera schachtii*) *versus* some alien genes for resistance in beet, *Fundam. Appl. Nematol.* 16:447.
Luedders, V.D., 1983, Genetics of the cyst nematode-soybean symbiosis, *Phytopathology* 73:944.
Luedders, V.D., 1987, A recessive gene for a zero cyst phenotype in soybean, *Crop Sci.* 27:604.
Luedders, V.D., 1989, Selection for zero cyst phenotypes with soybean, *Ann. Appl. Biol.* 114:509.
Luedders, V.D., 1990, A recessive soybean cyst nematode allele for incompatibility with soybean PI 88287, *Ann. Appl. Biol.* 116:321.
Medina-Fihlo, H.P., and Tanksley, S.D., 1983, Breeding for nematode resistance, *in*: "Handbook of Plant Cell Culture," Vol.I, I.D.A. Evans, Sharp, W.R., Ammirato, P.V., and Yamada, Y., eds., Macmillan, New York.
Mendoza, H.A., and Jatala, P., 1985, Breeding potatoes for resistance to the root-knot nematode *Meloidogyne* species, *in*: "An Advanced Treatise on *Meloidogyne*. Vol. I. Biology and Control," J.N. Sasser and Carter, C.C., eds., North Carolina State University Graphics, Raleigh.
Messeguer, R., Ganal, M., De Vicente, M.C., Young, N.D., Bolkan, H., and Tanskley, S.D., 1991, High resolution RFLP map around the root knot nematode resistance gene (*Mi*) in tomato, *Theor. Appl. Genet.* 82:529.

Mouchès, C., Pasteur, N., Bergé, J.B., Hyrien, O., Raymond, M., Robert de Saint Vincent, B., De Silvestri, M., and Georghiou, G.P., 1986, Amplification of an esterase gene is responsible for insecticide resistance in a California *Culex* mosquito, *Science* 233:778.

Müller, J., 1992, Detection of pathotypes by assessing the virulence of *Heterodera schachtii* populations. *Nematologica* 38:50.

Netscher, C., 1976, Observations and preliminary studies on the occurrence of resistance breaking biotypes of *Meloidogyne* spp. on tomato, *Cah. ORSTOM, Sér. Biol.* 11:173.

Netscher, C., 1983, Problems in the classification of *Meloidogyne* reproducing by mitotic parthenogenesis, *in*: "Concepts in Nematode Systematics," A.R. Stone, Platt, H.M., and Khalil, L.F., eds., Academic Press, London.

Netscher, C., and Taylor, D.P., 1979, Physiologic variation within the genus *Meloidogyne* ans its implication on integrated control, *in*: "Root-knot Nematodes (*Meloidogyne* species). Systematics, Biology and Control," F. Lamberti and Taylor, C.E., eds., Academic Press, London.

Nijboer, N., and Parlevliet, J.E., 1990, Pathotype-specificity in potato cyst nematodes, a reconsideration, *Euphytica* 49:39.

Noe, J.P., 1992, Variability among populations of *Meloidogyne arenaria*, *J. Nematol.* 24:404-414.

Parrott, D.M., 1981, Evidence for gene-for-gene relationships between resistance gene H1 from *Solanum tuberosum* ssp. *andigena* and a gene in *Globodera rostochiensis* and between H2 from *S. multidissectum* and a gene in *G. pallida*, *Nematologica* 27:372.

Parsons, P.A., 1980, Isofemale strains and evolutionary strategies in natural populations, *in*: "Evolutionary Biology," Vol. XIII, M. Hetch, Steere, W., and Wallace, B., eds., Plenum Press, New York.

Piotte, C., Castagnone-Sereno, P., Uijthof, J., Abad, P., Bongiovanni, M., and Dalmasso, A., 1992, Molecular characterization of species and populations of *Meloidogyne* from various geographic origins with repeated-DNA homologous probes, *Fundam. Appl. Nematol.* 15:271.

Prot, J.C., 1984, A naturally occurring resistance breaknig biotype of *Meloidogyne arenaria* on tomato. Reproduction and pathogenicity on tomato cultivars Roma and Rossol, *Rev. Nématol.* 7:23.

Riggs, R.D., and Winstead, N.N., 1959, Studies on resistance in tomato to root-knot nematodes and on the occurrence of pathogenic biotypes, *Phytopathology* 49:716.

Roninson, I. B., 1983, Detection and mapping of homologous, repeated and amplified DNA sequences by DNA renaturation in agarose gels, *Nucl. Acids Res.* 2:1911.

Sasser, J.N., 1979, Pathogenicity, host ranges and variability in *Meloidogyne* species, *in*: "Root-knot Nematodes (*Meloidogyne* species). Systematics, Biology and Control," F. Lamberti and Taylor, C.E., eds., Academic Press, London.

Shaner, G., Stromberg, E.L., Lacy, G.H., Barker, K.R., and Pirone, T.P., 1992, Nomenclature and concepts of pathogenicity and virulence, *Annu. Rev. Phytopathol.* 30:47.

Sidhu, G. S., 1987, Host-parasite genetics, *in*: "Plant Breeding Reviews," J. Janick, ed., Van Nostrand Reinhold, New York.

Sikora, R.A., Sitaramaiah, K., and Singh, R.S., 1973, Reaction of root-knot nematode-resistant tomato cultivars to *Meloidogyne javanica* in India, *Plant Dis. Reptr.* 57:141.

Slana, J.L., and Stavely, J.R., 1981, Identification of the chromosome carrying the factor for resistance to *Meloidogyne incognita* in tobacco, *J. Nematol.* 13:61.

Southards, C.J., and Priest, M.F., 1973, Variation in pathogenicity of seventeen isolates of *Meloidogyne incognita*, *J. Nematol.* 5:63.

Staskawicz, B.J., DahlbeckD., and Keen, N.T., 1984, Cloned avirulence gene of *Pseudomonas syringae* pv. *glycinea* determines race-specific incompatibility on *Glycine max* (L.) Merr., *Proc. Natl. Acad. Sci. USA* 81:6024.

Straus, D., and Ausubel, F.M., 1990, Genomic substraction for cloning DNA corresponding to deletion mutations, *Proc. Natl. Acad. Sci. USA* 87:1889.

Tanksley, S.D., Bernatzky, R., Laitan, N.L., and Prince, J.P., 1988, Conservation of gene repertoire but not gene order in pepper and tomato, *Proc. Natl. Acad. Sci. USA* 85:6419.

Taylor, D.P., 1975, Observations on a resistant and a susceptible variety of tomato in a field heavily infested with *Meloidogyne* in Senegal, *Cah. ORSTOM, Sér. Biol.* 10: 239.

Travis, G.H., Brennan, M.B., Danielson, P.E., Kozak, C.A., and Sutcliffe, J.G., 1989, Identification of a photoreceptor-specific mRNA encoded by the gene responsible for retinal degeneration slow (*rds*), *Nature* 338:70.

Triantaphyllou, A.C., 1985, Cytogenetics, cytotaxonomy and phylogeny of root-knot nematodes, *in*: "An Advanced Treatise on *Meloidogyne*. Vol. I. Biology and Control," J.N. Sasser and Carter, C.C., eds., North Carolina State University Graphics, Raleigh.

Triantaphyllou, A.C., 1987, Genetics of nematode parasitism on plants, *in*: "Vistas on Nematology," J.A. Veech and Dickson, D.W., eds., Society of Nematologists, Hyattsville.

Triantaphyllou, A.C., and Sasser, J.N., 1960, Variation in perineal patterns and host specificity of *Meloidogyne incognita*, *Phytopathology* 50:724.

Trudgill, D.L., 1991, Resistance to and tolerance of plant parasitic nematodes in plants, *Annu. Rev. Phytopathol.* 29:167.

Turner, S.J., 1990, The identification and fitness of virulent potato cyst-nematode populations (*Globodera pallida*) selected on resistant *Solanum vernei* hybrids for up to eleven generations, *Ann. Appl. Biol.* 117:385.

Turner, S.J., Stone, A.R., and Perry, J.N., 1983, Selection of potato cyst-nematodes on resistant *Solanum vernei* hybrids, *Euphytica* 32:911.

Van Der Plank, J.E., 1982, "Host Pathogen Interactions in Plant Disease," Academic Press, New York.

Van Kan, J.A.L., Van Den Ackerveken, G.F.J.M., and De Wit, P.J.G.M., 1991, Cloning and characterization of of cDNA of avirulence gene *avr9* of the fungal pathogen *Cladosporium fulvum*, causal agent of tomato leaf mold, *Mol. Plant-Microbe Inter.* 4:52.

Viglierchio, D.R., 1978, Resistant host responses to ten California populations of *Meloidogyne incognita*, *J. Nematol.* 10:224.

Weigel, R.J., and Nevins, J.R., 1990, Adenovirus infection of differentiated F9 cells results in a global shut-off of differentiation-induced gene expression, *Nucl. Acids Res.* 18:6107.

Whitehead, A.G., 1991, Selection for virulence in the potato cyst-nematode, *Globodera pallida*, *Ann. Appl. Biol.* 118:395.

Wieland, I., Bolger, G., Asouline, G., and Wigler, M., 1990, A method for difference cloning: gene amplification following substractive hybridization, *Proc. Natl. Acad. Sci. USA* 87:2720.

MOLECULAR APPROACHES TO AN UNDERSTANDING OF THE TRANSMISSION OF PLANT VIRUSES BY NEMATODES

Michael A. Mayo, Walter M. Robertson, Francesco J. Legorboru[1] and Karen M. Brierley[2]

Scottish Crop Research Institute
Invergowrie, Dundee DD2 5DA
United Kingdom

INTRODUCTION

In common with all parasites, viruses must have a means of moving from one host to another in order to survive. For the majority of plant viruses, this phase involves the assistance, either actively or passively, of a motile vector. For most viruses, this vector is an arthropod but a small number of viruses have other vectors, prominent among which are soil-inhabiting nematodes which feed on plant roots. However, although few viruses are involved, they can nevertheless cause some economically significant diseases.

The need to move between hosts is a potentially vulnerable phase in the life cycle of viruses and frequently this phase is where control measures are aimed. A detailed understanding of the mechanisms of virus transmission by vectors, including nematodes, is therefore worthwhile not only for its scientific interest but also as it may lead to new means of controlling important plant pathogens. This article reviews some recent advances in our knowledge of the molecular biology of nematode-transmitted viruses which may shed light on the mechanisms by which the transmission occurs.

---

[1] present address: Estacion de la Mejora de la Patata, Vitoria/Gastiez, Spain
[2] present address: Dept. of Agricultural Chemistry, University of Glasgow, UK

Advances in Molecular Plant Nematology
Edited by F. Lamberti *et al.*, Plenum Press, New York, 1994

THE VIRUSES AND THEIR VECTORS

The viruses transmitted by nematodes are of two distinct sorts. Those with rod-shaped particles are classified in the genus *Tobravirus* and those with isometric particles are classified in the genus *Nepovirus*. Taxonomically these genera are very different

Table 1. The nematode vectors of nepoviruses and tobraviruses

NEPOVIRUS VECTORS		TOBRAVIRUS VECTORS	
Longidorus spp.	**virus**	***Paratrichodorus spp.***	**virus**
L. apulus	AILV[1]	*P. allius*	TRV
L. arthensis	CRV	*P. minor*	TRV + PEBV + PRV
L. attenuatus	TBRV		
L. diadecturus	PRMV	*P. nanus*	TRV
L. elongatus	RRV + TBRV	*P. pachydermus*	TRV
L. fasciatus	AILV	*P. teres*	TRV + (PEBV)*
L. macrosoma	RRV	*P. tunisiensis*	TRV
L. martini	MRSV	*P. anemones*	(TRV + PEBV)
Xiphinema spp.	**virus**	***Trichodorus spp.***	**virus**
X. americanum	TobRV + TomRV + CRLV + PRMV	*T. viruliferus*	TRV + PEBV
		T. similis	TRV
		T. primitivus	(TRV + PEBV)
X. bricolensis	TomRV	*T. cylindricus*	(TRV)
X. californicum	TomRV + TobRV + CRLV		
X. diversicaudatum	ArMV + SLRV		
X. index	GFLV	* parentheses indicate that the evidence is not unequivocal	
X. italiae	GFLV		
X. rivesi	TomRV + TobRV + CRLV		

[1]**Abbreviations of virus names:** AILV – artichoke Italian latent virus; ArMV – arabis mosaic virus; CRV – cherry rosette virus; CRLV – cherry rasp leaf virus; GFLV – grapevine fanleaf virus; MRSV – mulberry ringspot virus; PEBV – pea early browning virus; PRMV – peach rosette mosaic virus; PRV – pepper ringspot virus; RRV – raspberry ringspot virus; SLRV – strawberry latent ringspot virus; TBRV – tomato black ring virus; TMV – tobacco mosaic virus; TobRV – tobacco ringspot virus; TomRV – tomato ringspot virus; TRV – tobacco rattle virus

(Mayo and Martelli, 1993). Nepoviruses are in the family *Comoviridae*, whereas tobraviruses, although not yet classified in a family, are on a distinct branch of the RNA viruses. The nematode vectors of these viruses are also of two distinct types. Tobraviruses are transmitted by nematodes of the genera *Trichodorus* (6 of 61 species) or *Paratrichodorus* (7 of 26 species) from the family trichodoridae whereas Nepoviruses (at least the 30% or so of the genus that are nematode-transmitted) are transmitted by the genera *Longidorus* (8 of 89 species) or *Xiphinema* (7 of 205 species) in the family longidoridae. These families are only distantly related as they have been put in different sub-orders of the order dorylaimida (Maggenti, 1991) or, in a recent proposal, in different orders (Hunt, 1993). Thus it is likely that both the ability of nematodes to transmit virus and the property of viruses to be transmitted by nematodes have arisen independently at least twice during evolution. Table 1 lists the currently accepted virus-vector pairs.

Both tobraviruses and nepoviruses are acquired when the appropriate nematodes feed on the roots of infected plants. Virus particles are bound to the mouthparts of the nematodes and can then be transmitted to further healthy plant hosts after as little as 15 to 60 minutes feeding (Harrison et al., 1974). Nematodes remain viruliferous for long periods of time, for example for 12 weeks in *Longidorus* sp., 1 year in *Xiphinema* sp. and more than 1 year in *Trichodorus* spp., but not after a molt (Harrison et al., 1974). There is no evidence for any multiplication of the viruses in the nematode vectors.

TOBRAVIRUSES

Electron microscopy of tobravirus retention by nematodes

Nematodes acquire tobravirus particles when they feed on the epidermal cells of young roots of infected plants. Virus particles can be detected by electron microscopy at sites in the lumen of the food canal along the length of the feeding apparatus of vector trichodorid species (Fig. 1a). Tobravirus particles are of two types which differ in modal length (Fig. 1b). Most long particles lie with their long axis parallel to the food canal whereas short particles are often found closely adpressed by one end to the lining (Fig. 1c) (Taylor and Robertson, 1970). The binding of virus particles appears to be a property of only certain regions of the nematode pharynx and the lining of the food canal in this region has been shown to stain for carbohydrate. This suggests the involvement of lectin-like structures in the binding reaction. Moreover, in high resolution micrographs it is possible to discern some material linking the lining of the food canal to the virus particle (Fig. 1d) (Robertson and Wyss, 1983).

Specificity of tobravirus transmission

In general, there is a correlation between the serotype of a particular strain of tobacco rattle virus (TRV) and the vector species which can act as a vector for that strain

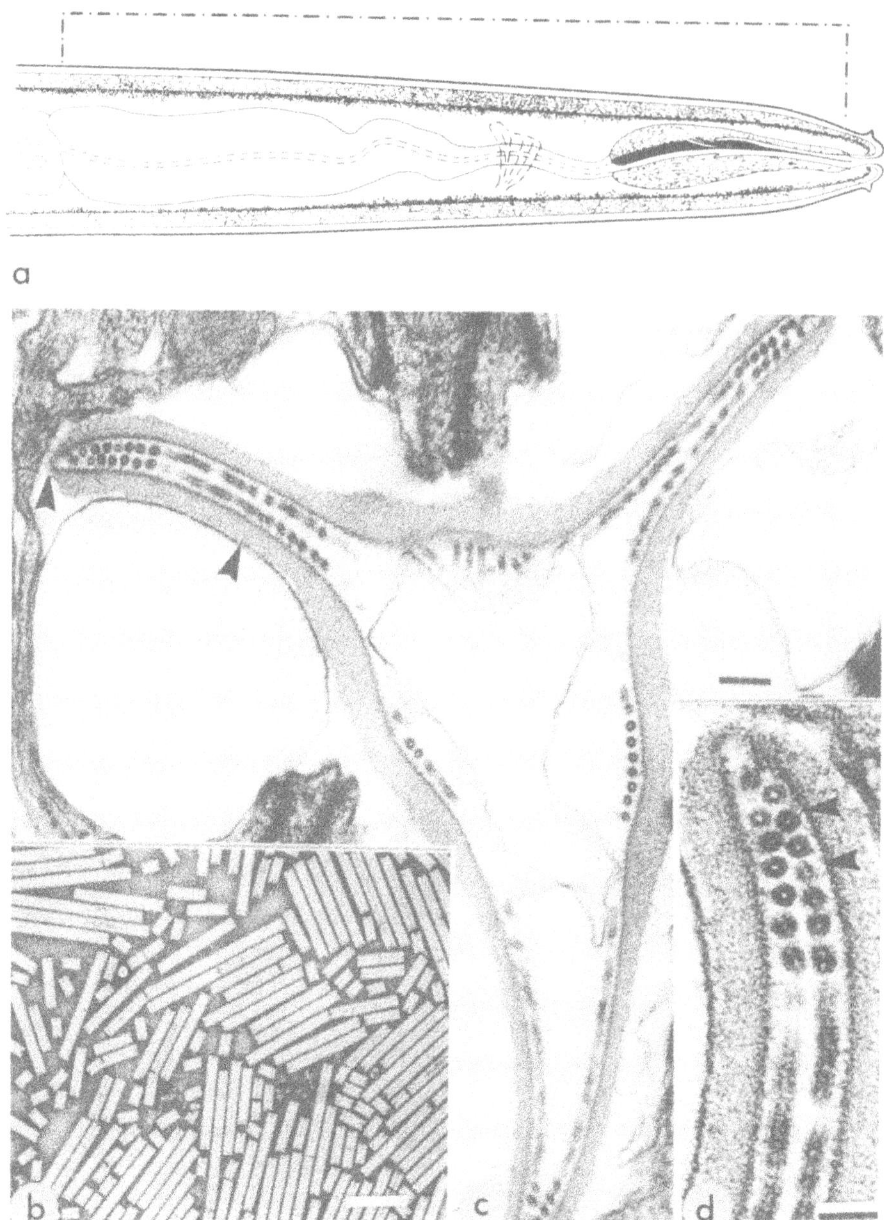

Fig. 1. Trichodorid vectors. (a) Diagram of feeding apparatus with sites of virus retention indicated by a bracket. (b) Electron micrograph of negatively stained particles of tobacco rattle virus (courtesy I.M. Roberts, SCRI). Bar = 100 nm. (c) Electron micrograph of a TS through the oesophagus of *Paratrichodorus pachydermus* exposed to a plant infected with TRV showing particles cut transversely and longitudinally and short particles adpressed end-on to the food canal wall. Arrows indicate region shown in (d). Bar represents 100 nm. (d) Part of (c) showing material (arrow) linking the particles to the wall. Bar = 50 nm.

(Ploeg et al., 1992). However, this specificity is more apparent in associations between *Paratrichodorus* spp. and tobravirus serotypes than in those involving *Trichodorus* spp. Thus the serotypes PRN and ORE were transmitted by *P. pachydermus* and *P. teres* respectively, whereas *T. cylindricus* transmitted the RQ and TCB2 serotypes of TRV as well as the SP5 serotype of pea early browning virus (PEBV) (Ploeg et al., 1992).

Location of determinants for transmission in tobravirus genomes

The genome of tobraviruses consists of two molecules of RNA. The larger RNA (approx. 6.8 kb) encodes proteins that are involved in RNA replication and virus transport; the smaller RNA (approx. 1.4 kb to approx. 4.5 kb) encodes the coat protein (Harrison and Robinson, 1978). In some tobravirus genomes other genes are present on one or other of the RNA species. Because of the differences in size between the RNA species and between the two types of virus particle which encapsidate each of the RNAs, it has proved to be relatively easy to separate the genome components of different tobraviruses. When plants were infected with artificial mixtures of genome parts from different viruses to make pseudo-recombinant strains, it was possible to assign phenotypic characteristics to each RNA species. The first evidence that the coat protein gene was in RNA-2 (Ghabrial and Lister, 1973) was obtained by such an experiment. When pseudo-recombinants were made between a nematode-transmissible and a non-transmissible strain, the transmission character was found to be determined by the strain contributing the RNA-2 (Ploeg et al., 1993). This suggests that the coat protein is involved in the transmission.

Structure of tobravirus particles

Tobravirus particles consist of a helical array of coat protein molecules which surround the RNA helix (Harrison and Robinson, 1978). The arrangement is very similar to that of the coat protein of tobacco mosaic virus (TMV) and the detail of this construction has been determined by X-ray crystallography (Namba et al., 1985). Goulden et al. (1992) showed that it is possible to align the amino acid sequences of tobravirus coat proteins with that of TMV coat protein in such a way that the residues which are thought to be critical for the folding of TMV protein to form a coat protein sub-unit (Altschuh et al., 1987) correspond in the two proteins. The resulting folding pattern for TRV protein shows that both the N-terminus and the C-terminus are located on the outside of the virus particle.

This model is supported by two lines of evidence. (1) When particles of TRV or pepper ringspot virus (PRV) were treated with proteases a small amount of the coat protein was removed from the C-terminus without affecting particle integrity or the infectivity of the virus preparations (Mayo and Cooper, 1973). (2) When TRV particles were used to raise a panel of monoclonal antibodies, several were obtained which reacted with the amino acid sequence at the extreme C-terminus of the coat protein (Legorboru, 1993). Moreover, analysis of the peptide-binding characteristics of rabbit polyclonal antiserum showed that the C-terminal sequence contained at least one major epitope (Legorboru, 1993; Legorboru et al., 1992).

Nuclear magnetic resonance (NMR) spectroscopy of tobravirus particles

Proton NMR spectroscopy is a method which examines molecules in solution and therefore in their native state. The peaks in the NMR spectra contain information about the chemical environments of the protons in the molecules. However, with virus particles many nuclei are in rigid parts of the particle and therefore only yield very broad peaks. Thus TMV particles do not give detectable spectra (Jardetsky et al., 1978). When peaks are detected these are associated with protons in nuclei in mobile parts of the virus protein. When particles of PRV were examined, a signal was obtained showing that these particles have a mobile component (Fig. 2a) (Brierley et al., 1993); similar results have been obtained with TRV particles (Mayo et al., 1993). Analysis of the signal from PRV particles by two-dimensional correlated spectroscopy (COSY) (Fig. 2b) assigned the individual signals to amino acid residues A,N,G,S,T and P. Examination of the amino acid sequence of PRV coat protein (Bergh et al., 1985) showed that these were the sole components only in the 38 C-terminal amino acids. It was concluded that the mobile element consists of the 11 amino acids at the extreme C-terminus together with some or all of the 27 amino acids to the N-terminal side of this position (see Fig. 3). This hypothesis was confirmed by NMR analysis of a peptide which was synthesised to have the same sequence as the C-terminal 17 residues of PRV coat protein. Fig. 3 shows an alignment of the C-terminal sequences of several tobraviruses and two tobamoviruses, which also have rod-shaped particles. The line on the left represents the aligned W or F residues of each sequence which in TMV protein is known to be a key structural component of the folded coat protein and located at the edge of the virus particle (Altschuh et al., 1987). Assuming, as discussed above, that tobravirus particles have a similar organization to those of TMV, residues to the right of this W or F will be on the outside of the virus particle.

Comparisons among tobravirus coat protein sequences

Fig. 3 shows that tobravirus coat proteins differ from those of tobamoviruses in having a longer 'protruding' C-terminal sequence. This, and the mobility of the sequence shown by the NMR experiments, suggests that this C-terminal sequence could be significant in the transmission of tobraviruses by nematodes. Further circumstantial evidence comes from sequence comparisons. The strains PLB and K20 of TRV are very similar serologically but differ in that PLB is not transmissible whereas K20 is (Ploeg et al., 1993). An alignment of the amino acid sequences of the coat proteins of these strains showed 15 mainly conservative changes in amino acid sequence except for the region of the C-terminus shown in Fig. 3 (Legoboru, 1993). In this region the sequences differ in several positions and in that the sequence of K20 is shorter. Because the C-terminal region of the coat protein is the most immunogenic on the TRV particle (Legorboru et al., 1992), Legorboru (1993) has proposed that the change from PIRPNP in K20 to AVRPNP in PLB at the extreme C-terminus is responsible for the change from transmissibility to non-transmissibility. This reinforces the suggestion that the C-terminal sequence plays

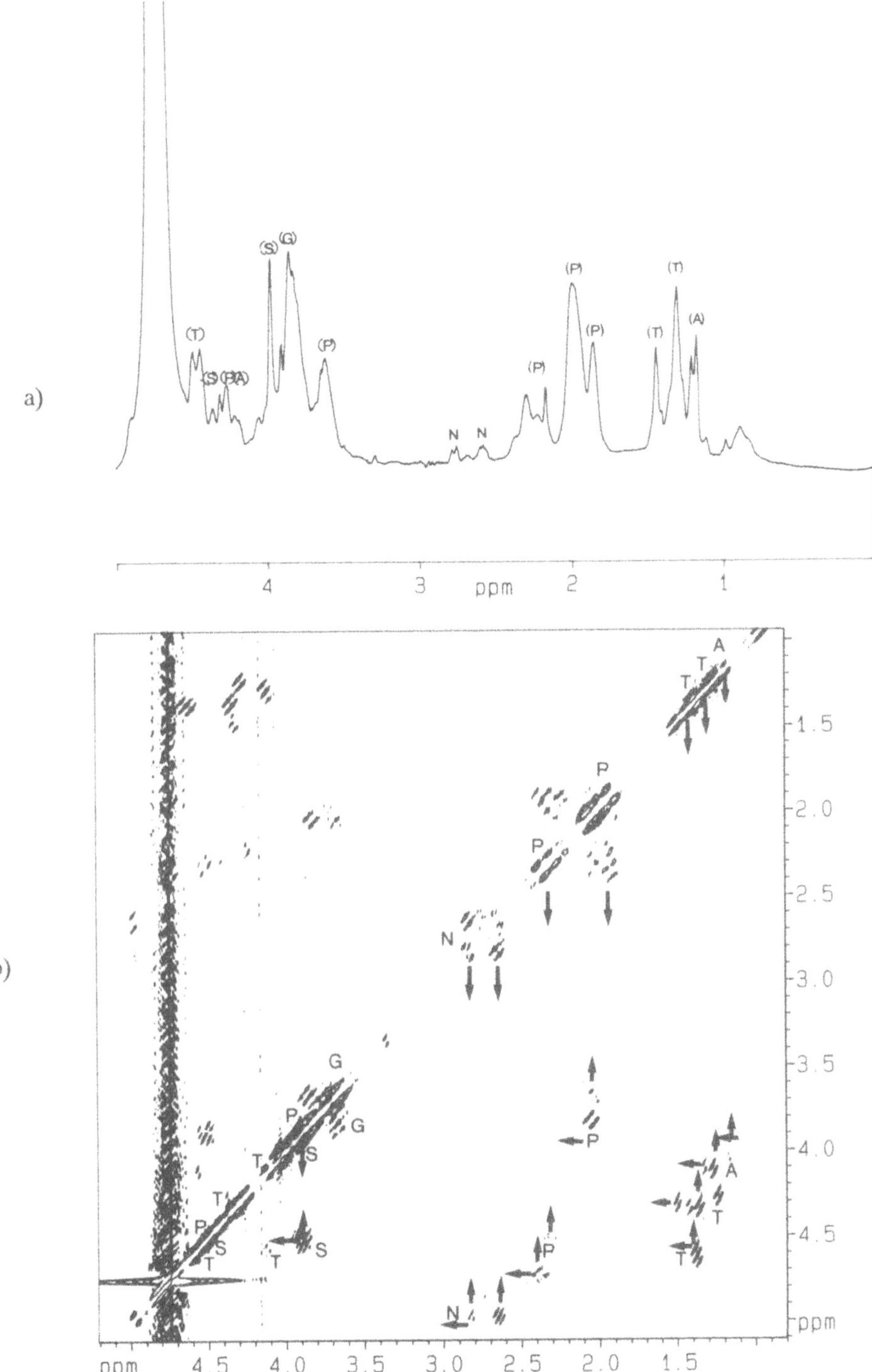

Fig. 2. Nuclear magnetic resonance spectra of particles of pepper ringspot virus. (a) One dimensional proton NMR spectrum; Amino acid assignments are N, asparagine and (tentatively) A, alanine; G, glycine; P, proline; S, serine; T, threonine. (b) Two-dimensional phase-sensitive COSY spectrum, assignments are unequivocal.

a role in transmission. It may also be significant that the amino acid composition of the terminal sequences of the tobravirus coat proteins are similar in that they are rich in G,P,A and S or T and also that the sequences contain several repeated sequences.

Recent observations

Several strands of evidence show that tobravirus particles have a surface–located relatively mobile element composed of the C–terminal sequence of the coat protein molecule. This sequence is readily accessible to proteases and is markedly immunogenic. And differences between the C–terminal sequences may explain differences in transmissibility between virus strains. Possibly the accessibility, mobility or immunogenicity of the region may be related to its reactivity in binding to structures in the nematode vector. If so, the susceptiblity to proteolysis of tobravirus particles could be a mechanism for the release of virus particles from the site at which they are bound to the

Tobraviruses		
PRV		GGSGSAPSGAPAGGSSGSAPPTSGSSGSGAAPTPPPNP
	K20	GGAAASSSAPPPASGGPIRPNP
TRV	PLB	GSASTPASGGSGATPPPASGGAVRPNP
	TCM	TAAAPVAAAGGTPPGGRSWTWNLV
PEBV	SP5	KETPQQQQNVTGPTVPATSSGGGKGPGVA
Tobamoviruses		
TMV		TSGPAT
CGMMV		SETTSKA

Fig. 3. Alignment of C–terminal sequences of coat proteins of tobraviruses and tobamoviruses. Sequences were from the EMBL database or Legorboru (1993). The vertical bar represents the conserved W/F residue described in the text.

nematodes. However there are complicating observations which suggest that this model may be overly simple. Close examination of TRV particles bound in nematode bodies suggests that the separation between virus particle and nematode cuticle is about 5 nm to

7 nm. This is much greater than the space that a peptide of about 25 amino acids could fill. Although much evidence implicates the virus coat protein in the transmission process, there is little to exclude the involvement of other proteins as well. Indeed recent experiments with PEBV show that (1) the coat protein of the nematode tranmissible strain A56 is virtually identical in sequence to that of the non–transmissible strain SP5 and (2) inserting the coat protein gene from the transmissible strain K20 into cDNA of a non–transmissible strain did not restore transmissibility to the progeny virus derived from transcripts of the cDNA (S. MacFarlane, pers. comm.). Thus a second virus gene product seems to be involved in nematode transmission. It might therefore be this protein which links the nematode surface to virus particles, perhaps by attaching to the carbohydrate–containing material and to the protruding mobile C–terminal part of the virus coat protein.

NEPOVIRUSES

Electron microscopy of nepovirus retention by nematodes

Nepovirus particles are acquired by *Longidorus* spp. or *Xiphinema* spp. when these feed on the young roots of infected plants. Virus particles are bound to surfaces in the alimentary tracts of the vectors but the sites at which particles bind differ between the two nematode genera. In *Longidorus* spp., particles are found on the odontostyle lumen or the guiding sheath whereas in *Xiphinema* spp., particles are bound to the cuticular lining of the odontophore and the oesophagus (Fig. 4). Indeed in *Xiphinema* spp. there is a sharp delimitation of the binding site in that particles are found attached to the wall of the odontophore but not to the immediately adjacent odontostyle.

There is also a difference between the two nematode genera in that virus particles seem to be separated from surfaces on the odontostyle of *Longidorus* spp. by a substance which lines the lumen but on the odontophore of *Xiphinema* spp. no such layer has been found (Fig. 4). The lumen of the odontostyle is also coated with a substance and in both genera the lumen binds cationized ferritin (Robertson, 1987). The lining of the odontophore of *X. diversicaudatum* has been found to have patches of carbohydrate associated with particles of arabis mosaic virus (ArMV) (Robertson and Henry, 1986). No carbohydrate has been found associated with the wall of the lumen of the odontostyle of *L. elongatus* or the substance which lines it.

Specificity of nepovirus transmission

For nepoviruses transmitted by *Longidorus* spp., there is a high degree of specificity between virus and vector. For example, serologically diverse strains of raspberry ringspot virus (RRV) are transmitted efficiently by *L. elongatus* or *L. macrosoma,* but not both (Harrison et al., 1974). However the specificity of transmission is not so strong with viruses transmitted by *Xiphinema* spp. *X. rivesi* can transmit cherry rasp leaf virus (CRLV), tobacco ringspot virus (TobRV) and tomato ringspot virus (TomRV) equally efficiently and most, but not all populations of *X. americanum* can transmit all three viruses to at least a limited extent (Jones and Brown, 1993). A further complication is that peach rosette mosaic virus can be transmitted by *X. americanum* as well as by *L. diadecturus* (Allen et al., 1984).

Location of transmission determinants in nepovirus genomes

Nepoviruses have genomes which consist of two RNA molecules of approx. 7.2 kb to 8.4 kb (RNA-1) and approx. 3.9 kb to 7.2 kb (RNA-2) (Murant, 1981). A fundamental feature of the family *Comoviridae*, to which the genus *Nepovirus* belongs (Mayo and

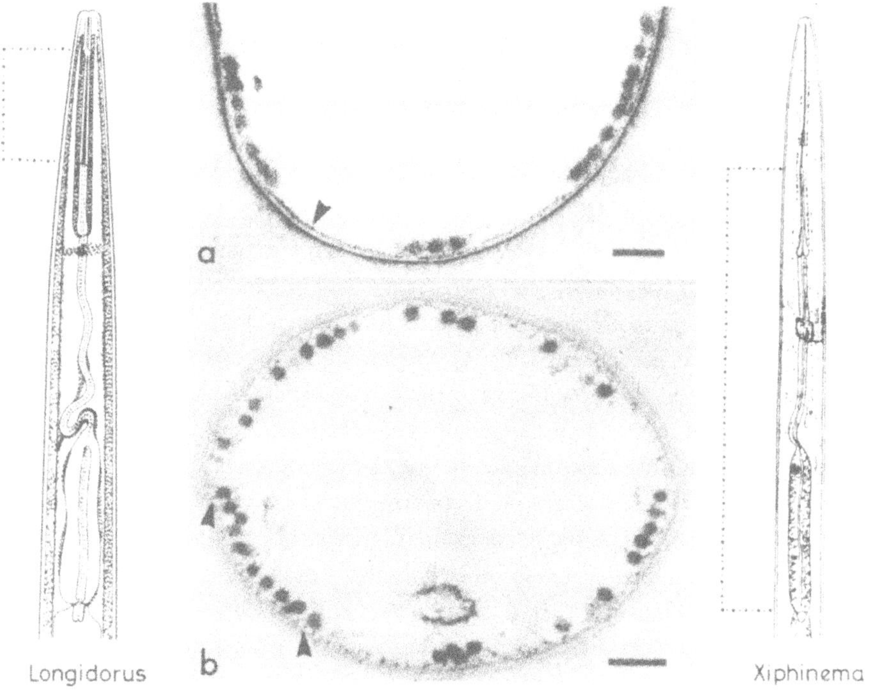

Fig. 4 Longidorid vectors. Diagrams show the feeding apparatus of *Longidorus* spp. and *Xiphinema* spp. with the sites of virus retention shown by the brackets. (a) Electron micrograph of an LS through the odontostyle of an *L. apulus* exposed to plants infected with artichoke Italian latent virus showing particles lining the lumen of the odontostyle. Arrow indicates material lining the lumen, diminishing in depth where particles are bound. (b) Electron micrograph of a TS through the anterior odontophore of a *X. diversicaudatum* which had fed on a strawberry latent ringspot virus-infected plant, showing virus particles lining the lumen of the food canal and material linking the particles to the wall (arrow). Bars = 100 nm.

Martelli, 1993), is that genome genome part is expressed as a single polyprotein. After synthesis in infected cells, the polyproteins are cleaved by viral proteases into the functional virus proteins. By separating the two RNA species from distinctive strains of either RRV (Harrison et al., 1972) or tomato black ring virus (TBRV) (Randles et al., 1977) and then inoculating heterologous mixtures, it was possible to prepare pseudo-recombinant strains of each virus to locate phenotypic characters to each RNA species. Such experiments have shown that the enzymes used in RNA replication are encoded by RNA-1 whereas the coat protein and the ability to be transmitted by specific nematodes is determined by RNA-2 (Harrison et al., 1974; Harrison and Murant, 1977).

Comparisons among nepovirus proteins

The degree of relationship between proteins can be represented by comparing their amino acid sequences when these are placed at right angles to each other. Fig. 5 shows the results of this test applied to the polyproteins encoded by the RNA-2 of some nepoviruses. When more than 17 amino acids were found to be identical in a stretch of 30 residues, a dot was made corresponding to the vertical and horizontal positions of the two stretches. Similarity between proteins is represented by the density of these points. In the plots shown in Fig. 5, the regions of the polyproteins which encode the coat proteins are to the right (horizontal) or the top (vertical) of the sequences. There were similarities between coat protein sequences in each of the pairs tested. However, it was in the regions of the polyproteins to the N-terminal side of the coat proteins that the most marked similarities were detected. This was so in the comparisons between TBRV and RRV and between grapevine fanleaf virus (GFLV) and TomRV. But in the comparison between GFLV and TBRV there was no similarity in this region. Thus this similarity correlates with the transmission of TBRV and RRV by *Longidorus* spp. and of GFLV and TomRV by *Xiphinema* spp. This was suggested by Blok et al. (1992) to reflect the use of this gene product in some part of the transmission mechanism. If this idea is correct, the plots also show that the virus sequences involved in the mechanism differed between the nematode genera.

Little is known about this gene product although it has been detected *in vivo* in TBRV-infected plants (Demangeat et al., 1992). However, the analogous gene product of cowpea mosaic comovirus, which, like nepoviruses, is in the family *Comoviridae*, is thought to be involved in virus transport (Wellink and Van Kammen, 1989) and is a component in the tubular structures that form in infected cells (Van Lent et al., 1991). Similar tubules are also found in cells infected by nepoviruses (Murant, 1981) and it is tempting to speculate that when feeding, nematodes may take up virus particles attached in some way to this tubule protein and it is this which becomes attached to the nematode. Tubular structures have been seen in the food canals of *L. apulus* that had fed on plants infected with AILV (Taylor et al., 1976).

Proteolysis of tomato black ring nepovirus particles

It is not known how the coat protein of nepoviruses is folded in the virus particle. However, the results of experiments with plants infected with TBRV (Demangeat et al., 1992) suggest an interesting parallel with the structure of tobravirus particles. TBRV coat

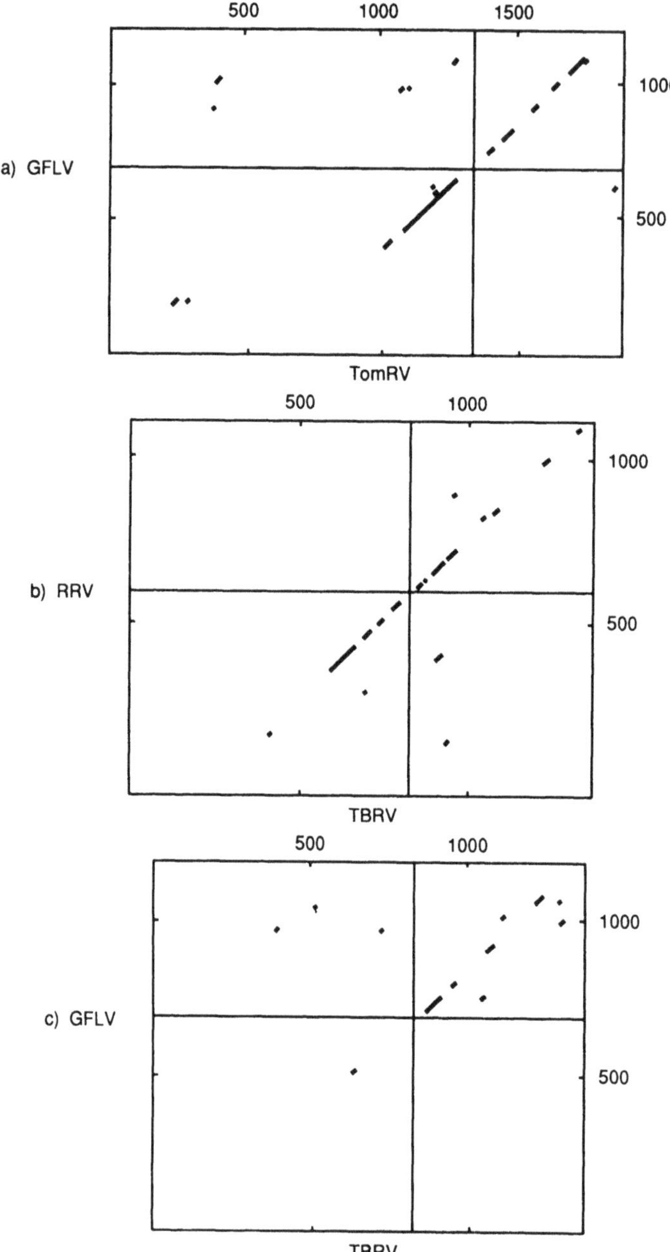

Fig. 5. Diagonal plots showing amino acid sequence similarities between polyproteins of (a) grapevine fanleaf virus (GFLV) and tomato ringspot virus (TomRV), (b) raspberry ringspot (RRV) and tomato black ring virus (TBRV), and (c) GFLV and TBRV. Plots were made using the COMPARE and DOTPLOT programs (Devereux et al., 1984). The numbers are the number of amino acids from the N-terminus which is at the bottom left; the lines show the positions of the N-terminal amino acids of the coat proteins of each virus. Sequences were taken from the EMBL database.

protein extracted from plants a few days after infection had an apparent molecular mass of 59,000. However, a few days later extracts of the same plants contained a smaller coat protein with an apparent molecular mass of 57,000. This protein corresponds in size to the coat protein which can be extracted from purified particles of TBRV. The C-terminal sequence of this protein showed that it is 9 amino acids short of the C-terminus of the predicted translation product of the RNA. It appears that the C-terminal 9 amino acids are removed by protease action while purified virus is stored in the infected plants or during virus purification. A tempting candidate for the structure responsible for the release of TBRV particles from the nematode is this C-terminal peptide. If such a structure were to prove to be involved in nematode transmission, these results suggest that virus particles may become non-transmissible during storage in infected plants and that purified virus may be non-transmissible.

Qualifying remarks

As with tobraviruses, the mechanism of transmission of nepoviruses is likely to involve separate binding and release steps. The binding is presumably efficient because often virus particles appear as close packed arrays with little available space left for further binding. But binding can occur which does not result in virus being transmitted. Thus large quantities of particles of the Scottish strain of RRV have been observed bound to the lumen of the odontostyle of *L. macrosoma* even though it does not transmit the virus (Taylor and Robertson, 1974; Trudgill and Brown, 1978). The release of virus particles is presumably relatively inefficient because nematodes remain viruliferous for many weeks.

Some caution is needed about simple interpretations particularly with respect to determinants on the coat proteins of the viruses because apparently very different virus particles can be transmitted by the same vector. SLRV particles have two coat proteins with molecular masses of approx. 43,000 and 27,000 (Mayo et al., 1974) and ArMV particles have only one coat protein with a molecular mass of 57,000 (Mayo et al., 1971) but both are transmitted by *X. diversicaudatum;* CRLV particles have three coat proteins with molecular mass of approx. 26,000, 23,000 and 21,000 (Jones et al., 1985) and those of TobRV have only one with a molecular mass of 57,000 (Mayo et al., 1971) but both are transmitted by the same *Xiphinema* spp. (Table 1).

Both these features suggest that it may be release rather than binding which is the main determinant of specificity and that a protein other than the coat protein may be involved. It would be particularly interesting to compare the polyprotein sequences upstream of the coat protein in pairs of viruses such as SLRV and ArMV, or CRLV and TomRV.

CONCLUSIONS

The mechanism of virus transmission by nematodes is of the same general type as that of the non-circulative transmission of many viruses by insects. That is, particles are bound from the ingested food and then released from the binding site during subsequent feeds. It is known that for both tobraviruses and nepoviruses, virus can be transmitted within an hour of a nematode feeding on a source but that the efficacy of the transmission is greater if the nematodes feed for longer on the source. In contrast, as nematodes remain viruliferous for many weeks or months the release mechanism would seem to be relatively inefficient. This fast binding/slow release mechanism is clearly a sound strategy for these viruses which often infect transient weed species which are not always available for nematodes to feed on and because nematodes move relatively slowly in soil.

A feature common to current speculations about the mechanisms of transmission is that a second virus gene product is involved. The evidence for this with PEBV is strong but that with nepoviruses is highly conjectural. Another common conjecture is that protease cleavage might be the way in which virus particles are released from their position in the nematode body and that this cleavage may involve loss of a few amino acids from the C-terminus of the coat protein. This idea is supported, albeit weakly, in that no other role has been suggested for this cleaved peptide and it seems unlikely that such a sequence would survive evolution when virus particles are fully infective after it has been removed.

However it is important to recognise the differences between the nematodes involved in transmission. The taxonomic distances between tobraviruses and nepoviruses and between the vectors of each group strongly suggest that each association has arisen by convergent evolution. Thus the more precise molecular details of the transmission mechanisms are likely to be distinct.

In summary, there is an increasing amount of information appearing about the molecular biology of viruses transmitted by nematodes. And it is becoming possible to conduct reverse genetics experiments with at least one nepovirus (GFLV: Viry et al., 1993) and two tobraviruses (TRV: Angenent et al., 1989; PEBV: MacFarlane et al., 1991). This means that it is feasible to determine the effect of changing individual amino acids in the coat protein or other virus proteins on virus transmission or other phenotypic characters. The challenge is now that the nematode side of the relationship be subjected to molecular scrutiny. Rapid progress has been made in the study of aphid transmission of viruses because the vector can be made viruliferous in the laboratory (Harrison and Murant, 1984). Such a laboratory-based artificial feeding system for the nematode vector species may yield equally rich rewards. Other opportunities would come from establishing a genetic basis for nematodes possessing or lacking the ability to be vectors followed by the exploitation of some of the new methods of molecular genetical analysis.

Acknowledgements

This work was supported by finance from The Scottish Office Agriculture and Fisheries Department, Edinburgh, UK and the Basque Regional Government, Spain. We are grateful to Dr S. A. MacFarlane for showing us data prior to its publication.

REFERENCES

Allen, W.R., Van Schagen, J.G. and Ebsary, B.A., 1984, Comparative transmission of the peach rosette mosaic virus by the Ontario populations of *Longidorus diadecturus* and *Xiphinema americanum* (Nematoda: Longidoridae), *Can. J. Plant Pathol.* 6:29.

Altschuh, D., Lesk, A.M., Bloomer, A.C. and Klug, A., 1987, Correlation of co-ordinated amino acid substitutions with function in viruses related to tobacco mosaic virus, *J. Mol. Biol.* 193:693.

Angenent, G.C., Posthumus, E. and Bol, J.F., 1989, Biological activity of transcripts synthesized *in vitro* from full-length and mutated DNA copies of tobacco rattle virus RNA 2, *Virol.* 173:68.

Bergh, S.T., Koziel, M.G., Huang, S-C., Thomas, R.A., Gilley, D.P. and Siegel, A., 1985, The nucleotide sequence of tobacco rattle virus RNA-2 (CAM strain), *Nucl. Acids Res.* 13:8507.

Blok, V.C., Wardell, J., Jolly, C.A., Manoukian, A., Robinson, D.J., Edwards, M.L. and Mayo, M.A., 1992, The nucleotide sequence of RNA-2 of raspberry ringspot nepovirus, *J. Gen. Virol.* 73:2189.

Brierley, K.M., Goodman, B.A. and Mayo, M.A., 1993, A mobile element on a virus particle surface identified by nuclear magnetic resonance spectroscopy, *Biochem. J.* 293:657.

Demangeat, G., Hemmer, O., Reinbolt, J., Mayo, M.A. and Fritsch, C., 1992, Virus-specific proteins in cells infected with tomato black ring nepovirus: evidence for proteolytic processing *in vivo*, *J. Gen. Virol.* 72:1609.

Devereux, J., Haeberli, P. and Smithies, O., 1984, A comprehensive set of sequence analysis programs for the VAX, *Nucl. Acids Res.*, 12:387.

Ghabrial, S.A. and Lister, R.M., 1973, Coat protein and symptom specification in tobacco rattle virus, *Virol.* 52:1.

Goulden, M.G., Davies, J.W., Wood, K.R. and Lomonossoff, G.P., 1992, Structure of tobraviral particles: A model suggested from sequence conservation in tobraviral and tobamoviral coat proteins, *J. Mol. Biol.* 227:1.

Harrison, B.D. and Murant, A.F., 1977, Nematode transmissibility of pseudo-recombinant isolates of tomato black ring virus, *Ann. appl. Biol.* 86:209.

Harrison, B.D. and Murant, A.F., 1984, Involvement of virus-coded proteins in transmission of plant viruses by vectors, *in*: "Vectors in Virus Biology", M.A. Mayo and Harrap, K.A., eds., Academic Press, London.

Harrison, B.D., Murant, A.F. and Mayo, M.A., 1972, Two properties of raspberry ringspot virus determined by its smaller RNA, *J. Gen. Virol.* 17:137.

Harrison, B.D., Robertson, W.M. and Taylor, C.E., 1974, Specificity and transmission of viruses by nematodes, *J. Nematol.* 6:155.

Harrison, B.D. and Robinson D.J., 1978, The tobraviruses, *Adv. Virus Res.* 23:25.

Hunt, D.J., 1993, "Aphelenchida, Longidoridae and Trichodoridae their systematics and Bionomics," CAB International, Wallingford.

Jones, A.T., Mayo, M.A. and Henderson, S.J., 1985, Biological and biochemical properties of an isolate of cherry rasp leaf virus from red raspberry, *Ann. appl. Biol.* 106:101.

Jardetzky, O., Akasaka, K., Vogel, D., Morris, S. and Holmes, K.C., 1978, Unusual segmental flexibility in a region of tobacco mosaic virus coat protein, *Nature* 273:564.

Jones, A.T. and Brown, D.J.F., 1993, Apparent lack of specificity in transmissionn of N. American nepoviruses by *Xiphinema* nematodes, Proc. 6th Int. Cong. Plant Pathol. (Montreal), p.304.

Legorboru, F.J., 1993, Immunogenic structure of tobacco rattle virus particles and its relation to vector transmission, Ph.D thesis, Univ. of Dundee.

Legorboru, F.J., Robinson, D.J., Torrance, L.T., Ploeg, A.T. and Brown, D.J.F., 1992, Surface features of the particles of tobacco rattle virus related to its transmissibility by trichodorid nematodes, Proc. EAPR Meeting, Vitoria-Gastiez, Spain, p. 47.

MacFarlane, S.A., Wallis, C.V., Taylor, S.C., Goulden, M.G., Wood, K.R. and Davies, J.W., 1991, Construction and analysis of infectious transcripts synthesized from full-length cDNA clones of both genomic RNAs of pea early browning virus, *Virol.* 182:124.

Maggenti, A.R., 1991, Nemata: higher classification, *in*: "Manual of Agricultural Nematology," W.R. Nickle, ed., Marcel Dekker, New York.

Mayo, M.A., Brierley, K.M. and Goodman, B.A., 1993, Developments in the understanding of the particle structure of tobraviruses, *Biochimie* 75:639.

Mayo, M.A. and Cooper, J.I., 1973, Partial degradation of the protein in tobacco rattle virus during storage. *J. Gen. Virol.* 18:281.

Mayo, M.A., Harrison, B.D. and Murant, A.F.,1971, New evidence on the structure of nepoviruses, *J. Gen. Virol.* 12:175.

Mayo, M.A. and Martelli, G.P., 1993, New families and genera of plant viruses, *Arch. Virol.* 133:496 .
Mayo, M.A., Murant, A.F., Harrison, B.D. and Goold, R.A., 1974, Two protein and two RNA species in particles of strawberry latent ringspot virus, *J. Gen. Virol.* 24:29.
Murant, A.F., 1981, Nepoviruses, *in*: "Handbook of Plant Virus Infections and Comparative Diagnosis", E. Kurstak, ed., Elsevier/North Holland, Amsterdam.
Namba, K., Caspar, D.L.D. and Stubbs, G.J.,1985, Computer graphics representation of levels of organization in tobacco mosaic virus structure, *Science* 227:773.
Ploeg, A.T., Brown, D.J.F. and Robinson, D.J., 1992, The association between species of *Trichodorus* and *Paratrichodorus* vector nematodes and serotypes of tobacco rattle tobravirus, *Ann. appl. Biol.* 121:619.
Ploeg, A.T., Robinson, D.J. and Brown, D.J.F., 1993, RNA-2 of tobacco rattle virus encodes the determinants of transmissibility by trichodorid vector nematodes, *J. Gen. Virol.* 74:1463.
Randles, J.W., Harrison, B.D., Murant, A.F. and Mayo, M.A., 1977, Packaging and biological activity of the two essential RNA species of tomato black ring virus, *J. Gen. Virol.* 36:187.
Robertson, W.M., 1987, Possible mechanisms of virus retention in virus-vector nematodes, Ann. Rep. Sco. Crop Res. Inst. for 1986, p.127.
Robertson, W.M. and Henry, C.E., 1986, An association of carbohydrates with particles of arabis mosaic virus retained within *Xiphinema diversicaudatum*, *Ann appl. Biol.* 109:299.
Robertson, W.M. and Wyss,U., 1983, Feeding processes of virus-transmitting nematodes, *in*: "Current Topics in Vector Research", K.F. Harris, ed., Praeger Publishers, New York.
Taylor, C.E. and Robertson, W.M.,1970, Location of tobacco rattle virus in the nematode vector, *Trichodorus pachydermus* Seinhorst, *J. Gen. Virol.* 6:179.
Taylor, C.E. and Robertson, W.M., 1974, Acquisition, retention and transmission of viruses by nematodes, *in*: "Nematode Vectors of Plant Viruses", F. Lamberti, Taylor, C.E., and Seinhorst, S.W., eds., London, Plenum, New York.
Taylor, C.E., Robertson, W.M. and Roca, F., 1976, Specific association of artichoke Italian latent virus with the odontostyle of its vector, *Longidorus attenuatus*, *Nematol. medit.* 4:23.
Trudgill, D.L. and Brown, D.J.F., 1978, Ingestion, retention and transmission of two strains of raspberry ringspot virus by *Longidorus macrosoma*, *J. Nematol.* 10:85.
Van Lent, J., Storms, M., Van der Meer, F., Wellink, J. and Goldbach, R., 1991, Tubular structures involved in movement of cowpea mosaic virus are also formed in infected cowpea protoplasts, *J. Gen. Virol.* 72:2615.
Viry, M., Serghini, M.A., Hans, F., Ritzenthaler, C., Pinck, M. and Pinck, L., 1993, Biologically active transcripts from cloned cDNA of genomic grapevine fanleaf nepovirus RNAs, *J. Gen. Virol.* 74:169.
Wellink, J. and Van Kammen, A., 1989, Cell-to-cell transport of cowpea mosaic virus requires both the 58K/48K proteins and the capsid proteins, *J. Gen. Virol.* 70:2279.

DISCUSSION

D. Baillie: I think that the proline-rich carboxy-terminal domain you described looks similar to a domain in let-653. This matches many carbohydrate-binding proteins in the SWISSPROT database. Do you think this might be relevant to the adhesion of tobravirus particles to nematode surfaces?
M. Mayo: This a very interesting observation that we had not noticed. It certainly could be that we have a carbohydrate-binding domain on the surface of tobravirus particles and we will look into the idea immediately.

L. Georgi: There are reports of nepoviruses that are transmissible by both *Longidorus* and *Xiphinema*. Do you plan to look at the sequence upstream of the coat protein in such a virus?

M. Mayo: Peach rosette mosaic virus is reported to be transmitted by both *L. diadecturus* and *X. bricolensis*. I would very much like to test the ideas presented here by sequencing the RNA-2 of this virus but I have no plans at present. I hope someone is stimulated to do it.

CHARACTERIZATION OF RESISTANCE GENES IN *BETA* SPP. AND IDENTIFICATION OF VIRULENCE GENES IN *HETERODERA SCHACHTII*

Joachim Müller

Biologische Bundesanstalt,
Institut für Nematologie und Wirbeltierkunde
D-48161 Münster

INTRODUCTION

The beet cyst nematode, *Heterodera schachtii* Schm., is a wide-spread pest in most areas of sugar beet cultivation. In agricultural practice the nematode is controlled by a crop rotation adapted to the infestation level, the use of resistant cruciferous green-manuring crops, or by the application of nematicides (where available). All these control strategies often are insufficiently effective or may have environmental or economical disadvantages. Therefore, the growing of resistant sugar beet is considered to be a valuable alternative or addition to these measures, and possibilities to establish resistant sugar beet cultivars have been studied in many research programmes.

SOURCES OF RESISTANCE IN *BETA* SPECIES AND BREEDING STRATEGIES

Resistance to *Heterodera schachtii* has never been found in cultivated beet (*Beta vulgaris* var. altissima Doll.). It occurs, however, in wild beet species, and efforts to transfer this resistance to sugar beet have been continued for more than 50 years. Resistance from *B. maritima* is based on a polygenic and recessive genetic system, providing only partial resistance, whereas that from *B. procumbens*, *B. webbiana* and *B. patellaris* is virtually complete and dominantly inherited. The latter type has been the more widely used even though transfer to cultivated beet has proved extremely difficult. Savitsky (1975), in North America, was the first to produce monosomic additions, i.e. plants with the diploid chromosome complement of *B. vulgaris* plus one extra chromosome of a wild beet species (*B. procumbens*) which contained the gene(s) for resistance. This achievement stimulated further research efforts in Europe, resulting in the production of many new monosomic additions, including some in which the resistance was derived from *B. webbiana* or *B. patellaris* (Speckmann and De Bock, 1982; Heijbroek et al., 1983; Löptien, 1984a). The history of this breeding and a survey of the current situation is given by Lange et al. (1990).

Further studies showed that up to three resistance genes can be distinguished within a single wild beet species (in *B. procumbens* and *B. webbiana*, but only one in *B. patellaris*). Löptien (1984b), Speckmann et al. (1985), and Jung et al. (1986) demonstrated that these genes are linked with morphological plant characters (a-, b-, and c-types). Isozyme markers were also successfully used to confirm the identity of the alien chromosome in the resistant additions (Van Geyt, 1986; Jung et al., 1986). Jung and Wricke (1987) found differences in the degree of resistance between a- and c-type additions from *B. webbiana*, while a- and b-type additions are considered to react identically. Lange et al. (1990) suggested that a single gene (or a group of closely linked genes) occurs in either one, two or three unlinked dosages in the donor species genomes, but they found no evidence that the genes of the various alien chromosomes have different effects, nor that per alien chromosome more than one gene might be involved in generating the resistance.

Fig. 1 shows schematically the concept for breeding nematode resistant sugar beet. Establishing an addition line was a long-lasting and difficult task. It was, however, only a first step. Crossing it with *B. vulgaris* is resulting in at least 50 % susceptibility in the offspring and due to the complete wild beet chromosome, this material has no marketable quality.

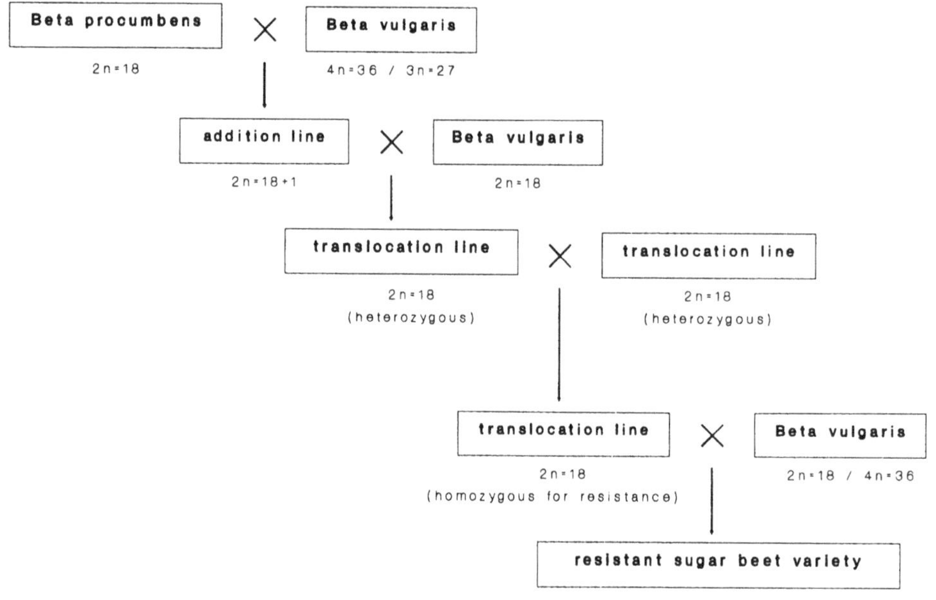

Figure 1. Strategy for breeding nematode resistant sugar beet.

Therefore, translocation lines carrying only a small part of the wild beet chromosome (including the resistance gene) in one of the 18 chromosomes of *B. vulgaris* have been developed. They are still heterozygous in resistance, but after testing their F_2-progeny translocation lines with homozygous resistance have been successfully identified.

Such material (named B883) has been released to breeding companies, and has been incorporated in breeding programmes. Unfortunately, transmission of the resistance gene was not complete in hybrids with *B. vulgaris* and this phenomenon is still the main problem in breeding resistant sugar beet cultivars.

DETECTION OF PATHOTYPES BY ASSESSING THE VIRULENCE OF *H. SCHACHTII* POPULATIONS

In contrast to the extensive literature on the genetic background of nematode resistance in the genus *Beta* nothing was known about virulence genes in beet cyst nematode populations. Breeders have usually considered *H. schachtii* to be a constant, homogeneous factor in the host parasite interaction. From an evolutionary point of view, however, this seems unlikely. Virulence genes complementary to the resistance genes in plants are likely to occur, albeit at a very low frequency, in some nematode populations. Repeated cultivation of resistant cultivars in infested soil should exert a selection pressure for nematodes with virulence genes, leading to the development of nematode populations with a high frequency of virulent individuals. These are called pathotypes if the plant-nematode interaction is based on a gene for gene relationship.

Müller (1992) demonstrated that virulence genes breaking the resistance derived from *B. procumbens* do exist in *H. schachtii* populations. Altogether 146 nematode populations of different geographical origin from Germany and 16 other European countries were collected and maintained on susceptible fodder rape. Juvenile suspensions of all these populations were separately inoculated to a beet genotype carrying homozygous resistance from chromosome 1 of *B. procumbens*. After the time necessary for one nematode generation only a few cysts had developed on some of the plants. However, after repeated transfer of these cysts to new resistant plants a remarkable

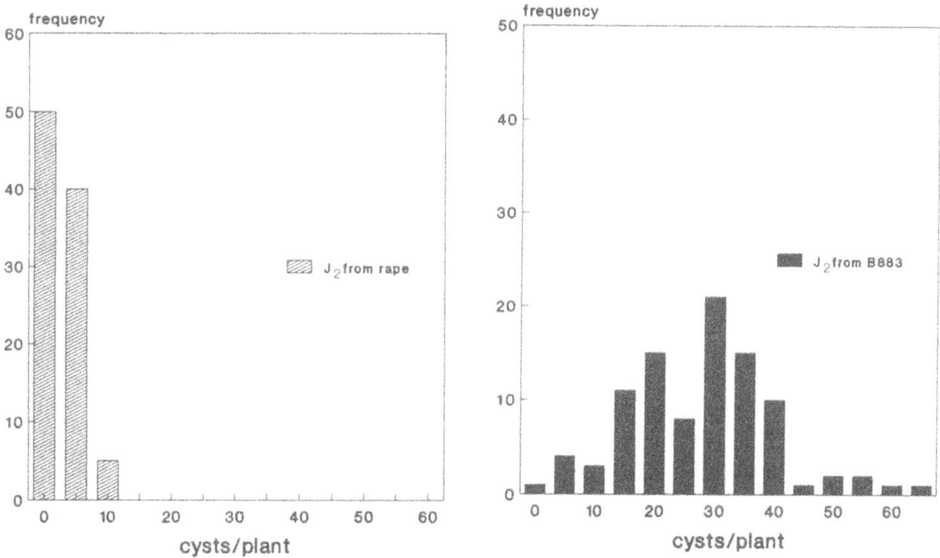

Figure 2. Cyst formation on a beet genotype carrying resistance from chromosome pro-1 with juveniles from rape (left side) and juveniles from resistant beet (B883 right side).

population increase was observed in some of the populations. These populations were maintained further on resistant beet and at the same time the original avirulent populations were reared on fodder rape.

Fig. 2 demonstrates the result of testing the beet genotype B883 (with resistance from chromosome 1 of *B. procumbens*) for its resistance by using nematode populations maintained on fodder rape (left side) or on resistant beet (right side). B883 is homozygous for resistance and no susceptible plants are expected. This holds true if it had been inoculated with the avirulent population from rape, whereas on almost all plants a considerable cyst production was observed when virulent juveniles from resistant beet had been used. The few cysts observed on plants inoculated with the nematode population from rape (left side) indicate that virulence genes do occur at a low frequency in the natural population.

CHARACTERIZATION OF RESISTANCE GENES WITH PATHOTYPES OF *H. SCHACHTII*

The above mentioned virulent pathotype was selected on a beet genotype with resistance from chromosome 1 of *B. procumbens*. The following experiments demonstrate the specificity of the virulence gene. Lange et al. (1993) tested several beet genotypes with different sources of resistance, two examples of which are presented here. In the first test, the addition line AN5 with heterozygous resistance from chromosome 1 of *B. patellaris* was used. The tested seed derived from a hybrid of this addition line with *B. vulgaris*, and as a consequence at least 50 % of the progeny was expected to be susceptible. Due to incomplete resistance transmission, the percentage of susceptibility was even higher, as Fig. 3A shows. On 71 of the test plants more than 25 cysts were counted per plant, and these plants are considered to be susceptible. The remaining 19 plants had no cysts or a maximum of 10 cysts per plant; they carry the resistance gene.

This differentiation in resistant/susceptible is only possible by using avirulent juveniles from rape. When the same seed material was tested with virulent juveniles from resistant B883, all plants produced cysts and there is no indication for resistance (Fig. 3V). This result supports the suggestion that the chromosomes pro-1 and pat-1 carry the same resistance gene and might be homologous.

In a second experiment the authors used the addition line AN101 which has heterozygous resistance from chromosome pro-7 of *B. procumbens*. Again the seed material was obtained from a hybrid with *B. vulgaris* and more than 50 % of the plants were expected to be susceptible. The result presented in Fig. 4A confirms this. Although resistant and susceptible plants are less clearly separated in the frequency distribution, some 12 % of all plants are considered to be resistant.

The same result was obtained when AN101 was inoculated with juveniles from resistant beet (Fig. 4V). The nematode pathotype was avirulent on this beet genotype and the about 14 % of plants carrying a resistance gene from chromosome pro-7 could be clearly distinguished from susceptible plants. It was concluded that either the resistance genes on pro-7 and pro-1 have different resistance mechanisms or the chromosome pro-7 carries an additional resistance gene or resistance generating factor.

Lange et al. (1993) proposed to use the symbol *Hs* for genes for resistance to *H. schachtii*. The different reactions to pathotypes will be shown by different numbers, whereas the origin of the alien genes is given in an additional superscript.

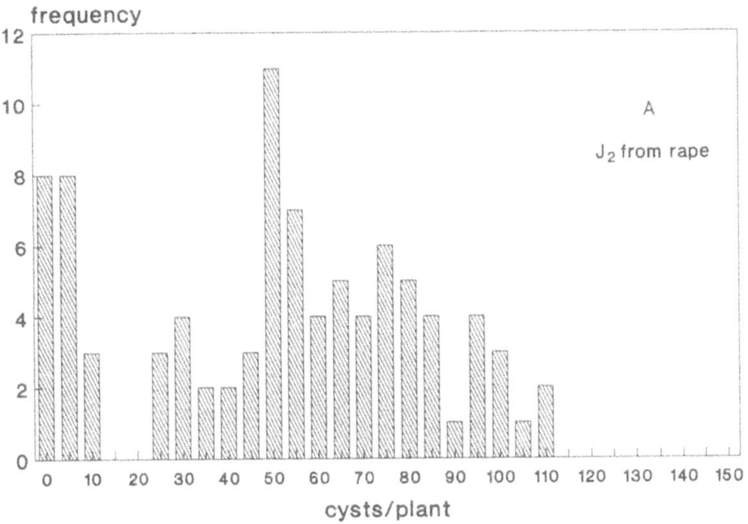

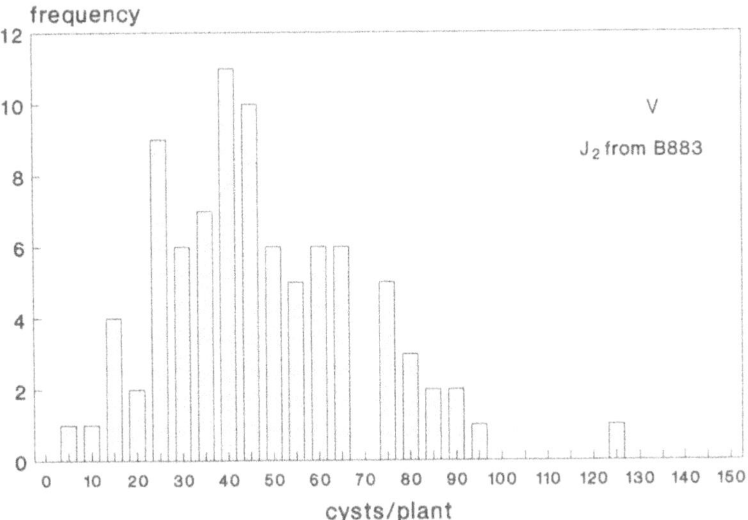

Figure 3. Cyst formation on a hybrid with beet genotype AN5 carrying resistance from chromosome pat-1 and infested with juveniles from rape (A = Avirulent) or from resistant beet (B883) (V = Virulent).

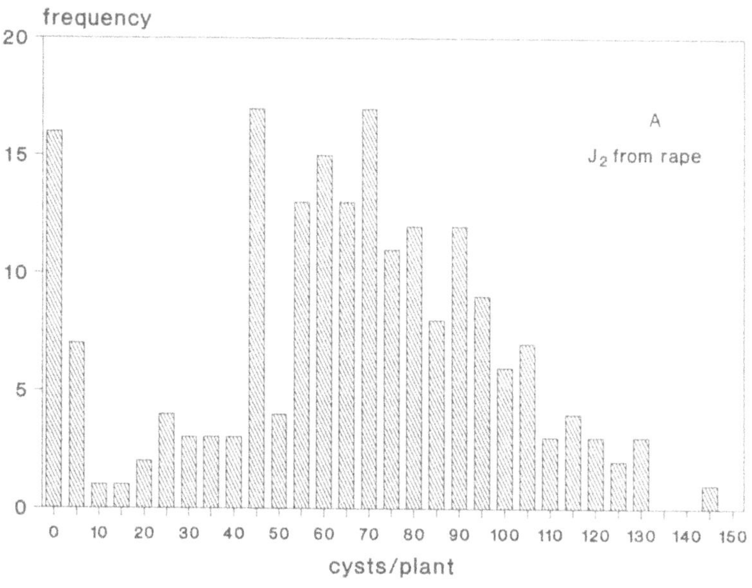

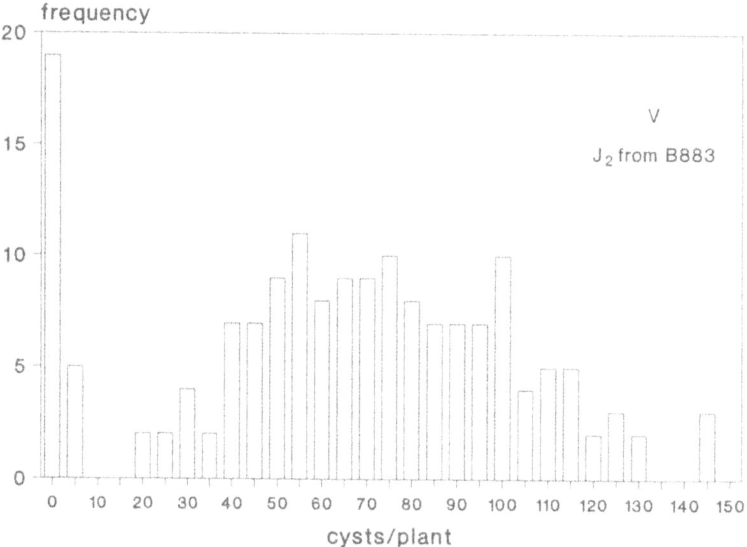

Figure 4. Cyst formation on a hybrid with beet genotype AN101 carrying resistance from chromosom pro-7 and infested with juveniles from rape (A) or from resistant beet (B883) (V).

The gene in the material of Savitsky (1975) and in B883 is called $Hs1^{pro-1}$. The one from *B. patellaris*, in accession AN5, which has a similar reaction to the new pathotype of *H. schachtii*, thus is called $Hs1^{pat-1}$. And the third gene, which occurs in AN101, originates from *B. procumbens*, and gives rise to a different reaction to the new pathotype is called $Hs2^{pro-7}$.

The three wild beet species *B. procumbens*, *B. webbiana* and *B. patellaris* exhibit complete resistance when inoculated with natural populations of *H. schachtii*. Only a very low percentage of all plants tested may produce a few single cysts. The same result was obtained when *B. patellaris* or *B. procumbens* were infested with the pathotype selected for virulence to $Hs1^{pro-1}$ (Fig. 5). As on the other hand the pathotype is able to break resistance derived from resistance gene $Hs1^{pat-1}$ (Fig. 3V), it was predicted that *B. patellaris* carries more than one resistance gene.

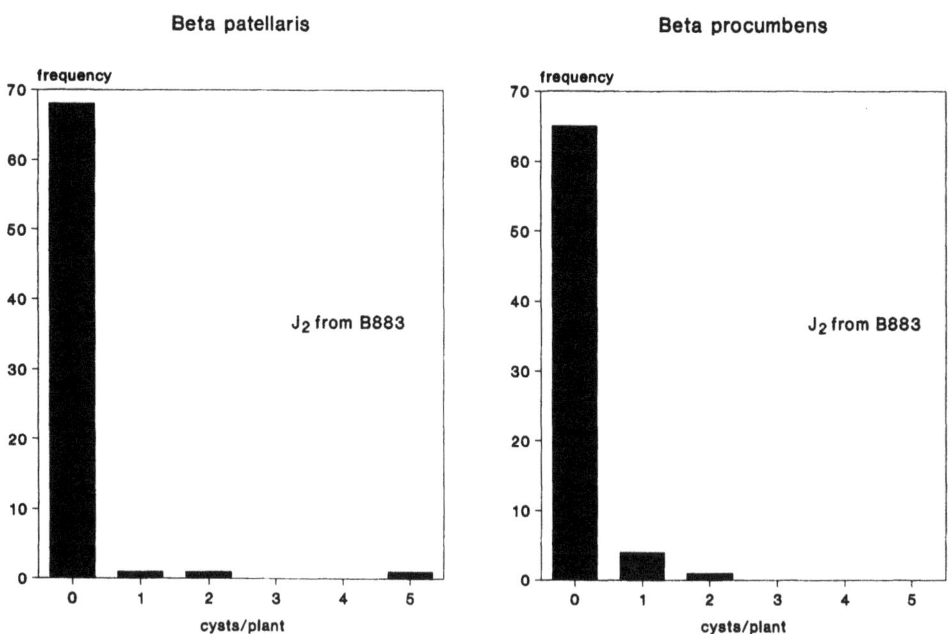

Figure 5. Cyst formation on the wild species *Beta patellaris* and *B. procumbens* with juveniles from resistant beet (B883).

CONCLUSIONS

Breeders demonstrated that up to seven resistance genes can be distinguished in the three wild beet species of the section *Patellares*. Five of them were considered to react identically and to give almost complete resistance to all *H. schachtii* populations. Only after selection of a virulent nematode pathotype was it possible to differentiate between these sources of resistance. The levels of cyst formation in the wild species *B. procumbens* and *B. patellaris* were the lowest of all resistant plant materials tested, which is in concordance with earlier studies (e.g. Hijner, 1951; Yu, 1984; Müller, 1992). This means that the resistance in B883, AN101 and AN5 in fact is incomplete as compared to the original species. The high level of resistance in *B. procumbens* might be explained by the combined action of at least the postulated genes *Hs1pro-1* and *Hs2pro-7*. The high level of resistance in *B. patellaris* justifies the expectation that other genes for resistance than *Hs1pat-1* might occur in this species.

REFERENCES

Heijbroek, W., Roelands, A.J., and De Jong, J.H., 1983, Transfer of resistance to beet cyst nematode from *Beta patellaris* to sugar beet, *Euphytica* 32: 287.

Hijner, J.A., 1951, De gevoeligheid van wilde bieten voor het bietecystenaaltje (*Heterodera schachtii), Meded. Inst. rat. Suikerprod.* 21: 1.

Jung, C., Wehling, P. and Löptien, H., 1986, Electrophoretic investigations on nematode resistant sugar beets, *Plant Breeding* 97: 39.

Jung, C. and Wricke, G., 1987, Selection of diploid nematode-resistant sugar beet from monosomic addition lines, *Plant Breeding* 98: 205.

Lange, W., Jung, C. and Heijbroek, W., 1990, Transfer of beet cyst nematode resistance from *Beta* species of the section *Patellares* to cultivated beet, *Proc. 53th Winter Congr. int. Inst. Sugar Beet Res., Brussels, 14-15 Feb., 1990*: 89.

Lange, W., Müller, J. and De Bock, T.S.M., 1993, Virulence in the beet cyst nematode versus various alien genes for resistance in beet (*Beta vulgaris*), *Fund. Appl. Nematol.* 16: 447.

Löptien, H., 1984a, Breeding nematode-resistant beets. I. Development of resistant alien additions by crosses between *Beta vulgaris* L. and wild species of the section Patellares, *Zeitschr. für Pflanzenzücht.* 92: 208.

Löptien, H., 1984b, Breeding nematode-resistant beets. II. Investigations into the inheritance of resistance to *Heterodera schachtii* Schm. in wild species of the section Patellares, *Zeitschr. für Pflanzenzücht.* 93: 237.

Müller, J., 1992, Detection of pathotypes by assessing the virulence of *Heterodera schachtii* populations, *Nematologica* 38: 50.

Savitsky, H., 1975, Hybridization between *Beta vulgaris* and *B. procumbens* and transmission of nematode (*Heterodera schachtii*) resistance to sugar beet, *Canad. J. Genet. Cytol.* 17: 197.

Speckmann, G.J. and De Bock, T.S.M., 1982, The production of alien monosomic additions in *Beta vulgaris* as a source for the introgression of resistance to beet root nematode (*Heterodera schachtii*) from *Beta* species of the section *Patellares, Euphytica* 31: 313.

Speckmann, G.J., De Bock, T.S.M. and De Jong, J.H., 1985, Monosomic additions with resistance to beet cyst nematode obtained from hybrids of *Beta vulgaris* and wild *Beta* species of the section *Patellares*. I. Morphology, transmission and level of resistance, *Zeitschr. für Pflanzenzücht.* 95: 73.

Van Geyt, J., 1986, The use of an isozyme marker system in sugar beet genetics and breeding, *PhD Thesis, Vrije Universiteit Brussel.*

Yu, M.H., 1984, Resistance to *Heterodera schachtii* in *Patellares* section of the genus *Beta*, *Euphytica* 33: 633.

PARTICIPANTS

PIERRE ABAD: INRA. Centre de Recherches d'Antibes. Laboratoire de Biologie des Invertébrés. 123, bd Francis Meilland BP 2078 06600 Antibes Cédex, France

ALBERT ABBOTT: Dept. of Biological Sciences. 132 Long Hall. Clemson University. Clemson, SC 29634-1903 U.S.A.

HOWARD J. ATKINSON: University of Leeds. Centre for Plant Biochem. and Biotechnology. Leeds, LS2 9JT U.K.

DAVID L. BAILLIE: Simon Fraser University. Institute Molecular Biology and Biochem. Burnaby, British Columbia Canada V5A 1S6

THOMAS J. BAUM: Dept. of Plant Pathology. University of Georgia. Athens, GA 30602-7274 U.S.A.

JAAP BAKKER: Dept. of Nematology. Agricultural University Wageningen. P.O. Box 8123/6700 ES Wageningen, The Netherlands

PAOLO BAZZICALUPO: Istituto Internazionale di Genetica e Biofisica.Via Marconi 12. 80100 Napoli, Italy

DAVID BERTIOLI: Entomology and Nematology Dept. Rothamsted Exp. Stn. Harpenden, Herts AL5 2JQ U.K.

DAVID McK. BIRD: Dept. of Nematology. University of California. Riverside, CA 92521 U.S.A.

VIVIAN BLOK: Dept. of Zoology. Scottish Crop Research Institute Invergowrie. Dundee, DD2 5DA U.K.

PHILIPPE CASTAGNONE-SERENO: INRA. Centre de Recherches d'Antibes. Laboratoire de Biologie des Invertébrés. 123, bd Francis Meilland BP 2078 06600 Antibes Cédex, France

DAVID J. CHITWOOD: Nematology Lab. USDA. Bldg. 011A, BARC-W. Beltsville, MD 20705 U.S.A.

ERIC DAVIS: Dept. of Plant Pathology. N.C. State University. Raleigh, NC 27695-7616 U.S.A.

CARLA DE GIORGI: Dip. di Biochimica e Biol. Molecolare. Università di Bari 70100 Bari, Italy

FRANCESCA DE LUCA: Dip. di Biochimica e Biol. Molecolare. Università di Bari 70100 Bari, Italy

MAURO DI VITO: Istituto di Nematologia Agraria. CNR. Via Amendola 168/5. 70126 Bari, Italy

SEAN R. EDDY:Medical Research Council Centre. Hills Road. Cambridge, CB2 2QH U.K.

ADRIAN A. F. EVANS: Imperial College Silwood Park. Department of Biology. Ascot, Berks SL5 7PY U.K.

MIREILLE FARGETTE: Laboratoire de Nématologie. CIRAD, BP 5035. 34032 Montpellier Cedex 1, France

CARMEN FENOLL: Dep. de Biologia. Universidad Autonoma de Madrid. 28049 Madrid, Spain.
MARIELLA FINETTI SIALER: Dip. di Biochimica e Biol. Molecolare. Università di Bari. 70100 Bari, Italy
LAURA GEORGI: Dept. of Biological Sciences. 132 Long Hall. Clemson University. Clemson, SC 29634-1903 U.S.A.
NICOLA GRECO: Istituto di Nematologia Agraria. CNR. Via Amendola 168/5. 70126 Bari, Italy
FLORIAN M.W. GRUNDLER: Institut für Phytopathologie. Universität zu Kiel. Hermann Rodewald Strasse 9. 24118 Kiel, Germany
RICHARD S. HUSSEY: University of Georgia. Dept. of Plant Pathology. Julian H. Miller Plant Sciences Building. Athens, Georgia 30602-7274 U.S.A.
MICHAEL G.K. JONES: Murdoch University. School of Biological and Environmental Sciences. Perth, Western Australia 6150
DAVID T. KAPLAN: USDA-ARS. 2120 Camden Rd. Orlando, FL U.S.A.
HINANIT KOLTAI: ARO. The Volcani Center. Institute of Plant Protection. P.O.B. 6. Bet-Dagan, 50250. Israel
FRANCO LAMBERTI: Istituto di Nematologia Agraria. CNR. Via Amendola 168/5. 70126 Bari, Italy
JOHN W. MARSHALL: The New Zealand Institute for Crop and Food Research LTD. Private Bag 4704. Christchurch, New Zealand
MICHAEL A. MAYO: Scottish Crop Research Institute Invergowrie. Dundee, DD2 5DA U.K.
MAURICE MOENS: Rijksstation voor Nematologie en Entomologie. Burg. van Gansberghelaan 96. B-9220 Merelbeke, Belgium
JOAKIM MÜLLER: Biologische Bundesanstalt. Institut für Nematologie und Wirbeltierkunde. Toppheideweg 88. D-48161 Münster, Germany
ANDREAS NIEBEL: Universiteit Gent. Laboratorium Genetika. Ledeganckstratt 35 .B-9000 Gent, Belgium
SEAN O'LEARY: Genetics Lab. Biology Dept. St. Patrick's College. Maynooth, Co Kildare, Ireland
CHARLES H. OPPERMAN: Dept. of Plant Pathology. North Carolina State University. Box 7616, Raleigh, 27695-7616 U.S.A.
MARK S. PHILLIPS: Scottish Crop Research Institute Invergowrie.Dundee, DD2 5DA U.K.
THOMAS O. POWERS: University of Nebraska. Dept.Plant Pathology. 406 Plant Sciences. Lincoln, NE 68583-0722 U.S.A.
PATRICK QUÉNÉHÉRVÉ: Orstom. Laboratoire de Nématologie.BP 8006-97259 Fort de France Cedex Martinica (French West Indies)
CELESTE RAY: University of Georgia. Dept. of Plant Pathology. Athens, Georgia 30602-7274 U.S.A.
ALEX P. REID: Imperial College Silwood Park. Dept. of Biology. Ascot, Berks SL5 7PY U.K.
WALTER ROBERTSON: Scottish Crop Research Institute Invergowrie. Dundee, DD2 5DA U.K.
ANN M. ROSE: Dept. of Medical Genetics. 226-6174 University Blvd. University of British Columbia.Vancouver, B.C. V6T 1W5 Canada.
ARJEN SCHOTS: Dept. of Nematology.Agricultural University Wageningen. P.O. Box 9060. GW Wageningen, The Netherlands
PETER C. SIJMONS: MOGEN. International nv Eisteinweg 97. 2333 CB Leiden, The Netherlands

YITZHAK SPIEGEL: The Volcani Center. ARO. Institute of Plant Protection. P.O.B. 6. Bet -Dagan, 50250 Israel
RENZO TACCONI: Osservatorio per le Malattie delle Piante. Via di Corticella 133. 40129 Bologna, Italy
JAN J. VAN DE HAAR: RZ Research BV P.O. Box 2. 9123 ZR Metslawier,The Netherlands
THIERRY C. VRAIN: Agriculture Canada Research Station. 6660 N.W. Marine Drive. Vancouver, B.C. V6T 1X2 Canada
VALERIE M. WILLIAMSON: Dept of Nematology. University of California. Davis, CA 95616 U.S.A..
CAROLIEN ZIJLSTRA: Research Institute for Plant Protection. (IPO-DLO). P.O. Box 9060. NL 6700 GW Wageningen, The Netherlands.

INDEX

www.ingramcontent.com/pod-product-compliance
Ingram Content Group UK Ltd
Pitfield, Milton Keynes, MK11 3LW, UK
UKHW051130260726
13967UKWH00010B/2968

* 9 7 8 1 4 7 5 7 9 0 8 1 8 *